SOVIET CIVIL LAW

SOVIET CIVIL LAW

O. N. Sadikov, Editor

With an introduction by William B. Simons

M. E. Sharpe, Inc.
ARMONK, NEW YORK
LONDON, ENGLAND

Available in the United Kingdom and Europe from M. E. Sharpe, Publishers, 3 Henrietta Street, London WC2E 8LU.

Originally published as *Sovetskoe grazhdanskoe pravo* by "Iuridicheskaia literatura," Moscow. © 1983 by O. N. Sadikov. Russian text translated by arrangement with VAAP, the USSR Copyright Agency.

This translation first appeared in *Soviet Statutes & Decisions*, Vol. XXI, Nos. 1, 2, 3, 4 and Vol. XXIII, No. 2, © 1984, 1985, 1987 by M. E. Sharpe, Inc. Translated by Lucy Cox (Part I) and Arlo Schultz (Part II).

Library of Congress Cataloging-in-Publication Data

Sadikov, O. N. (Oleg Nikolaevich)
Soviet civil law.

Translation of: Sovetskoe grazhdanskoe pravo.
Includes index.
1. Civil law—Soviet Union. 2. Domestic relations—Soviet Union.
I. Title.
LAW 346.47 87-4518
ISBN 0-87332-429-3 344.706

Printed in the United States of America

Table of Contents

PART I. CIVIL LAW

General Provisions

The Law of Property

The Law of Obligations

Copyright and the Law of Inventions

The Law of Succession

PART II. FAMILY LAW

Table of Citations

Introduction

This volume is an unabridged translation of the textbook *Soviet Civil Law*, originally published in 1983 under the auspices of the USSR Ministry of Justice. Edited by Professor O. N. Sadikov, the work includes contributions from nine Soviet legal scholars.

There is a substantial body of Western literature devoted to Soviet law and its various branches, including a number of works that digest and interpret Soviet law and practice for the reader who cannot consult Soviet legal sources in the original. Only occasionally are Soviet legal treatises or law school textbooks translated into English. This is particularly true with regard to Soviet civil and family law, the two related branches of law examined in the present volume. By contrast, laws, edicts, and normative acts and regulations from virtually all levels of the Soviet political and governmental structure are available in English translation.

Soviet legal literature in translation fulfills an important role in providing the Western reader an avenue of *direct* access to Soviet legal thought and doctrine. Yet, this literature needs to be understood and appreciated in its proper context.

The picture of Soviet civil and family law that is presented in this volume—while perhaps adequate and complete for the Soviet reader for whom it was originally intended—will not be fully accessible to readers who are not already familiar with the Soviet sociopolitical, economic, and legal system. Accordingly, I would like to highlight several points that the reader of this volume should keep in mind.

First, the published reports of Soviet judicial decisions are *not* a formal source of law. In this respect, the USSR is similar to the other major continental civil law countries of Europe (indeed, in general, the law of prerevolutionary imperial Russia, as well as of postrevolutionary Soviet Russia, evidences a noticeable influence—and on occasion a part-and-parcel borrowing—of French and German legal institutions and doctrines). Nevertheless, the decisions of Soviet courts and tribunals are published and disseminated on a periodic basis both in the Soviet Union and abroad. Soviet sources are usually forthright in acknowledging the influence that such publication has on the legal system as a whole, even if it falls far short of the precedent established through judge-made case law in common law countries. Furthermore, the USSR Supreme Court from time to time formulates and publishes "guiding explanations" covering broad issues of Soviet law which *are* formally binding on all Soviet courts.

Second, unlike some other Eastern European countries which have a separate and formal source of *economic* law, there is at present no economic code in the USSR. However, there have been recurring debates among Soviet legal scholars and practitioners on the proposition that this area should be separated into a distinct branch of Soviet law. Thus, Soviet civil law deals with property and other relationships where Soviet (or for that matter, foreign) citizens are involved on at least one side as well as those relationships involving only Soviet (or foreign) organizations, institutions, or enterprises.

Third, while the civil codes (*kodeksy*) of the Soviet republics, and the federal principles (*osnovy*) of civil legislation (on which the former are based) are the *primary* source of Soviet civil law, they by no means form the total body of Soviet civil law legislation. There exists an entire gamut of subordinate rules and regulations contained in the legislation of the supreme soviets of the USSR and the union republics and autonomous republics, as well as their governments, ministries, departments, and agencies, extending down to the level of local soviets of people's deputies and their executive committees. The same also holds true for the republican codes and federal principles devoted to marriage and the family.

Fourth, as mentioned above, although Soviet legislation evinces the influence of French, German, and other continental civil codes, there are a number of aspects of Soviet civil and family law that do not have any direct parallel in nonsocialist (i.e., noncommunist) civil law countries or even in prerevolutionary Russian law. Some examples are as follows: the all-encompassing nature of planned economic relations in Soviet law and society; the provision that civil rights are protected by Soviet law only when they are exercised in conformity with the purposes of a socialist society in the process of establishing communism; the requirement that in exercising their rights and performing their duties, Soviet citizens and organizations must comply not only with the laws but also with the rules of socialist community life and the moral principles of a socialist builder of communism; the mandate that parents in bringing up their children are expected to raise worthy members of a socialist society; and the proclaimed task of Soviet family law to strengthen the Soviet family based on the principles of communist morality.

Finally, the reader should be aware that the relationships that fall into the general categories of civil and family law are seldom if ever regulated *solely* by civil or family legislation. This is especially true in the USSR, given the predominance of state ownership and control and Party guidance of the economy through comprehensive planning and its leadership of society as a whole. In addition to extralegal influences or controls, Soviet administrative law affects civil and familial relationships. It is only partially codified and—perhaps more than any other branch of Soviet law—is characterized by the prevalence of *unpublished* rules and regulations. Some contemporary Soviet legal scholars have suggested that the remainder of Soviet administrative law should also be codified. However, Soviet law provides that laws, edicts, and decrees of a nonnormative character need not be published, but only communicated to the relevant addressee. It is important to be aware of this problem, although it is difficult to gauge its effect on the day-to-day operation of Soviet civil and family law.

William B. Simons

Part I

CIVIL LAW

General Provisions

Chapter 1. Concepts, Principles, and System of Soviet Civil Law

1. *Civil law as a branch of Soviet law.* Soviet civil law is one of the basic branches of Soviet law. Among the various social relationships, the determinant ones are the economic relationships. It is the task of civil law to regulate those which are conditioned by commodity-money forms. Therefore, civil law holds an important place in Soviet society, and its object is a wide range of social relationships.

In the first place, civil law regulates relationships connected with the solving of the tasks of economic development. The economic relationships between socialist organizations are arranged mainly on the basis of plan contracts and obligations; moreover, the improving of the industrial mechanism, the more intensive utilization of economic levers and stimuli, such as economic accountability [*khozraschet*], credit and profit, involve a wider application of civil law and a strengthening of its role.

Secondly, civil law regulates relationships meant to satisfy the material and spiritual needs of citizens, relationships which are linked to the development of creative activity. Thus, it actively helps in the solving of the basic economic task of socialism – the raising of the material and cultural level of the life of the people.

In the third place, the norms of Soviet civil law are of great significance in the regulation of relationships between citizens which concern their exercising of the right to personal property, inheritance, and other rights. In this way, the norms of civil law are employed in different areas to regulate diverse relationships.

A more exact definition of civil law necessitates the clarification of two basic questions: precisely what relationships does civil law regulate, that is, what is the object of civil law regulation, and how does the law regulate it, that is, what are the most general legal devices it utilizes or in what lies the method of civil law?

2. *The object of Soviet civil law.* In characterizing the object of Soviet civil law, the law defines three groups of relationships regulated by it. Among these relationships, Article 1 of the Principles of Civil Legislation of the USSR and the Union Republics (hereinafter to be cited by the abbreviation Principles) lists property relationships, personal non-property relationships which are connected with property relationships, and, in instances stipulated by law, other personal property relationships. Analogous positions are stipulated in Article 1 of the Civil Code of the RSFSR. Let us look at each of these groups of relationships.

The first group comprises those property relationships which are the fundamental object of civil law. The concept of property relationships does not coincide with the concept of economic relationships; they are a part of the latter, a narrower category. Property relationships are relationships based on property, that is, on material objects and other things of economic value. Thus, these are relationships based on concrete material objects and among concrete subjects.

Relationships among concrete subjects always have inherent in them a voluntary character, and this trait constitutes an important characteristic of property relationships. It is precisely because of the strength of this voluntary character that property relationships can be the subject of legal regulation, for the law exercises its regulatory function by influencing the will of the participants in the relationships.

Thus, property relationships are those economic relationships which are formed among concrete subjects, have a voluntary character and whose object is material objects and other things of economic value, for instance, services or other results of activity, even if they do not acquire material forms, e.g. inventions, trademarks, objects of literary creativity.

However, not all property relationships make up the object of civil law. According to the Principles, Soviet civil law regulates property relationships which are determined by the use of commodity-money form in communist construction. In other words, we are talking of property relationships which are linked with economic activity, its prerequisites and results.

In the main, it is the law of property which appears as the prerequisites and result of economic activity, and economic activity itself is conducted through the utilization of various contracts: purchase-sale, property-hire, carriage, etc. Property relationships regulated by civil law usually are reciprocal, although sometimes they are not (inheritance, gifts).

Those property relationships which are not connected with economic activity or with the commodity-money form are regulated not by civil but by other branches of law. For example, tax relationships, although they exist in monetary form, are not connected with economic activity and are regulated by financial law. Property relationships within the sphere of state administration are the object of administrative law.

The second group of relationships which makes up the object of civil law is that of personal non-property relationships connected with property relationships. Most frequently, these types of non-property relationships arise in connection with objects of creative activity. To this group belong, for instance, the creator's right to the work of research, literature, and art or to an invention or rationalization proposal. Here also are included some rights which belong only to socialist organizations, for instance, the right to the firm name. Under this name, the organization enters into economic activity with its products and services, and it is in the final analysis connected with property rights.

The third group of relationships which the law places within the sphere of civil law consists of some "other personal non-property" relationships, that is, non-property relationships not directly connected with property relationships. These relationships, which express interests of a moral nature, are the individualization of the subject in society, his individuality, the worth of his personality, and data concerning his personal life. These personal, non-property rights, not connected with property rights, form part of civil law when this is expressly specified by law. With these belong the protection of honor and dignity (Art.7 Civil Code), the right to one's own image (Art.514 Civil Code), and several others.

When the issue is about relationships which form the object of civil law, its norms act independently from the composition of the parties in these relationships. The subjects of civil law may be citizens, as well as organizations (only citizens, or only organizations; as well as both citizens and organizations). However, the regulation of civil-law relationships among citizens and among organizations have several, sometimes quite significant characteristics.

3. *The method of Soviet civil law.* If the object of regulation answers the question, what relationships are subject to the influence of civil law, then the method defines how this influence is exercised. It conveys the means by which the law influences regulated relationships, that is, manifests itself through basic devices which make up the content of that regulation.

The civil law method of regulation has significant characteristics in comparison with other branches of law, and this is plainly evident if civil law is compared with administrative law. In civil law the relationships between parties arise for the most part on the basis of their agreement (contract), while in administrative law they arise on the basis of the obligatory order of one of the parties (administrative act); in civil law, a right is defended, as a rule, through the method of bringing a suit in court or *arbitrazh*, while in administrative law through the method of filing a complaint to the competent organ.

The first characteristic of the method of civil law is connected with the status of its subjects. Economic activity assumes the equality of its participants, the impossibility that one of them force his will onto the other. This equality is a necessary condition of the basic relationship of economic activity and exchange. Therefore, a characteristic feature of the method of civil law is the equality of the parties involved in the relationships regulated by civil law. This does not mean that the parties involved in a specific relationship are allotted the same rights. On the contrary, rights may belong to only one side, while the other may bear only obligations. But the origin of the legal relationship and its substance depend in equal measure on both sides.

The second characteristic of the method of civil law is determined by the fact that economic activity demands the allocation to its participants of a certain degree of independence, of freedom of choice and maneuverability. This trait of the method, which is termed "dispositivity", consists in the presence of certain possibilities in the disposition of the law, in the taking of decisions. Civil law sets up limits, within which the parties themselves may define their own actions. However, it is possible that there are circumstances when the realization of a right in a civil law relationship may give rise to an obligation in an administrative law relationship, and the discretion of the subject [*sub"ekt*] of civil law is limited to the defined boundaries which are expressed, in part, in the imperative norms of law.

The principle of "dispositivity" takes many forms. Civil law contains a large number of dispositive norms from which the parties may deviate. There exists a large number of civil rights, of which the law does not contain an exhaustive enumeration, just as in the law there is not an exhaustive listing of the grounds for the development of civil legal relationships (Art.4 Civil Code).

It is important to note that the principle of dispositivity applies not only to citizens but also to socialist organizations. Precisely because of this, the role of contractual relationships between them is very important. Their significance was emphasized during the XXV and XXVI Congresses of the CPSU. The party has repeatedly noted the importance of economic activity and initiative.

The third characteristic of the method of civil law is determined by the character of sanctions, e.g. the means of enforcement. When the rights of a participant in a civil law relationship are violated, it is, above all, his property interests which suffer. Hence the problem of restoring the property situation to the position which would have existed had there been no violation. The restoration of the property situation must be carried out at the expense of the violator. Therefore, the basic form of the restoration of property damage is the claiming of damages (Art.219 Civil Code), and, as a rule, their compensation must be complete.

Finally, the fourth characteristic of the method of civil law is linked with the carrying out of the enforcement. As a general rule, civil rights are protected by independent institutions which are not linked to any of the parties, whose special function consists in the settling of disputes – courts, *arbitrazh*, or arbitration tribunals (Art.6 Civil Code). The defense of civil rights by other organs in administrative procedure occurs only as an exception, in cases specified by law.

4. *The delimitation of civil law from contiguous branches of law.* The delimitation of civil law from contiguous branches of law is made possible by the characteristics of object and method. First of all, there is delimitation according to the object of the corresponding relationships, e.g. property relationships are separate from non-property relationships. Thus, family law regulates basically non-property relationships, and it follows that family relationships fall outside the borders of civil law. In addition, delimitation is also made by method, as there exist a number of branches which regulate property relationships. It is precisely through the criterion of method that it is possible to delimit, for instance, financial law, which is characterized by relationships of authority and subordination, from civil law. Method is a supplementary criterion for the division of the branches of law. Those branches of law which are contiguous to the civil branch of law, whose delimitation is particularly important, are specifically listed in paragraphs 3 and 4, Article 2 of the Civil Code. They are administrative, financial, family, labor, land

and collective farm law, and others. Among these branches one must separate, first of all, those which regulate property relationships not through the method of equality but through the method of authority and subordination – administrative, financial, and land law. In these branches, property relationships do not have a commodity-money character, and they employ methods of authority in regulation.

Property relationships which are part of family, labor, or collective farm law also are not commodity-money relationships, and they do not serve economic activity, as is characteristic of civil law. The property relationships of family members (the payment of alimony) are not primary and reflect their personal ties; compensation for damages as a measure of liability does not play a role here.

As far as labor law is concerned, it regulates relationships in the work process which, under the conditions of socialism, do not have a commodity character. The same holds true of collective farm law characterized by intra-cooperative relationships (an important role among these being held by labor relationships) which do not have a commodity character. However, in the realization of the results of cooperative farm production, relationships having a commodity character do arise and are regulated by civil law.

Sometimes norms of law are promulgated to regulate a specific area of activity which go beyond the limits of one branch of law, for instance, capital construction or consumer services. These acts may form branches of legislation which combine the norms of various branches of law and therefore are called complex acts. This happens, for instance, in economic legislation which regulates relationships between socialist organizations in the sphere of economic activity and contains norms of civil, administrative, financial, and other branches of law. Treatises and texts may be devoted to these complex branches of legislation, and they may become the subject of special educational courses. However, in such cases new branches of law are not formed.

Thus, Soviet civil law may be defined as the totality of norms regulating through its own particular methods, under the conditions of a planned economy, property relationships which are dependent on the utilization of the commodity-money form, personal non-property relationships which are connected to these, and, in circumstances which are stipulated by law, other personal non-property relationships of socialist organizations and citizens, as well as among citizens and socialist organizations.

5. *The leading role of the CPSU in the development of Soviet civil law.* An examination of the object of civil law shows how substantial its significance is in the resolution of problems facing the country. Civil law aids in the raising of the material and cultural level of the life of the people, the growth of social productivity, and the heightening of its effectiveness. This explains the importance which the party has attributed to civil law as a means of solving economic and social problems ever since the first years of the existence of the Soviet state.

The elaboration of the first civil code, the 1922 Civil Code of the RSFSR, was carried out under the direct guidance of V.I. Lenin, who gave instructions on the fundamental, most basic, of its positions. His instructions on denying the protection of civil rights which are exercised in conflict with their social-economic purpose (we do not acknowledge anything private in the economic sphere), and on the importance of the principle of *khozraschet*, which he called commercial accounting, are well known.[1]

Later on, the Party also devoted significant attention to the development of Soviet civil law. The decisions of the Party Congresses and plenums, and the decrees of the Central Committee of the CPSU are the basis for the preparation and adoption of the most important civil law acts. In this connection, it is important to note the significance which is accorded to the strengthening of socialist legality and the legal order [*pravoporiadok*]. The importance of this trend is specifically noted in the Program of the CPSU.

1. V.I. Lenin, *Complete Collected Works*, vol.44, pp.398-399; p.220.

In 1981, additions and changes, relating to the principles which were expressed in conformity with decisions of the party organs in the Constitution of the USSR, were made in the Principles of Civil Legislation of the USSR and the Union Republics.

As regards the regulation of economic activity, civil law legislation is being developed and improved. The XXVI Congress of the CPSU proposed "the perfection of the structure of *khozraschet* relationships and of the mutual economic interests and responsibility for the fulfillment of the tasks set by the plan and contractual obligations between the supplier and user, as well as between the client and contractor". It is envisaged to elevate the role of finance-credit levers in the intensification of production, the consolidation of *khozraschet*, and the strengthening of the economic regime; to strengthen the effectiveness of economic sanctions for the non-fulfillment of plan quotas and contractual obligations, or the irrational expenditure of material and financial resources.[2] The Central Committee of the CPSU and the Council of Ministers of the USSR have adopted a number of resolutions specifically devoted to economic legislation.[3]

The Party also directs the development of those institutions of Soviet civil law which are connected with the interests of citizens. Thus in agreement with directives of the Party, the Fundamentals of Housing Legislation of the USSR were adopted and work is going on toward publication of housing codes for the union republics. In this fashion, the development of civil law in all basic, principal questions is carried out under the direct and unmediated guidance of the Communist Party.

6. *The principles of Soviet civil law are the fundamental premises expressed in civil law.* It is necessary to distinguish them from the legal devices for the regulation of social relationships, which are part of a given branch of law, e.g. from

2. *Materials of the XXVI Congress of the CPSU*, Moscow 1981, p.199.
3. *Collection of Decrees of the USSR* 1971 No.1 item 1; 1975 No.6 item 98; 1979 No.18 item 118

the method of legal regulation. The principles usually are not directly formulated in the law, but they find expression in many of its norms, they are subject to application in the absence of a direct indication in the norms of law, and are the basis for the interpretation and application of legal norms. It is possible to distinguish the following principles of Soviet civil law:

a) The most important principle of Soviet civil law is the fundamental significance of socialist ownership. That is to say, it constitutes the basis of the economic system of the USSR and defines the character of the entire Soviet economy (Article 10, USSR Constitution). The basis for the personal property of citizens consists of earned income in the socialist economy; the material and spiritual needs of citizens are satisfied, for the most part, as a result of social production. The fundamental significance of socialist ownership finds many manifestations in civil law. For instance, special measures are provided for the protection of socialist ownership which serve for its safeguarding and productive use;

b) The basis of the plan is characteristic of socialist civil law. This principle, which is formulated in general form in Article 16 of the Constitution of the USSR, reflects the nature of the socialist economy. Under the conditions of socialism, the activity of commodity-money relations is managed by centralized state planning. The planned aspect of the economy exerts a definite influence on the content of the legal regulation of commodity-money relationships, in other words, on civil law. The influence of the plan basis is particularly apparent in civil law relations between socialist organizations. The acts of the plan are one of the important grounds for the origin of civil law relations between organizations and determine many rights and obligations of the parties.

But it would be a mistake to link the significance of the plan basis only to relationships among socialist organizations; the plan aspect also appears in civil relationships in which citizens participate. First of all, this is related to the relationships of citizens with socialist organizations, where many contractual terms are defined by acts of the plan. An administrative act in this area may be the basis of the origin of civil relations (an order for the contract for the rental of living accommodations). The plan element also exerts an important influence on those civil law relations between citizens which are formed under the influence of relationships in which socialist organizations participate;

c) The protection of social interests in the realization of civil rights. This principle is expressed in the law in general in Article 39 of the Constitution of the USSR, which declares that the realization of civil rights must not cause harm to the interests of society and state. This principle is expressed in Article 5 of the Principles and Article 5 of the Civil Code, which provides for the protection of civil rights, except in those instances when they are exercised in conflict with the purpose of these rights in a socialist society. The great significance of this principle is underscored by the fact that V.I. Lenin frequently pointed out the necessity of its inclusion in the Soviet civil code.[4] In this way, the correct correlation between personal and social interests is ensured;

d) The protection of personal interests. Article 56 of the Constitution provides for the protection of the personal life of citizens, and Article 57 states that respect for the individual is the duty of all government organs, social organizations, and officials. Civil law carries out the protection of the individual to an ever greater degree, and not only in regard to property rights. To the sphere of influence of civil law also belong personal rights of citizens not connected with property rights, which nevertheless require protection and may be defended through means of civil law, e.g. through the recognition and restoration of the situation existing prior to the violation of a right, the prevention of actions violating a right, and so forth. The process of the widening of the protection of such rights through the methods of civil law will continue;

e) Taking into account, in the realization of civil rights, the interests and legal rights of the other party. This is the

4. V.I. Lenin, *Complete Collected Works*, vol.44, p.398, 412; vol.54, p.169.

expression in civil law of a general principle secured in Article 39 of the Constitution of the USSR. The effect of this principle appears in many provisions of civil law. Thus, Article 168 of the Civil Code stipulates that each party must fulfill its obligations by means most economical to the socialist economy, and to render to the other party all possible assistance in fulfilling of obligations. In accordance with this principle, the degree of liability of the debtor may be lessened if the creditor intentionally or negligently contributed to an increase in the amount of damage inflicted upon him by the other side, or did not take measures to reduce it (Article 224, Civil Code).

7. *The system of Soviet civil law is its structure, the composition of its individual institutes and norms in their defined sequence.* The system of civil law exists objectively, as it reflects the actual social relationships which constitute the subject of that branch. The system of law receives its expression in legislation, primarily in acts of codified nature, such as the Principles or the Civil Code. It is also expressed in the discipline of law and in the educational process, which adhere to this system, making easier the study and teaching of civil law.

Soviet civil law includes two large groups of institutions and norms: the general and special (particular) parts. The general part contains provisions which have significance for all or many of the other institutions of the special parts of civil law.

The general part contains the institutions and norms which define the subject of civil law and the means of defending civil rights. Here also are entered such norms as are applicable to the participants in civil law relationships – the subjects of civil rights (citizens and juridical persons); to an important source of the origin and termination of civil relations – transactions. Further comes the institution used to complete transactions: representation and the most widely used of its bases – letters of authority. Finally, general significance is held by rules on periods of limitation, in part for the enforced carrying out of subjective rights [*sub"ektivye prava*] through the court, *arbitrazh* or an arbitration tribunal.

The special part of civil law includes a large number of institutions. Here is included, first of all, the law of ownership, which appears as the premise and result of many other legal property relationships, above all contractual ones.

The next division of the special part is the law of obligations, which expresses the dynamics of property relationships. The law of obligations is divided into several general provisions (the origin and termination of obligations, and performance of obligations and securing performance, and responsibility for their violation). Further come the individual types of obligations, the majority of which originate in contract, but some – from unilateral transactions which cause harm or from other grounds. This is the broadest section of civil law.

The special part also includes a group of norms which is related to the objects of creative activity (copyright, the law of discovery, the law of invention). These norms are directed toward the increase and wide use of things of spiritual importance, the protection of the rights of authors, inventors and rationalizers, and, in the final analysis, they serve to round out the development of personality, to raise the level of the economy, and to educate mankind.

Finally, the last group of civil law norms deals with the legal consequences connected with the death of a citizen, when questions about the succession to property and of the other rights of the deceased occur (the law of inheritance).

It is necessary to differentiate the system of civil law from the inner structure of particular normative acts. The inner structure of the most important codified acts of civil law – the Principles and civil codes – reflects the system of civil law. But the structure of particular normative acts may have special characteristics. For instance, the Civil Codes of the union republics are richer in their content than the Principles, which is most of all apparent in the section on the separate aspects of obligations. There also exist some structural differences between the civil codes of the union republics.

The system of civil law as a whole is wider than the structure of the codified acts. It includes institutions which do not exist in the codified acts, for example, contracts for the carrying out of scientific-research and experimental-construction work in the areas of design and energy-supply. This shows the breadth of civil law and the process of its development. However, these differences do not affect the unity of the system of Soviet civil law, which is expressed in its totality in codified acts: The Principles of Civil Legislation and the civil codes of the union republics.

Chapter 2. Sources of Soviet Civil Law

1. *Concept and aspects of sources.* After a study of the object, principles, and system of civil law, it is essential to elucidate where and in what form those legal norms, which in their totality form civil law, receive their expression. It is customary to designate the forms of expression of civil law norms by the term "*sources of civil law*". There are two sources of civil law: normative acts[1] and custom. In the USSR custom is not widely applied and has the character of a subsidiary source of civil law.

The USSR is a federal state which consists of sovereign states – the union republics. Therefore, the sources of Soviet civil law consist both of all-union as well as republic normative acts. In their totality they constitute the entire body of Soviet civil legislation.

The development and improvement of civil law in our country is directed by the Communist Party. In accordance with the decisions of the XXV Congress of the CPSU, work on a *Digest of Laws* is being carried out which will make our laws more accessible to Soviet citizens. One of the chapters of the *Digest of Laws* will contain the basic normative acts in the area of civil law. At the XXVI Congress of the CPSU it was noted that the revision of Soviet legislation, carried out on the basis of the Constitution, was of great use and work on the improvement of our legislation will continue.[2]

A special place among the sources of civil law is held by the fundamental laws of the Soviet state – the Constitution of the USSR and the constitutions of the union republics. In them are secured: the leading role of socialist ownership of the means of production in the economic system of the USSR; protection by the state of the personal property of citizens based on earned wages; the carrying out of the management of the economy on the basis of state plans for economic and social development; the equality of citizens of the USSR in all

1. Normative acts are also called by another term – legislative acts, although the former is more prevalent.
2. See: *Materialy XXVI s"ezda KPSS*, p.64.

spheres of life. All normative acts in the area of civil law are based on these constitutional principles and adhere to them strictly.

Civil law includes normative acts adopted by various organs of state authority and administration. Here are included the Laws of the USSR and the Edicts of the Presidium of the Supreme Soviet of the USSR, the laws of the union republics and the edicts of the Presidia of the Supreme Soviets of the union republics, decrees of the Council of Ministers of the USSR; decrees of the Councils of Ministers of the union republics, orders and instructions of all-union and republic ministries and departments, and decisions of executive committees of the Soviets of people's deputies. Each state organ is allocated a specific sphere of influence within which limits it has the right to promulgate normative acts related to its authority.

The number of normative acts within the area of civil law is quite large, giving rise to the question of how to attain unity within such a broad area of legislation. In law there exists the principle of coordination, in light of which an act of a lower state organ may not contradict an act of a higher organ. In addition to this, there is being carried out systematic work on the improvement of civil legislation and the publication of internally-linked systematized normative acts which are called codified acts. This ensures the unity and agreement, both in the content as well as in the form, of all-union and republic acts of civil legislation.

2. *Laws of the USSR and edicts of the Presidium of the Supreme Soviet of the USSR, laws of the union republics, and edicts of the Presidia of the Supreme Soviets of the union republics.* The most important laws in the area of civil law are the Principles of Civil Legislation of the USSR and the Union Republics, as well as the civil codes of the union republics. These are acts of a codified nature, containing the system of norms of all the basic institutions of civil law.

The Principles were adopted on 8 December 1961 and went into effect on 1 May 1962. They contain a preamble, which explains the significance and tasks of Soviet civil law, and 129 articles. The Principles consist of eight parts: I) General Provisions; II) Law of Ownership; III) Law of Obligation; IV) Copyright; V) Law of Discovery; VI) Law on an Invention, Rationalization Proposal, and Industrial Model; VII) Law of Inheritance; and VIII) Legal capacity of foreign citizens and stateless persons, and Application of civil laws of foreign states and international treaties.

The Principles give the definition of the object of civil law and establish its system (see Chapter 1); there are also provisions for the competence of the Soviet Union and the union republics to promulgate normative acts in the sphere of civil law. Questions of competence are discussed in Article 3 and in a number of other articles in the Principles.

In accordance with Article 3 of the Principles, all-union legislation regulates relations among socialist organizations with regard to the delivery of goods, capital construction, the procurement of agricultural products, transportation (except by motor-car), accounting and credit, relationships with regard to state insurance, inventions, and several others. Other relationships are regulated by legislation of the Soviet Union and of the union republics, and in some instances, by civil legislation of the republics.

In addition to this, the Principles contain the basic positions on all the most important institutions of civil law. These positions of the Principles are developed in the Civil Codes of the union republics and in other normative acts of civil law.

The first Civil Code of the RSFSR was adopted in 1922. In the wake of the development of the Principles, new civil codes were adopted in all the union republics in the years 1963-1964. At the present time in the RSFSR the Civil Code is in effect which was adopted on 11 June 1964 and went into effect on 1 October of that year.[3]

All the civil codes of the union republics are arranged on the basis of the system established by the Principles. They

3. Henceforth, all references to the Civil Code (CC) will be to the Civil Code of the RSFSR. If there are essential differences in the civil codes of the other union republics, they will be specially referred to.

reproduce to the fullest extent the norms of the Principles and, at the same time, supplement and develop them. For instance, in the Principles there are 5 articles on the contract of purchase-sale, while in the Civil Code of the RSFSR there are 18 articles; on the work contract [*dogovor podriada*] there are 3 and 18 articles, and so forth.

In addition to this, in the civil codes of the union republics there are contained norms on those institutions of civil law whose regulation, according to the Principles, is within the competence of the union republics. For instance, in the civil code are contained norms on the contract of motor transport, on contract of barter, of gift, of loan, of agency, of deposit for safe custody, as well as of public competition. Therefore, the civil codes of the union republics contain a significantly greater number of articles than do the Principles. Thus, the Civil Code of the RSFSR contains 569 articles. Approximately the same number of articles is found in the civil codes of the other union republics.

The system of the Civil Codes of the union republics and their contents is fundamentally the same. However, the Civil Codes of several republics contain some differences both in the system of the structure, as well as in the contents of various norms. These differences are not significant and may be explained by the geographic particularities of the various republics and their traditions, as well as by a diverse approach to the regulation of specific civil law relationships.

Important sources of Soviet civil law are the recently adopted laws on housing law: the Principles of Housing Legislation of the USSR and the Union Republics of 24 June 1981, and the housing codes of the union republics, which were promulgated on the basis thereof.

Among edicts, sources of civil law are the Merchant Shipping Code of the USSR, approved by the the Presidium of the Supreme Soviet on 17 September 1968, and the edict of the Presidium of the Supreme Soviet of 2 October 1981 on the compulsory state insurance on property owned by citizens, etc.

3. *Normative acts of the Council of Ministers of the USSR and the Councils of Ministers of the union republics.* The Council of Ministers of the USSR is the highest administrative organ and in accordance with Article 131 of the USSR Constitution carries out the management of the national economy and of social-cultural development in our country. It has adopted a large number of important normative acts in the area of civil law.

The normative acts of the Council of Ministers of the USSR are issued in the form of decrees [*postanovleniia*], and acts on private matters in the form of orders [*rasporiazheniia*]. The most important acts are adopted jointly by the Central Committee of the CPSU and the Council of Ministers of the USSR.

Normative acts adopted by the USSR Council of Ministers which have general significance are usually in the form of statutes, rules, or charters [*ustavy*]. For example, the USSR Council of Ministers approved the Statutes on the socialist state production enterprise (1965), on production associations (1974), and on industrial associations (1973). Among other normative acts approved by the USSR Council of Ministers one may include the Statutes on contracts of delivery (1981), and the Charter of the railroads of the USSR (1964), etc.

Thus, the normative acts of the USSR Council of Ministers are concerned, first of all, with the regulation of civil law relationships between socialist organizations. However, many of its acts are devoted to the regulation of relationships in which citizens participate, for instance, the Statute on discoveries and inventions (1973).

The Councils of Ministers of the union republics, which carry out the management of the state economy and the social-cultural development of their specific republics, adopt normative acts on questions of civil law which are under the competence of the republic. They issue normative acts on contracts of motor transport, carriage, contracts on the hire of property, on housing construction cooperatives, and on a number of other issues.

4. *Normative acts of the ministries and departments of the USSR and union republics.* The ministries and departments issue normative acts in the form of orders and instructions on questions of civil law which lie within their competence. This competence is established by statutes of the pertinent ministry (department), by laws or decrees of the USSR Council of Ministers and the Councils of Ministers of the union republics.

The normative acts of a number of ministries of the USSR (Ministry of Finance, Ministry of Trade, Ministry of Communications, ministries of transportation and several others) have general applicability and must be observed by all citizens as well as by organizations which are not part of the system of these ministries. Examples of this are the rules for carrying out of merchandising operations in stores, and rules on the transport of freight and passengers. Normative acts of the majority of all-union departments also have general applicability: [the acts] of the State Planning Committee of the USSR (*Gosplan*), of the State Central Construction Office of the USSR (*Gosstroi*), of the State Committee on the Supply of Technical Materials of the USSR (*Gossnab*), and of the State Committees on Science and Technology and on Inventions and Discoveries.

The state banks of the USSR, which issue instructions on questions of credit and accounting (the State Bank [*Gosbank*], the Construction Bank [*Stroibank*], Foreign Trade Bank [*Vneshtorgbank*]), form a special group of all-union departments. These instructions are applicable to all organizations and citizens when they carry out the corresponding banking operations.

Some ministries and departments of the USSR issue normative acts which are not generally applicable but are rather applicable only to organizations and employees of that ministry.

Normative acts in the area of civil law are also issued by ministries and departments of the union republics if the resolution of certain questions falls within their competence. For instance, the Ministry of Housing and Communal Services of the RSFSR issues a number of normative acts on the contract for the rental of housing, and the Ministry of Motor Transport of the RSFSR issues acts on the contract of motor carriage.

5. *Normative acts of other organs.* Normative acts of several other organs also may be sources of common law. Thus, executive committees of the Soviets of people's deputies may adopt decrees on questions within their competence, for instance, on the procedure for the use of housing accommodations, on conveying freight in city transport, and several others.

In instances established by law, organs of cooperative and social organizations may also issue normative acts: such organizations as the Soviet Central Union of Consumer Societies (*Tsentrosoiuz*), the Central Committee of the All-Union Order of the Red Flag Volunteer Society of the Army, Air Force and Navy of the USSR (*DOSAAF*), the All-Union Central Trade Union Council (*VTSPS*) and others. This type of normative act usually regulates civil law relationships arising within the framework of a given cooperative system or social organization.

6. *Custom and rules of socialist communal life.* Custom as a source of civil law is a rule of behavior which is formed as a result of lengthy usage and has received the protection of the state. Among those social customs which have spread into our way of life and are part of socialist morality, legal custom distinguishes itself by the fact that it is protected by the state and failure to observe it results in the right to turn to a court or other state organ for redress.

Under the conditions of a socialist state, the wide use of custom as a source of law is not necessary, and legal custom is rare in the USSR. For example, in accordance with Articles 134 and 151 of the Merchant Shipping Code of the USSR, the time limit during which freight must be loaded into and unloaded from a vessel is decided by agreement among parties, or, in the absence of such an agreement, the time limits generally observed in the port where the loading and unloading takes place. These time limits are set down in the Collection of customs of the specific port.

In Article 59 of the USSR Constitution and in a number of articles of the Principles and the Civil Code, reference is made to the duty of citizens and organizations to observe the rules of socialist community life. In accord with Article 5 of the Civil Code, "in the exercise of rights and the performance of duties, citizens and organizations must observe the laws and respect the rules of socialist community life and the moral principles of a society building communism". The rules of socialist community life are not sources of civil law; they are part of socialist morality.

The rules of socialist community life play an important role in applying civil law acts: they are used in the evaluation of the conduct of people and organizations and, at the same time, aid in the correct application of legal norms.

7. *Significance of court and* arbitrazh *practice.* When civil law disputes involving citizens or cooperative farms arise, they are decided by courts, but disputes between organizations are reviewed by organs of *arbitrazh.*[4] What is the legal significance of the decisions of the court or *arbitrazh* and are they a particular source of civil law?

The last question must be answered in the negative. In the first place, the task of courts and of *arbitrazh* lies in resolving civil law disputes if they cannot be regulated by the parties themselves rather than to pass new norms of law. Secondly, each dispute has its specific character and is different from other disputes, even though they may bear a superficial resemblance to each other. From this it follows that is is impossible to be guided by the decision concerning one specific dispute in settling other, albeit analogous disputes.

Nevertheless, the work of the court and *arbitrazh* organs

4. The activities of courts and *arbitrazh* are studied in the course "Soviet Civil Procedure".

has important significance. It is important above all for the understanding it gives of the norms of civil law and their correct application. The work of the court and *arbitrazh* tribunals also supplies needed material for the improvement of the norms of legislation in force, and aids in the preparation of new normative acts.

Especially significant are the decrees of the plenums of the Supreme Court of the USSR and the Supreme Courts of the union republics. The plenums of these courts give guiding explanations to courts on questions concerning the application of legislation in the review of court cases. The State *Arbitrazh* of the USSR and the union republics are required by law to give instructive orders on questions of the applicability of legislation in the decisions of disputes between organizations.

In some instances, the legislation in force charges the Supreme Courts and State *Arbitrazh* to affirm rules of a general nature. For instance, the State *Arbitrazh* of the USSR affirms the rules on the method for accepting delivered goods. This type of document is a form of normative act and is, of course, a source of civil law.

8. *Effect of norms in time, space, and on range of persons.* All normative acts have defined limits of effect, of which it is important to have knowledge for their correct application. Especially significant is the effect in time of a civil act. This is the period from the time the law goes into effect to the time when it loses force.

Sometimes, the date of its effectiveness appears in the act itself, and it is usually fixed at the beginning or middle of the month, which aids in the practical implementation of given act. If such a date is not established, laws of the USSR, and edicts and decrees of the Presidium of the Supreme Soviet of the USSR of a general nature enter into force over the entire territory of the USSR at the same time, ten days after their publication in the *Gazette of the Supreme Soviet of the USSR*[5]

5. Edict of the Presidium of the Supreme Soviet of the USSR, adopted 19 June 1958, in the redaction of 6 May 1980. (*Gazette of the Supreme Soviet of the USSR* 1980 No.2 item 374).

or in *Izvestiia*. The same sequence is established by the union republics for analogous republic acts. The ten-day period gives the time necessary for the study of a newly-adopted act.

If the date of implementation is not given, normative acts of the Council of Ministers of the USSR go into effect from the moment of their adoption.[6] The same rule holds true for normative acts of the Council of Ministers of the union republics, of ministries and departments, as well as of executive committees of local Soviets.

As a rule, civil law normative acts take effect only in future time, that is, they do not have retroactive force. However, in rare circumstances, a law may have retroactive force. For example, when the Civil Code of the RSFSR went into effect in 1964 it was decided that its norms would be applied to relationships which had arisen earlier.

As a general rule, normative acts are in force until their repeal or until the norms are replaced by norms contained in newly-adopted normative acts. In this latter situation, the earlier act is not always repealed and it may not be clear which norm is in effect. This vagueness is resolved through the interpretation of legal norms. (See section 10 of this Chapter.)

The effect of civil law norms in space is the effect of legal norms within a specific territory. This is determined by the organ which adopts the act.

Normative acts adopted by all-union organs are effective over the entire territory of the USSR, acts of republican organs over the territory of a specific republic, and acts of executive committees of local Soviets within the limits of the specific province (*oblast'*), district (*raion*), or city. However, sometimes the sphere of effect of a normative act may be limited to other boundaries; for instance, some acts concerning the contract for the rental of housing apply only to the districts of the Far North.

6. Decree of the Council of Ministers of the USSR of 20 March 1959 (*Collection of Decrees of the USSR* 1959 No.6 item 37).

In accordance with Article 74 of the USSR Constitution, in instances of a discrepancy between a law of a union republic and an all-union law, the law of the USSR prevails. Instances are also possible when civil relationships originate on the territory of one union republic and are carried out on the territory of another republic.

The question of which republican legislation prevails in these circumstances is decided in Article 18 of the Principles. In part, in relationships which arise from the law of property, the applicable law is that of the place where the property is located; for obligations in transactions, the law of the place where the transaction was concluded; for inheritance, the law of the place where the inheritance is opened.

The effect of civil law acts on a circle of persons is contained in the definition of the subject of civil law whose relationships fall within the range of a given act. The majority of norms of the Principles and the Civil Code have general significance and extend over the relationships of organizations as well as of citizens. However, some of their norms are designated only for citizens (for instance, norms on the work agreement for capital construction). Some normative acts have a more limited range of effect and apply, for instance, only to commercial or transportation organizations.

9. *Imperative and dispositive norms.* The norms contained in normative acts may be imperative (mandatory) or dispositive (discretionary). This division of norms is of great practical significance.

The rules of imperative norms may not be changed by the agreement of citizens and organizations. An example of this type of norm is the rule of limitation of actions which provides, according to Article 80 of the Civil Code, that "no change in the periods of limitations or the calculation of such periods is permitted by the agreement of the parties".

In contrast, dispositive norms allow for a change in the norms by agreement of the parties, although if such an agreement did not exist, the rule contained in the dispositive norms is applied. For instance, according to Article 284 of the Civil Code, "the lessor is obliged to carry out at his own expense those capital repairs on the article hired unless other provisions are made by law or contract".

There are a significant number of dispositive norms in civil law. This may be explained by the multiplicity of economic activity and the various possibilities open to its participants. Under these circumstances, the parties in a civil law relationship may have the right – while keeping in mind the peculiarities of the specific economic operation as well as their own possibilities and requirements – to define their mutual rights and obligations concretely within the limits of the law.

However, sometimes it is difficult to define whether a given norm is imperative or dispositive, since the text of the norm may not include any instructions on this matter. In these situations, one must turn to the interpretation of the norm.

10. *Interpretation of civil law norms.* In order to apply legal norms it is necessary to understand their true meaning. In juridical language this is known as *interpretation*. The necessity for interpretation lies in the fact that legal acts are in force over a period of many years and during that period the social relationships which are regulated by them may undergo changes. The interpretation of legal norms may exist in various forms.

In terms of juridical force, there is a distinction between *obligatory* and *non-obligatory* interpretation. Obligatory interpretation is rendered by a state organ authorized to do so. If the interpretation is carried out by the organ which adopted the given norm, this interpretation is called *authentic* [*autentichnyi*]. Non-obligatory interpretation is that given by courts in the decision of concrete disputes, as well as scholarly interpretation (in literary sources, reports, etc.).

Depending on the method, interpretation may be grammatical, logical, systematic, or historical. Grammatical interpretation is an explanation of norms by the general rules of grammar. Logical interpretation utilizes in its aims of interpretation the rules and arguments of logic. If, for instance, a possible understanding of a norm leads to a clearly inapplicable result, that understanding of it may not be applied.

In a systematic interpretation, the meaning of a norm is established by comparing it with other norms and explaining the general and particular features of the various norms. Historic interpretation is based on knowledge of the history of the adoption of the norm by listing the factors resulting in the introduction of the norm, and a comparison of it with earlier, analogous norms.

As a rule, all the methods listed are utilized in the interpretation of a norm, because interpretation is a complex and responsible procedure, and in order to attain a correct interpretation it is necessary to utilize all [possible] measures.

As far as results are concerned, one may distinguish between the literal, the restrictive, and the extended (broad) interpretation. In the majority of cases, the interpretation of a normative act shows that the underlying meaning of the norm agrees with its verbal formulation to the fullest extent. This ought to be the case because the inner substance of the law and its verbal formulation ought to coincide. This interpretation of legal norms is termed the literal interpretation.

However, there are instances when the verbal formulation of a legal norm is shown to be wider than its underlying meaning. Here the restrictive interpretation comes to aid, since with its help it is possible to conclude that the legal norm must be understood and applied in a restrictive fashion.

For instance, a cursory acquaintance with Article 256 of the Civil Code, containing the definition of a contract for a gift, may give the impression that this contract is applied not only in agreements among citizens and between citizens and organizations, but also among socialist organizations.

However, in considering, by means of logical and systematic interpretation, that the activity of socialist organizations is based on reciprocity, it is necessary to conclude that the contract for a gift may not be applied between socialist organizations. This is an example of a restrictive interpretation.

There also occur circumstances when the interpretation of a norm leads to the conclusion that its real content is wider than the verbal formulation. For instance, Article 48 of the

Civil Code states that "a transaction which does not correspond with the provisions of the law is invalid." Logical and systematic interpretation of this norm leads to the conclusion that under the circumstances, by the term "law" is meant not only laws adopted by the Supreme Court but also other normative acts.

In spite of the multitude of sources of civil law, there occur situations which do not have proper legal regulation. These are usually new features of life or unusual and rare situations for which there are no provisions in legislation. However, even in these circumstances the rights of the affected parties must receive proper legal protection.

With this aim, Article 12 of the Principles of Civil Procedure of the USSR and Union Republics provides that "in the absence of law regulating disputed relationships, the court shall apply the law regulating similar relationships and in the absence of such a law the court shall proceed from the general principles and meaning of Soviet legislation". This rule also guides the organs of *arbitrazh*.

The resolution of a dispute on the basis of an act which regulates similar situations is commonly known as analogy of law, and the resolution of disputed questions arising from the general principles and meaning of Soviet legislation is termed analogy of legislation. Analogy of law and legislation occur rather infrequently in the practice of the application of law.

11. *Publication of the sources of civil law*. The lawyer must have quick access to the relevant legal norms necessary to decide problems occurring in life. For this, it is necessary to know where and how normative acts are published.

There exist official publications in which are found: Laws of the USSR and the Edicts of the Presidium of the Supreme Soviet of the USSR – *The Gazette of the Supreme Soviet of the USSR*; Laws of the RSFSR and Edicts of the Presidium of the Supreme Soviet of the RSFSR – *The Gazette of the Supreme Soviet of the RSFSR*. Decrees of the Council of Ministers of the USSR are issued in the publication titled *Collection of Decrees of the Government of the USSR* (abbreviated as *SP SSSR*). The decrees of the Council of Ministers of the RSFSR are published in *Collection of Decrees of the Government of the RSFSR* (abbreviated as *SP RSFSR*). Analogous publications exist also for the other union republics.

Normative acts of the ministries and departments are published in the monthly *Bulletin of Normative Acts of Ministries and Departments of the USSR*. The ministries of the USSR put out the *Collection of rules for transportation and tariffs* for the corresponding form of transport (railroad, sea, river).

The basic sources of civil law are published periodically in the form of collections arranged according to the system of civil law, which facilitates their use. The most recent to make its appearance is the collection *Civil Legislation: Collection of Normative Acts* (Moscow 1974). There are also collections of legislation on the contract of supply, on work contracts for capital construction, contracts for the rent of dwelling space, and a number of other issues.

Chapter 3. Civil Legal Relationships

1. *The concept of civil legal relationships.* The category of civil legal relationships is very important for an understanding of civil law and the mechanism of its action. If a norm of law is abstract and of a general nature, then a legal relationship is always concrete. It exists in that form in which the norm of law is realized. *A legal relationship* is a concrete legal connection between concrete subjects: it is expressed in the establishment of their mutual rights and obligations.

An example of a legal relationship is the contract of purchase and sale. It involves two parties – the seller and the buyer. The seller is obliged to give to the buyer the article bought and has the right to receive money for it; the buyer has the right to receive this article and has the obligation to pay money for it. There are special features with regard to the legal relationship concerning ownership, in which only one party, the owner of the article, is clearly designated. He has the right to possess the article, to use and to dispose of it. All other persons are under the obligation not to violate the right of the owner; this leaves undefined the range of liable parties.

What are the specific features of civil legal relationships? *In the first place*, a social relationship is the subject of a civil legal relationship – this is a property or personal non-property relationship. *Secondly*, the regulation of the relationships mentioned above is accomplished through the norms and methods of civil law.

As a rule, for the other branches of law the content of the legal relationship is fully prescribed in the norms of law. In civil law, the content of the legal relationship also depends on the will of the parties if the norm is dispositive in nature and does not prohibit the parties from making concrete their mutual rights and obligations or establishing them through its provisions. The content of civil law may also be defined by an administrative act, primarily an act of planning, which is obligatory for participants in a legal relationship.

Thus, *a civil legal relationship is a property or personal non-property relationship regulated by a norm of civil law which establishes or sanctions the mutual rights and obligations of concrete subjects – the parties of the legal relationship (or at least of one subject).* A civil legal relationship contains several elements, each of which demands consideration. Among those elements are the subjects, content, and object of a legal relationship.

The participants in a legal relationship may only be *subjects* of civil law. A special feature of the subjects of civil law is that they may be not only individual citizens, but also organizations which are called juridical persons when they enter into civil relationships. The state is a special subject of civil law. The state may act as a party in a civil legal relationship, for instance, in the inheritance of property. Several persons (plurality of persons) may appear on one side of a legal relationship (in the capacity of subjects of rights or obligations). The issue of the subjects of civil law (legal relationships) will be examined later in more detail (see Chapters 4-6).

The content of civil legal relationships is subjective rights and legal obligations. Furthermore, in legal relationships they are usually grouped in such a way that subjective rights belong to one party, while the other party bears the obligations corresponding to those rights. Let us examine the juridical essence of civil laws and obligations.

A subjective right is a right belonging to a specific subject; it must be distinguished from the norms of a law which embody a general rule of conduct not connected with a concrete subject. A civil right consists of the fact that the subject of the right, the participant in a legal relationship, is empowered by law to act in a certain manner. In other words, he either realizes or does not realize his subjective right at his own discretion. It sometimes happens that the realization of a subjective right may make the subject liable under another legal relationship (for instance, a claim for damages may be a right under a civil legal relationship but a liability under an administrative relationship), but this does not change the character of subjective rights.

The conduct of the bearer of a subjective right may consist of performing positive actions (for instance, the claiming of articles which belong to him under the law of ownership), demanding specific behavior on the part of the persons under obligation (for instance, the transfer of articles according to a contract) or in turning to competent organs (a court or *arbitrazh*) to take forcible measures against the person under obligation. However, the law does not only help the bearer of a subjective right, but also places him within defined limits beyond which it is not possible to go. A subjective right always has limits which are outlined by law.

Thus, a subjective right may be defined as a *means of possible behavior by a person empowered by law to act.* This behavior includes these possibilities: performing a positive action, demanding specific behavior on the part of a liable person, and turning to competent organs to take forcible measures against a liable person.

The legal obligations of the other participant in a civil legal relationship represent *the means of obligatory behavior of that person* (for instance, the buyer in an agreement of purchase and sale is obliged to pay money; a contractor is obliged to perform the work). In the event of a violation of this obligation, measures of state enforcement may be taken (a debt may be claimed forcibly, or work may be carried out at the expense of a careless contractor).

As a rule both sides in a legal relationship have rights as well as obligations, and the right of one side corresponds to the responsibility of the other. These types of relationships are called bilateral. An example of this is that of purchase and sale by which one side, the seller, is obliged to transfer an article and has the right to receive money, while the other side, the buyer, has the right to receive the article but is obligated to pay money. Unilateral agreements, under which one side has rights and the other obligations, occur more rarely. For example, in loan agreements, one side, the lender, has the right to demand the return of money, while the other side, the borrower, is obliged to return it. However, in thus classifying legal relationships it is necessary to keep in mind that in every legal relationship, unilateral as well as bilateral, there are never fewer than two participants. It is not possible to have a legal relationship consisting of only one party.

The subjective composition and distribution of the rights and obligations of the participants are divided into *absolute* and *relative*. In an absolute relationship only the side empowered by law is defined, and all other persons must refrain from actions violating the rights of the side empowered by law. On the other hand, in a relative legal relationship, all participants are defined, and the persons under obligation more often than not must perform positive actions. Examples of absolute legal relationships are the law of ownership and the law of copyright; examples of relative legal relationships are the rights and obligations in purchase and sale, property-hire, carriage, and other obligations.

An analysis of a legal relationship would be incomplete without examining the element of *object* [*ob"ekt*] on the grounds of which it is based. There are various objects of legal relationships, which are commonly divided into: 1) *articles* (constituting the object of the law of ownership, of the contract of purchase and sale, etc.) and other property (for example, the right to demand under a loan by contract); 2) *actions, results of work performed* (the carriage of goods or repairs on radio receivers under a work agreement); 3) *products of creative activity* (for example, a literary work or an invention); 4) *personal non-property valuables* (name, honor, and dignity).

The classification of objects has great practical significance because the character of the legal relationship and the extent of the right of the participants in a legal relationship depend on the type of object. Thus, the object is a criterion for dividing legal relationships into property and non-property relationships. Personal rights of a non-property nature, in contrast to property rights, are not assignable to another party; when they are violated, compensation is not in the form of property; limitation of actions is not applied in claims for the defense of personal rights.

2. *Juridical facts and their classification.* A concrete legal relationship arising on the grounds of a norm of law occurs only under specified conditions. Only if they are present do concrete parties have mutual rights and obligations and thus establish the legal relationship. The conditions by which the law links the existence, change, or termination of a legal relationship are called *juridical facts.*

Article 4 of the Civil Code lists those juridical facts which are most important in civil law. First of all, juridical facts are divided into *events* [*sobytiia*] and *actions* [*deistviia*]. *Events* are circumstances whose occurrence does not depend on the will of the people. They include events which are entirely independent of the human will such as natural occurrences (earthquakes, floods, and other natural catastrophes) as well as occurrences of a social character (epidemics, war). Human death also falls into the category of events.

Actions, on the other hand, consist of happenings whose occurrence is a result of human will. It is important to note that the category of actions also includes inaction, insofar as this is also indicative of will. Thus, applying Article 509 of the Civil Code, an author who concludes a contract for the use of his work by a publisher does not have the right to give his work to another organization to be used in the same fashion.

The division into events and actions holds important practical significance. In accordance with Article 4, only those events which are directly specified by law may be considered juridical facts in civil law. In considering actions as juridical facts, the law gives only an approximate, rather than an exhaustive, listing. Other actions for which there are no provisions in law may also give rise to civil rights and obligations should this develop from the general grounds and meaning of legislation. The absence of an exhaustive listing of actions as juridical facts is an important feature of civil law.

Actions are extremely varied. First of all, they may be divided into lawful and unlawful. *Unlawful actions* which break the requirements of law may also give rise to civil law consequences. Among such unlawful actions are those causing harm to another person (torts); they also include the acquisition or retention of property at the expense of someone else without proper grounds. *Lawful actions* are divided into *legal conduct* and *legal acts. Legal conduct* is a legal action which is not directed at achieving legal results but does in fact lead to that according to law. Thus, the creation of works of literature, science, or art results in copyright although the creator did not strive for this. It is a characteristic of legal acts that they lead to legal results.

Legal acts are divided into *administrative acts* and *transactions. Administrative acts* are lawful actions directed at legal results but emanating from a government functionary. Thus, the juridical facts in legal relationships concerned with housing emanate from an order given out by the executive committee [of a local Soviet]. In civil legal relationships among socialistic organizations, the administrative acts which give rise to civil legal relationships are the acts of planning. For instance, the monthly plan for the carriage of goods by rail is a juridical fact. The role of acts of planning is defined by the plan character of the socialist economy.

A special place in the system of legal acts belongs to court and arbitration decisions. Legal relationships arise, change, and terminate on the basis of these decisions, in part during the process of enforced execution [of an action].

When lawful actions, which are meant to result in certain civil legal relationships, emanate from the participants in civil legal relationships they are called transactions. Transactions are the most prevalent bases for the origin, change, and termination of civil legal relationships. The most important variant of transaction is contract, e.g. an agreement of will of two or more people. A detailed classification of transactions is given in Chapter 7.

Usually only one juridical fact is sufficient for the origin, change, or termination of a juridical fact. But quite often such results occur as a result of a definite combination of juridical facts. Such a combination is known as a factual composition, which includes the totality of juridical facts necessary to give rise to legal consequences. Thus, the right of an author of a literary work to compensation results from the factual composition which includes, first of all, conduct (the creation of the work), secondly, transaction (the contract for the use of the work), and, thirdly, actions concerning the use of the work (publication, performance on stage, etc.).

3. *Realization of civil rights and their guarantee. Execution of obligations.* The realization of a civil right consists in carrying out those actions constituting the content of the subjective right which has arisen. For this it is necessary that the other side, that is, the liable party, execute its obligations either voluntarily or under compulsion.

The realization of subjective rights is supported by a number of legal guarantees established by civil legislation. By way of example of an important legal guarantee, Article 6 of the Civil Code provides for various methods for the protection of civil rights, to which the bearer of a right may resort if an obligation is not executed voluntarily and his rights have been infringed (see section 4).

In addition to this, the actions of a person empowered by law in the realization of his subjective rights are placed within fixed limits which are provided for in Article 5 of the Civil Code. This norm establishes that it is not possible to have subjective rights which contradict the interests of socialist society and the tasks which it sets for itself. It is also not permissible to realize rights which violate the rights and lawful interests of another party, the principle of socialist cooperation, or the mutual benefit of the parties. In other words, misuse of law is not allowed.

The character of the actions in the realization of rights is evaluated by taking into account the rules of socialist community life and the moral principles of our society (para. 2, Article 5 Civil Code), that is, taking into account not only the legal but also the moral norms reflecting the foundations of Soviet society. For instance, in deciding some housing disputes, the observance by tenants of the rules of socialist community life are taken into consideration.

Besides Article 5 of the Civil Code, there are a number of special norms having the same aim. Thus, it is not permitted to use personal property to gain unearned income (para 3, Article 104 Civil Code), or to fail to take proper care of a house or cultural treasures (Articles 141, 142 Civil Code), which under these circumstances may be given over to [the care] of the state. Article 5 of the Civil Code is applied only in the absence of such special norms.

The performance by the liable party of those actions representing his legal obligation, which are in the interest of the party empowered by law, or refraining from action, constitute the execution of liability. The liable party should execute his liabilities voluntarily. In instances where liabilities are not discharged voluntarily, the party empowered by law may turn for defense of his right to a court or *arbitrazh* which will make a decision whether to compel the liable party to execute his obligation or to subject him to other compulsory measures.

4. *Procedure and means of the defense of civil rights.* The procedure for the defense of civil rights is defined by the consideration of the equality of the parties in civil legal relationships, and disputes are reviewed by an organ before which both parties have equal standing. In accordance with Article 6 of the Civil Code, the defense of civil rights is carried out by a court, *arbitrazh*, or arbitration tribunal and only in instances especially set forth in law may such defense be assigned to an administrative organ.

The court considers disputes in which the participants are citizens, collective farms, and inter-collective farm and state collective farm associations, organizations, and their associations, disputes in which foreign citizens are participants, and certain other special categories of disputes which are set forth in law (Article 25 Code of Civil Procedure). In instances designated in the Statute on Comrades' Courts, disputes between citizens are reviewed by comrades' courts.

In accordance with Article 9 of the Law on State *Arbitrazh* of 30 November 1979,[1] disputes between socialist organizations

1. *Official Gazette of the Supreme Soviet of the USSR* 1979 No.49 item 844.

(with the exceptions noted above) are considered by *arbitrazh* (state or departmental). The decisions of *arbitrazh* have an obligatory character and are subject to compulsory execution. In order to prevent unnecessary trials in *arbitrazh* and to regulate relationships peacefully, it has been provided that before a suit is brought to *arbitrazh* the claims must be shown to the other side.

Citizens may transfer civil law disputes between them to a chosen arbitration tribunal, which is formed under the supervision of the parties and which consists of one or more judge. If a decision of a chose arbitration tribunal is not executed voluntarily, it may be subject to execution under compulsion. The By-law on the Chosen Arbitration Tribunal comprises Addendum No. 3 of the Code of Civil Procedure.

The defense of civil rights by administrative procedure is allowed only in instances especially set forth in law. In accordance with the Edict of the Presidium of the Supreme Court of the USSR of 14 March 1955, disputes between state, cooperative (except collective farms), and other public organizations involving sums up to 100 rubles are decided by administrative procedure.

Article 6 of the Civil Code lists several methods for the defense of civil rights:

a) the violation of a right may consist of the negation of its existence or the assertion that the right belongs to another person. Under these circumstances, the establishment of the existence of the right is a necessary precondition of its realization. Therefore, the law lists as the first means of defense the *declaration* of the right by the organs for the defense of rights which were noted above. An example might be a declaration of authorship in the case of a dispute concerning the authorship of an invention or literary work;

b) another means of the defense of a civil right lies in the *restoration of the same situation as existed before the violation.* An example of this is the return to the owner of property which is illegally in someone else's possession or the repair of a damaged article by the liable party;

c) the restoration of a previously existing situation leading to the *recovery of damages* caused by the violation of a right. In some instances, the reimbursement of damages represents the restoration of the situation prior to the violation. When such restoration is not possible (for instance, when an article is ruined), the situation is resolved through monetary reimbursement for the harm caused. To this group of measures is related the recovery of forfeiture (in instances set forth in law or by contract), which is also a form of monetary compensation;

d) the meaning of the next measure of defense – *the specific performance of an obligation* – depends on the fact that under the conditions of the socialist economy a duty must be performed as such, and to substitute monetary compensation for it does not result in the solving of the problems of the nation's economy or the satisfaction of the interests of its creditors.

The specific performance of a duty is set forth, for example, in Article 217 of the Civil Code, when the obligation consists in the handing over of a specific, defined article. However, application of this measure is not always possible, for instance, when the obligation consists in the execution of a specific job or in providing a particular service to a citizen;

e) if the violation is of continuous character, the method of the defense of the right is the *termination of the actions* violating the right. This type of defense may be undertaken when an owner faces difficulty in using his property (the entrance to his garage is obstructed);

f) in some instances the method of defense of a right consists in *terminating a legal relationship or changing it (modifying it)*. Thus, Article 246 of the Civil Code provides that in the sale of a specified article the buyer may either abrogate the contract (terminate the relationship) or pay a lesser price (change in the legal relationship).

The law may also provide other methods for the defense of a right. Thus, Article 284 of the Civil Code establishes the right of a tenant to repair the property himself and to charge the expense to the landlord if the latter does not carry out capital repairs for which he is liable.

The most universal method of defending civil rights is the claiming of damages. This method is set forth in Article 219 of the Civil Code in case of the non-performance or improper performance of any obligation by the liable party. Article 221 of the Civil Code is in accord with this, stating that compensating for damages does not free the liable party from the specific performance of the obligation. All other methods of the defense of civil rights are applied only in instances especially set forth in law or contract.

5. *Defense of the honor and dignity of citizens and organizations.* The defense of the honor and dignity of citizens or organizations, which represent a non-property value not linked with property relationships, is realized by a special method. This method of defense depends first of all on the special character of the object of defense itself.

An introduction to the civil law protection of honor and dignity is defined by the significance accorded to the defense of the personality in all its manifestations. Criminal law defense is not sufficient; the condition for its application, in accordance with Article 130 of the Criminal Code of the RSFSR (Crim.C), which makes provisions for responsibility for slander, is the "deliberate" spreading of false, malicious fabrications about someone, that is, with "deliberate intent". But the defense is also concerned about non-intentional and even innocent circulation of information which is ruinous to honor and dignity. In addition to this, criminal law does not provide for the retraction of the circulated information and this is the basic concern of the defamed party.

According to Article 7 of the Civil Code, the following conditions have to be present for the defense of honor and dignity: a) circulation of information which has factual, not value, character; b) is ruinous to honor and dignity in its essence; c) does not correspond to reality.

Information is considered to be circulated if it is disseminated by a citizen or organization orally or in writing (including in print) to an undefined range of persons, several persons or one person (but not only to the defamed party). Information is defamatory which brings harm to a citizen or organization in the opinion of society or in the opinion of several individuals from the point of view of law, the rules of socialist community life, and the principles of communist morality.

The burden of proving the truth of the circulated statements lies on the parties which circulated it, e.g. the presumption exists that they do not correspond to reality. If this is not proved, the court must make a decision on the *retraction* of the false information. This is the essence of the civil law method for the defense of honor and dignity, and it is of restorative character.

The court establishes the method of retraction, but if the information is disseminated in print, it must also be retracted in print. If the information is contained in a document emanating from organizations, this document must be changed. In case of non-fulfillment of the court's decision on the retraction of defamatory information, the court has the right to impose a fine on the offender recoverable to the benefit of the state. But the payment of a fine is not grounds for relief from carrying out the decision of the court.

Chapter 4. Citizens as Subjects of Civil Law

1. *Concept and content of the civil legal capacity of citizens. Civil legal capacity is general legal capacity, the juridical capacity to have civil rights and obligations.*

According to Article 9 of the Civil Code, the capacity to have civil rights and obligations (legal capacity) obtains for all citizens of the RSFSR and the other republics of the USSR in equal measure. Every citizen of a union republic is a citizen of the USSR (Article 33 USSR Constitution). The equality of civil legal capacity of all Soviet citizens reflects the principle affirmed in Articles 34 and 35 of the USSR Constitution concerning the equality of all citizens of the USSR regardless of origin, social or property situation, racial or national affiliation, sex, education, language, religious affiliation, type and character of work, place of residence, or other conditions. The equality of citizens of the USSR is secured in all areas of economic, political, social, and cultural life (para. 2, Article 34 USSR Constitution).

The content of the legal capacity of citizens is generally defined in Article 10 of the Civil Code: citizens may in accordance with the law possess personal property, use living quarters or other property, inherit and make wills, choose an occupation and place of residence, enjoy rights of copyright to a work of science, literature or art, discovery, invention or rationalization proposal; and also have other property and personal rights. This formulation of law does not give an exhaustive, but only an approximate, listing of those rights and their corresponding obligations which a citizen may possess. At the same time, strictly speaking, the rights listed here go beyond the limits of civil law and its object. Thus, the right to the choice of living quarters and the freedom to move are regulated by administrative law, and the choice of an occupation and the realization of labor activity are regulated basically by labor law. However, both these elements of legal capacity also have great significance for the origin and realization of civil rights, such as those to personal property arising on the basis of earned income or savings, both the personal and property rights of an author or inventor, the right to use living quarters, etc., and this explains their inclusion within the framework of civil legal capacity.

At the basis of civil legal capacity are those rights and freedoms of citizens which are provided for in the USSR Constitution and legal capacity is called upon to aid in their realization. Articles 13 and 17 of the USSR Constitution have special significance. According to Article 13, the basis of personal property of citizens is earned income. Under personal property are included articles of everyday use, personal use and comfort, for use in a subsidiary household plot, a dwelling, and savings from labor. The personal property of citizens and the right to inherit it are protected by the state. Citizens may have the use of plots of land, granted by legal procedure, for the managing of a subsidiary household (including the keeping of cattle and fowl), gardens or orchards, and also for individual housing construction. Property under the individual ownership of citizens or under the personal use by citizens may not be used to derive unearned income or to the detriment of society. According to Article 17 of the USSR Constitution, the labor activity allowed by law in the USSR lies within the sphere of crafts and trade, agriculture, and serving the everyday needs of the population, as well as other forms of activity based exclusively on the personal labor of citizens and members of their families. The state regulates individual labor activity, securing its use for the interests of society.[1]

Articles 13 and 17 of the USSR Constitution permit making more precise the limits of civil legal capacity and defining the borders for the realization of the right to personal property and other civil rights. They serve as orientation in judicial practice in deciding on questions of which concrete civil rights and obligations may or may not belong to citizens in

1. See: Statute on Handicraft and Artisan Trade of Citizens, affirmed by decree of the Council of Ministers of the USSR of 3 May 1976 (*Collection of Decrees of the USSR* 1976 No.7 item 39), and the instruction of the Ministry of Finance of the USSR on the procedure for its application of 18 June 1976 (*Bulletin of Normative Acts of Ministries and Agencies of the USSR* 1977 No.1).

this or that concrete circumstance, and whether they coincide with the aims and purpose of civil legal capacity in a socialist society.

The most important characteristic feature of civil legal capacity in the USSR is the fact that its practical realization is guaranteed not only through juridical means – by the Constitution and other laws – but also by the material conditions of life in a socialist society; by its economic system, by socialist ownership of the means of production which is free from the exploitation of labor, the realization of the principle of socialism, "from each according to his ability, to each according to his work", and by the absence of unemployment.

In the USSR, civil legal capacity is indivisible from the personality of a person. It originates at the birth of an individual, belongs to him throughout his life, and terminates with his death. It does not depend on the physical or psychological state of health of a person nor on the fact whether or not a person is able to realize it himself. The newly born, the mentally ill, or the retarded possess civil legal capacity to the same degree as does an adult healthy person; like the healthy adult, they may become subjects of various civil rights (the right to inheritance, right of property, right to use of housing, copyright, etc.). It is another matter that they may not be able to realize their rights themselves, but this will be dealt with later.

In some instances the law establishes the protection of future, as yet not existing legal capacity, the protection of the interests of a yet not existing subject of law. Thus, in accordance with Article 530 of the Civil Code, children conceived during the lifetime of an individual leaving an estate but born after his death may be named as heirs. Sometimes, as an exception, the law may link the origin of defined spheres of civil legal capacity with the attainment of a definite age: one may become a member of a housing-construction cooperative only at age 18 (para. 6 of the Model Charter of a Housing-Construction Cooperative).

The indivisibility of legal capacity from the personality who bears it appears, in part, in the fact that a person may not of his own will limit his own legal capacity. Any transaction directly or indirectly aimed at limiting legal capacity is invalid (Article 12 Civil Code). This means that if someone carries out a transaction according to which he is bound not to make a will, or not to change an already existing will, or never to take ownership of a house or any other rights or obligations, the transaction is invalid.

No one may be limited in legal capacity other than under circumstances and by procedure established by law (Article 12 Civil Code). Full deprivation of a person's legal capacity is generally not provided for in law. However, in exceptional circumstances, the deprivation of certain elements of a person's legal capacity is allowed; for instance, for persons who by the decision of a court are undergoing punishment for committing a crime (Article 8 Principles of Corrective Labor Legislation of the USSR and the Union Republics). The deprivation of freedom means not only the loss of freedom of movement and choice of place of residence for a period of time established by the court, but at the same time subjects him to a specified regimen established by law. These persons are limited in the right (legal capacity) to lay claim to property for personal ownership or for use regardless of the place of the loss of freedom; to choose a type of work and in several other property and non-property rights. In addition to this, criminal law provides for such forms of punishment as exile, banishment, assignment to corrective labor tasks without deprivation of freedom, conditional sentence to deprivation of freedom with mandatory assignment to labor, deprivation of the right to occupy specific offices or to engage in specific activities (Articles 21 and 24-2 Criminal Code) which lead to the limitation of freedom to choose a type of work and place of residence. These limitations may be established only for a specific time set by law.

It is necessary to distinguish the limitation of legal capacity from the deprivation of a citizen's specific, concrete right belonging to him both in the procedure for confiscating property under a judgment of a court in a criminal matter (Article 35 Criminal Code) as well as in instances provided for by law in decisions of a court in civil matters (Articles 49, 58, 109-111, para. 4, Article 473 Civil Code). Here also is included the deprivation of the right of a citizen to inherit in circumstances provided in Article 531 of the Civil Code (for more on this, see Chapter 41). This type of deprivation of some kind of concrete right belonging to a citizen does not limit his civil legal capacity – there exists the possibility of using analogous or other rights.

2. *Dispositive capacity of citizens and conditions of its limitation. Civil dispositive capacity is the term for the possibility that a person may, through his own actions, acquire civil rights and create for himself civil obligations.* Civil dispositive capacity is attained to full extent with the advent of adulthood, i.e. the age of 18 (Article 11 Civil Code).

Minors up to the age of 15 (Article 14 Civil Code) have the right to enter into petty, everyday transactions, e.g. activities which are connected with everyday needs and are usually carried out by children themselves or at the request of adults – the purchase of goods in small quantities in a store, market, or school cafeteria, the giving of articles for minor repairs, acquiring tickets on city transport, for children's theater, etc. They also have the right to place deposits into a savings bank under their own name and to dispose of these deposits (para. 20 Statute of the State Worker's Savings Bank of the USSR).

All other legally significant acts within the sphere of civil law are executed in the name of minors under 15 years of age by their parents, adoptive parents, or guardians (see Chapter 46 on guardianship and curatorship), and those persons have civil (property) responsibility for any harm caused by the minor (see Chapter 35).

Upon attaining the age of 15 years, the dispositive capacity of minors widens significantly. They may not only carry out by themselves petty everyday transactions but may also dispose of their earnings or stipends and also realize their copyright and patent rights. They may also on their own dispose of deposits placed into a savings bank in their name by other persons (Article 13 Civil Code). They themselves are responsible for any harm caused by them on general grounds, and parents or persons who take their place have only auxiliary responsibility (Article 451 Civil Code). They have the right to execute all other transactions by themselves, but only with the consent of their parents, adoptive parent, or curators. In the presence of sufficient grounds, the organs of guardianship and curatorship may on their own initiative or through the intercession of interested persons restrict or deprive a minor from 15 to 18 years of age of the right to dispose of his earned wages or stipend on his own cognizance.

In instances when the law allows for entry into marriage before 18 years of age (see Chapter 43), a citizen not having attained the age of 18 may claim dispositive capacity to the fullest extent at the time of marriage (para. 2, Article 11 Civil Code).

A citizen who, as a result of mental illness or mental retardation, e.g. due to psychological disturbance, is not able to understand the significance of his actions or to control them, may be declared by a court as no longer having dispositive capacity. A guardian is appointed and all juridical actions in his name are executed by the guardian; the citizen himself is not responsible for any harm which he may cause. In instances of a return to health, the court recognizes him as capable and dismisses the guardian (Article 15 Civil Code).

The dispositive capacity of a person may be restricted by decision of a court in instances when he subjects his family to burdensome material difficulty as a result of the abuse of alcoholic beverages or narcotic substances. A curatorship is established over such a person. Only with the agreement of the curator may he carry out transactions involving the disposal of property, as well as receive earned wages, a pension, or other types of income or dispose of them, with the exception of petty everyday transactions. With the cessation of the abuse of alcoholic beverages or narcotic substances, the court lifts the restrictions on dispositive capacity and accordingly the curatorship which was established is also lifted (Article 16 Civil Code; Article 62 RSFSR Criminal Code).

Guardians carry out transactions in the name of their wards and are their lawful representatives (see Chapter 8 on representation). Curators render assistance to the persons under their curatorship and give consent for the execution of those actions which they may not execute on their own. If the transaction goes beyond the limits of the petty, then the guardian has a right to conclude it, while a curator has the right to conclude it only with prior decision of an organ of guardianship and curatorship. The conclusion of contracts of gift in the name of a person under curatorship is usually not permitted (Article 133 RSFSR Code on Marriage and the Family). Parents and adoptive parents fulfill the functions of guardians and curators without special designation (see Chapter 46 for guardianship and curatorship).

3. *Place of residence and its juridical significance. By the place of residence of a citizen is meant the place where he permanently lives* (para. 1, Article 17 Civil Code). Freedom of choice as to a place of residence enters into the content of civil legal capacity (Article 10 Civil Code). A place of residence holds juridical significance in the realization and protection of civil rights.

Article 174 of the Civil Code contains dispositive norms which define the place of performance of obligations according to the place of residence of one of the parties. The place of opening of an inheritance is defined as the last place of residence of the person leaving the inheritance (Article 529 Civil Code). Guardianship is established according to the place of residence of the ward or the place of residence of the guardian (Article 120 Code on Marriage and the Family), and supervision of the activity of guardians and curators is carried out at the place of residence of the ward (Article 136 Code on Marriage and the Family). The place of residence of the parties has decisive significance in determining the jurisdiction of civil matters by courts (Articles 117-118 Code of Civil Procedure).

According to law, the place of residence of persons under fifteen years of age, as well as those having attained majority who are under guardianship, is the place of residence of their parents, adoptive parents, or guardians although it may not coincide with the place of actual residence of the ward (para. 2, Article 17 Civil Code).

4. *Declaration of a citizen as missing and declaring him dead.* The protracted disappearance of a citizen, when there is no information as to him or his whereabouts, and when it is generally unknown whether he is alive, creates uncertainty in regard to the rights of not only that citizen himself but also of other citizens linked with him through one or another form of legal relationship. Those interested in eliminating this uncertainty are persons who are close to the missing person, those who have the right to receive alimony from him, his possible heirs, dependents having the right to a pension in the event of the death of the bread-winner, persons who are insured as beneficiaries of the missing person, etc. Creditors of the missing person interested in receiving that which is due them may not bring suits and receive satisfaction. In order to eliminate the created uncertainty, Soviet law permits [the following]:

a) A citizen may be declared missing in a juridical proceeding by petition of interested parties if in the course of one year there is no information as to his whereabouts at the place of his permanent residence (para. 1, Article 18 Civil Code).

Declaring a person missing leads to a series of legal consequences. Under Article 19 of the Civil Code, a guardian is placed over his property and from this property maintenance is paid to those citizens whom the missing person was lawfully obliged to support and debts on his other obligations are also discharged. By way of an exception, an organ of guardianship and curatorship may, by a petition of interested persons, designate a guardian to protect the property of the missing person prior to a declaration by a court that he is missing. The dependents of the missing person acquire the right to receive pensions (para. 60 Statute on the Procedure for Granting and Paying State Pensions).The effect of a contract of agency is terminated where one of the participants was a person who has been declared missing (Article 401 Civil Code) as is the effect of any letters of authority issued by him or in his name (Article 69 Civil Code).

In case of the appearance or the discovery of the whereabouts of a missing person, the court by petition of interested parties annuls the decision declaring him missing. On the basis of the decision of a court, the guardianship over his property is annulled (Article 20 Civil Code); an annulled marriage may be re-established upon the joint application of the spouses if the spouse of the person declared missing has not entered into a new marriage (para. 2, Article 42 Code on Marriage and the Family).

b) A citizen may through judicial procedure be declared dead if in the place of his permanent residence there has been no information as to his whereabouts in the course of three years, or after six months if he disappeared without any information under dire circumstances or under conditions giving grounds for believing that he died during some specific, disastrous event, for instance, the wreck of a ship on which he was located or a natural calamity in the place of his whereabouts. A person serving on military duty or a citizen who disappears without information in connection with military service may be declared missing by judicial procedure not sooner than two years from the day of the cessation of military operations.

The day of death of a citizen declared missing is considered to be the day when the decision of the court declaring him dead is legally effective. In the instance of the declaration as dead of a citizen who disappeared without a trace under dire circumstances or under conditions giving cause for belief that he perished under a specific disastrous occurrence, the court may declare the day of the death of that citizen to be that of the presumed disaster (Article 21 Civil Code).

In the presence of the conditions provided for by law as mentioned above, a person may be immediately declared to be dead without a prior declaration thereof by a court that he is missing.

Declaring a person dead leads to the same juridical consequences as physical death: the opening of inheritance, termination of marriage, the proper persons obtain the right to receive payments which are linked to the death of the citizen, etc.

In instances of the appearance or discovery of the whereabouts of a citizen declared dead, the relevant decision is annulled by the court upon petition of interested parties. Irrespective of the time of his appearance, the citizen may demand the return of his property from the various persons to whom this property was given free of charge (heirs, people to whom it was handed over, etc.). The re-appearing citizen has the right to demand the return of property from someone to whom it passed for consideration (for instance, it was bought by him) only in the event that the acquirer knew that the citizen declared dead was, in fact, alive. If the property of the citizen declared dead went by right of inheritance to the state and this was realized, then after the annulment of the declaration of death the citizen will be reimbursed the sum of money received upon realization (Article 22 Civil Code). In the last instance, there is no right to the return of the property in kind. The terminated marriage of such a person is considered re-established if the spouse has not entered into a new marriage (para. 1, Article 42 Code on Marriage and the Family).

It is necessary to distinguish between declaring a citizen dead and establishing the fact of his death according to the judicial procedure set forth in the Code of Civil Procedure (para. 8, Article 247). The necessity for the judicial establishment of the fact of death arises in circumstances when there is proof that the given party actually perished at a specific time and under specified circumstances, but the organ of the Registry of Acts of Civil Status (*ZAGS*) for one reason or another refuses to register the fact of death (for instance, if there is no medical statement). In this instance, according to judicial procedure, there is established that the juridical fact actually took place and the rules of Articles 18-22 of the Civil Code are not applied.

5. *Legal capacity and dispositive capacity of foreign citizens and stateless persons.* In the USSR, foreign citizens and stateless persons[2] enjoy civil legal capacity equally with Soviet citizens with some particular exceptions which may be established by law of the Soviet Union (Articles 562, 563 Civil Code). For instance, they may be forbidden to visit some locations or to live there permanently. Beside this, Article 562 of the Civil Code (Article 122 Principles) states that the Council of Ministers of the USSR may establish corresponding restrictions on citizens of those countries which place special restrictions on the legal capacity of Soviet citizens. This type of corresponding restriction is called retortions.

Civil dispositive capacity of a foreign citizen is defined by the law of the country of which he is a citizen, and for a stateless person, by the law of the country in which he has permanent residence. However, in regard to transactions concluded in the USSR and to obligations resulting as a consequence of causing harm in the USSR, dispositive capacity is always defined by Soviet law (Article 563-1 Civil Code).

2. Persons living on the territory of the USSR who are not citizens of the USSR and do not have proof of being citizens of a foreign state are considered stateless persons (Article 9 Law on Citizenship of the USSR of 1 December 1978).

Chapter 5. Juridical Persons

1. *Concept, significance, and characteristics of a juridical person.* No only individual citizens, but also socialist organizations may act in the capacity of a participant in a civil legal action: state enterprises and institutions, collective farms and other cooperative organizations, trade unions, volunteer societies, and other public organizations (Article 23 and 24 Civil Code). These subjects of civil law are called *juridical persons.*

Enterprises performing economic activity carry out specific production or engage in its preparation, dispose of their production and obtain raw materials and supplies, and perform economic services for other enterprises and citizens. According to socialist methods of production, these kinds of relationships between specific enterprises, organizations and citizens are commodity-money relationships, have an equivalent-reciprocal character, and assume the form of civil legal relationships (see Chapter 1). Consequently, these enterprises and organizations would not be able to carry out the economic relationships necessary to them if it were not possible for them to be subjects of civil rights and obligations.

However, not every organization is considered a juridical person. A shop or department of an enterprise or institution, or a division of a state farm are not juridical persons. Only those organizations which are independent participants in commodity-money exchange may be subjects of civil legal relationships. For this, the organization must possess specific characteristics, which are listed as follows:

1. *Organizational unity.* A juridical person must be organized in a specific manner as a unit, the internal structure of which should answer the goals and tasks of its activity, and it must have organs which carry out its dispositive capacity. The organizational unity of a juridical person usually is expressed and embodied in the charter of the juridical person or in a special statute therefor.

2. *The presence of separate property.* In order to be a participant in civil legal relationships, a subject of civil law,

an organization must possess property which is separate from the property of other organizations. The degree to which this property is separate varies with different types of juridical persons, and this separateness itself receives various organizational embodiment. Thus, the separateness of the property of collective farms and other cooperative organizations is expressed by the fact that they are the owners of the property belonging to them. For state enterprises which are not owners of the state property given over to them (for more on this see Chapter 11), as well as for enterprises created by cooperative and social organizations and inter-cooperative enterprises, the degree of the separateness of property is expressed in the allocation to them of capital and circulating assets and the fact that they have an independent balance.[1] For state budgetary institutions this type of feature is indicated by their having an independent estimate and the allocating to the manager of the institution of rights to disburse credits (Article 24 Civil Code). For more details on this see below.

3. *Independent property liability.* The creditors of a juridical person may turn only to the juridical person itself with claims arising from duties and other obligations of the juridical person; actions arising on the basis of these demands may be put only against the separate property of this juridical person.

In accordance with this general principle, which is expressed in Article 32 of the Civil Code, the state is not liable for the debts of those individual state organizations which are juridical persons, nor are these organizations liable for obligations of the state (Article 33 Civil Code). In the same manner a state, cooperative, or social organization is not liable for obligations of enterprises subsidiary to it if the latter are

1. Independence (completeness) of bookkeeping balance consists in the fact that in it are reflected all expenditures and sources of income of a given organization. In the bookkeeping account of a shop or other sub-division of a juridical person, a number of expenditures which constitute what is called general factory expenditures without which the given sub-division would not be able to operate, are not expressed, even if this sub-division is allotted a separate (but not independent) balance. It is important to distinguish the independent balance from combined balances, consisting of economic systems formed by a number of juridical persons. Likewise, a combined balance is not an organizational-bookkeeping expression of the separateness of the property of a juridical person.

themselves juridical persons, with exceptions specifically provided for in law (Article 34 Civil Code). Cooperative farms which form an inter-cooperative organization, as well as participants of a state-collective farm or other state-cooperative organization which is a juridical person, are not liable for the debts of such an organization, just as this organization is not liable for the obligations of its institutions (Article 36 Civil Code). Sometimes, however, auxiliary liability of other persons is permitted (para. 3, Article 34; Article 35; para. 2; Article 36 Civil Code). Thus, by a special law or charter of a cooperative organization, provision may be made for auxiliary liability of the members of this organization for its debts to the extent of their multiple share (para. 1, Article 36 Civil Code, see also Article 35 and para. 3, Article 34 Civil Code).

4. *Entering into civil intercourse in one's own name.* With this feature the economic-operative independence of the juridical persons receives ultimate expression. Arising from the basis of property (second and third features) and organizational (first feature) pre-conditions, this feature is the final, external expression of the independence and the civil legal subjectivity of the juridical person. A juridical person under its own name, independently (within the limits established by law and by plan tasks) disposes of its own property and acquires civil rights and obligations, in part concludes contracts and enters into obligations, and bears independent liability under them.

Keeping all this in mind and basing oneself on law (Article 23 Civil Code), one may give the following definition of a juridical person: *A juridical person is an organization which possesses separate property, may acquire civil rights under its own name and bear liabilities, and has the right to act in the capacity of plaintiff and defendant in court, in arbitrazh, or in a chosen arbitration tribunal.*

2. *Charter (statute), legal capacity, and dispositive capacity of juridical persons.* Subsidiaries and representation. A juridical person acts on the basis of its charter or statute; in

some instances a general statute on organizations of a given type may be substituted for an individual charter (statute) (Article 25 Civil Code). This is the case, for instance, with railroad transport enterprises.

The charter defines the tasks and range of activity of a given organization, its subordination, internal structure, the organs of the juridical person and their competence, the sites and name of the juridical person, as well as other necessary stipulations.

For some types of organizations, typical or model charters are approved. The juridical significance of such charters consists in the fact that charters of individual organizations of a certain type may not contradict the model (typical) charter, and if an individual organization for some reason does not have its own individual charter, then the model charter, as stipulated in Article 25 of the Civil Code, has for it the significance of the general statute on organizations of that given type.

Juridical persons are created under the control of the state in order to carry out specific activities. Therefore, in contrast to the principle of the equal, the so-called general legal capacity of citizens, juridical persons act under the principle of *special legal capacity*: each juridical person possesses civil legal capacity only in accordance with the established aims of its activity (Article 26 Civil Code). This means that it may acquire only those civil rights and obligations, and therefore, perform only those economic operations which correspond with the aims of its activity, with the tasks under its charter. Transactions which overstep the boundaries of its special legal capacity are invalid (Article 50 Civil Code).

Thus, the principle of special legal capacity by the methods of civil law limits the sphere of activity of juridical persons and excludes the possibility that it might deviate from the assigned tasks. Special legal capacity promotes the strict and orderly carrying out of the planning basis of socialist economic organizations. The correct application of this principle helps to solve the problem of the development of specialization and of the efficient cooperation of enterprises – one of the most important conditions of technical progress and the rational organization of the labor of society. The extent of special legal capacity of a juridical person is defined by existing legislation and by the charter of the juridical person, as well as by relevant acts of the plan.

A juridical person acquires civil rights and incurs civil obligations, e.g. exercises its dispositive capacity through its organs (Article 28 Civil Code), whose structure and competence is defined by law either by statute or the charter of the juridical person. In their capacity as organs of state juridical persons, the heads (director, chief, manager) and their deputies usually act alone. Cooperative and social organizations usually have collegial organs, and not only one but several (the highest, the general assembly of the members of the organization, or an assembly of representatives, or a congress; subordinate to the highest organ are the board of directors or presidium; normally besides the board of directors and an individual chairman there is also an organ which has a specific competence assigned by the charter).

An organ is a structural part of the juridical person, and all actions performed by it within the limits of its competence are considered actions of the juridical person and give rise to corresponding rights and obligations for the latter. Actions of an organ which overstep the limits of its competence do not create rights and obligations for it. Thus, if according to the charter, the right to conclude specific contracts belongs to the highest organ of the organization – the general assembly – then the conclusion of such a contract by the chairman acting alone or even by the board of directors does not engender rights and obligations for that organization.

It is necessary to distinguish between the conclusion of transactions, e.g. actions directed at acquiring rights and obligations, and actions directed towards carrying them out. Juridical persons realize their obligations and rights through the efforts of the entire collective of their workers; therefore, the corresponding actions of any one worker, performed by him in connection with the performance of his official (labor) duties are considered the actions of the juridical person itself and result in legal consequences for the latter. For instance, the damaging of objects of a client by the careless actions of the craftsman of a workshop is considered to be an act of the juridical person and results in its civil liability before the person suffering the damage (this does not exclude subsequently transferring this liability to the guilty worker through the procedure and within the limits established by labor legislation). At the same time, actions which may be outwardly linked with [the performance of] work, but which are not executed in the process of performing official duties, are not considered to be acts of the juridical person. For instance, if the workers of a repair and construction office go on a drunken spree in a repair shop and as a result of this ruin the property of a customer, this type of act may not be considered to be an act of the juridical person (the office) and does not engender the civil (property) liability of the latter; this liability falls on the guilty parties.

In law this principle is formulated in conformity with liability for causing harm (Article 445 Civil Code). However, it is also applied in instances of responsibility for obligations based on contract, as well as for the performance of acts by workers in executing other rights and obligations of the juridical person.

The situs of a juridical person is the place where its permanent operating organ is located (Article 30 Civil Code).

Juridical persons may establish *branches and representative offices*. As a rule, branches are established to carry out activities of the juridical person in another location (for instance, branches of higher educational institutions, scientific and design organizations) or to carry out this activity to a lesser extent (for instance, branches of production combines of department stores). Representative offices are formed to perform auxiliary operations in a location different from that of the juridical person (for instance, the representative offices of heavy industrial enterprises in areas where primary suppliers or consumers are located). Branches and representative offices are not themselves subjects of civil law; they are part of the juridical person (Article 31 Civil Code). The legal status and authority of the officials of the branches (representative offices, departments, sectors, etc.) are defined either by law or by the statute or charter of the juridical person. If the law or charter of the juridical person assigns to an official of a branch (or representative office) the right to conclude juridical acts in the name of the juridical person, then this official is an organ of the juridical person. In the absence of such authorization by law (charter), the officials of branches and representative offices may act in the name of the juridical person on the basis of authority granted them by letters of authority or through other means provided in law (see Chapter 8).

3. *Origin and termination of juridical persons.* It is possible to distinguish three methods or three procedures for the origin and termination of juridical persons: by order, by permission, and by normative without prior permission.

In the procedure of an order, a juridical person originates as the result of a direct order of an organ of state administration. State juridical persons originate through this procedure. Thus, ministries, state committees, and departments of the USSR establish, reorganize, and liquidate enterprises, organizations, and institutions under the authority of union subordination. Enterprise, institutions, and organizations of union republic and local subordination are established, reorganized, and liquidated by the Councils of Ministers of the union republics or by procedure established by them (see Statute on the Procedure for Establishing, Reorganizing, and liquidating Enterprises, Associations, Organizations, and Institutions, approved by decree of the Council of Ministers of the USSR, 2 September 1982).[2] In the RSFSR such a right was given to the ministries and departments of the RSFSR, to the Council of Ministers of Autonomous Soviet Socialist Republics, the executive committees of territorial, provincial, and the Moscow and Leningrad city Soviets of People's Deputies.

In the procedure by permission, the initiative for forming a

2. *Collection of Decrees of the USSR* 1982 No.25 item 130.

new juridical person belongs to the initiators who are citizens or juridical persons. Government organs participate in the organization of this type of juridical person by granting permission, and by verifying in every instance the advisability of its formation. Through the procedure by permission originate voluntary sport, study, and other societies, residential housing, and vacation-home construction cooperatives (see, for instance, the Model Charter of the Residential Housing Construction Cooperative, approved by a decree of the Council of Ministers of the RSFSR, 2 October 1965), and several consumer cooperative organizations. Sometimes the government transfers its right to grant permission for the formation of a juridical person to certain social organizations. Thus, organizations of the trade unions (except for the trade-union committee and the Central Committee of the trade unions) originate with the permission of higher-level trade union agencies.

The essence of *the normative without prior permission procedure* is reflected in its name: certain types of juridical persons, in circumstances provided by norms of law originate without prior permission, i.e. neither an order nor permission of governmental agencies is required for their formation; such permission in regard to certain types of juridical persons was granted earlier by a norm of law, and therefore they may freely be established by initiators who are either citizens or organizations. A government organ is competent only to verify that the newly established juridical person exists according to the requirements set forth by law and that the procedure regarding its establishment is observed. By the normative without prior permission procedure are established collective farms, mutual-insurance funds on collective farms (see para. 3 of the Model Charter for these funds, approved by decree of the Council of Ministers of the RSFSR, 6 January 1958[3]), trade union committees of enterprises, institutions, organizations, and the Central Committees of trade unions.

The law sets forth the procedure by which the various

3. *Collection of Decrees of the RSFSR* 1958 No.6 item 68.

types of juridical persons are established. Social organizations whose procedure of formation is not defined by law are formed in accordance with their charter or statute (Article 27 Civil Code).

The activity of a juridical person may be *terminated either by order or by voluntary procedure.* State juridical persons may terminate their activity only by order of that organ by which they were formed (Article 38 Civil Code), i.e. by the procedure of order.

The activity of cooperative and social organizations may be terminated either by procedure of order or through voluntary procedure: in the first instance, by a decree of the proper organ of state management, or sometimes by order of a higher link in the cooperative system; in the second situation, by decree of a higher organ of the juridical person being liquidated.

A juridical person may be dissolved either through the method of *liquidation* or by *reorganization* (Article 37 Civil Code).

Reorganization of the juridical person, i.e. dissolution without complete liquidation of its affairs and property, may be carried out in the form of *merger, division*, or *accession*. Through a merger the independent existence of the merging organizations of juridical persons is liquidated, and a new juridical persons is formed in their place, which continues their activity and to whom is passed their property (active and passive, i.e. rights and obligations). In division, new juridical person arise in place of the liquidated juridical person. In accession the liquidated juridical person joins another juridical person to which is passed its property and which continues its activity (see paras. 106-107 of the Statute on the Socialist State Production Enterprise).[4] In all instances of reorganization there takes place the so-called universal legal succession, i.e. the transfer of the entire complex of civil rights and obligations which belonged to the liquidated

4. The rules on termination contained in this Statute are in fact also applied in instances of the termination of other juridical persons.

juridical person in accordance with the transfer balance. The property passes on the day of signature of the transfer balance unless otherwise provided for by law or decree of reorganization (Article 37 Civil Code).

The reorganization of cooperative (including collective farms), state-cooperative (including state-collective farms), and social organizations is allowed only by decision of the general assemblies of their members (participants) or assemblies of their delegates (Article 39 Civil Code).

In the *liquidation* of a juridical person, the organ making the decision on the liquidation creates a liquidation commission, which within a set period of time presents claims to other persons and accepts claims brought against the given juridical person [being liquidated] and terminates all its affairs. According to the Statute on the State Socialist Production Enterprise (para. 109) this time limit may not be less than one month, and the management of operations up to the liquidation may be given to the director of the enterprise being liquidated rather than to the liquidation commission (para. 108).

In accordance with Article 39 of the Civil Code, the grounds on which cooperative and state-cooperative (including state-collective farm) organizations may be liquidated must be indicated in law or in the charter (Statute) of the given organization. Thus, the law makes provision for the enforced liquidation of a cooperative organization in instances of a decline in the number of its members to a limit lower than the one mandated by law or charter or upon the expiration of the period for which it was formed, as well as upon performing actions detrimental to the interests of the state. Voluntary liquidation is carried out by decision of the general assembly. Social organizations, according to Article 39 of the Civil Code, are liquidated on grounds given in their charters (Statutes).

Supervision over the observance of the existing procedure for the formation, reorganization, and liquidation of juridical persons is carried out by the State Bank of the USSR during the opening and closing of deposit and current accounts, and also by the organs of the Ministry of Finance. For some organizations a special registration is established (thus, collective farms must register with the executive committees of district Soviets of People's Deputies: housing and vacation-home construction cooperatives, with the executive committees of the local Soviet which gave permission for the organization of the cooperative); in these instances, supervision is also exercised by the organs which carry out the registration.

4. *State juridical persons.* The law (Art.24 Civil Code) lists the following types of state organizations which may act as juridical persons: a) state enterprises and other state organizations operating on the basis of economic accountability, having capital and circulating assets and an independent budget (for instance, production associations); b) institutions and other state organizations financed out of the state budget and having independent accounts whose managers have the right of disposal of credits; c) state organizations financed out of other sources and having an independent account-balance. Let us examine these types of juridical persons more closely:

A) The majority of state juridical persons *operate on the basis of economic accountability.* This means that at the formation of the organization it was given certain property – capital and circulating assets – from the unified fund of state socialist ownership and by using this it carries out its economic activity on the basis of self-repayment, i.e. it acquires raw materials, obtains supplies, uses services, etc., and sells its products or services for money, covering in this way its costs and creating specified savings.

The necessity of economic accountability arises from the economic principles of socialism as the first phase of Communist society, on the strength of which the output which is produced and other results of economic activity function as merchandise and are sold for money. Under these circumstances economic categories such as cost price, cost, and profit have great economic significance, and the organization of state enterprises on the basis of economic accountability gives the possibility of acting through these categories in the enterprises in the direction of a more rational and economic use by them of material and labor resources and of the increasing of labor productivity and the raising of profitability, insofar as the financial conditions and the material security of an enterprise on economic accountability depends on the results of its own activity. By the same token economic accountability allows for the application of the principle of material incentive in the result of one's own activity not only toward individual workers but for entire collectives – enterprises and similar organizations.

State enterprises operate on the basis of the Statute on the Socialist State Enterprises, which was approved by decree of the Council of Ministers on 4 October 1965. At first its effect was felt only by industrial, construction, agriculture, transportation, and communication enterprises, but later it gradually spread to other branches of the economy, and now there are almost no enterprises which are not based on this Statute. The activity of an enterprise is based on a combination of central planning and management with the economic independence and initiative of the enterprise itself (para. 1 Statute). It has a charter ratified by its superior organ, while in instances provided in the legislation of the USSR, it may act on the basis of the general statute on enterprises of a given type (para. 7 Statute). The management of an enterprise is carried out on the basis of one-man management (para. 4 Statute). It is headed by a director (chief, manager) who organizes the work of the enterprise and has full responsibility for its state and operations, and he acts in the capacity of an organ of the juridical person (paras. 89, 90 Statute). Deputies of the director may also act in the capacity of organs of the juridical person of the enterprise; their competence is established by the director.

The present period is characterized by a concentration in industry, and more and more frequently *industrial associa-*

tions (combines)[5] are being formed on the basis of various enterprises. They consist of a series of factories, plants, scientific-research, experimental-construction, and other production units for which the production association strengthens its capital and circulating funds and which have more independence than a shop. But these production units are not transformed into enterprises and are not considered enterprises by legislation. For a production enterprise, financial operations involving the budget or suppliers and customers, credit relationships, and supervision by banks do not as a rule depend on the location of its production units. From this it follows that only the combine as a whole, and not the production units, is the subject of property relationships which originate through the use of the commodity-money form. Therefore, production units are not juridical persons; only the combine as a whole is a juridical person. All legal acts which are performed by production units (by their managers and other employees) are executed in the name of the production association and engender the property (civil) liability of the production association as a whole, even if the production unit has its own bank account.

Production associations (combines) include such well-known industrial organizations as the "ZIL" production association in Moscow, the "Elektrosila" production association in Leningrad, and many more. Scientific-production associations (SPO), are organized in a fashion analogous to production associations (combines) and function on the basis of their Statute which was confirmed by a decree of the Council of Ministers of the USSR in 1975.

In accordance with the Statute on Enterprises (para. 10), a combine, trust, firm, or other economic organization, to which

5. Statute on the Production Association (Combine), confirmed by decree of the Council of Ministers of the USSR, 27 March 1974 and applies only to industries (*Collection of Decrees of the USSR* 1974 No.8 item 38). By decree of the State Construction Agency of the USSR, with the concurrence of the State Planning Committee of the USSR, the State Committee on Labor of the USSR, the Ministry of Finance, and the Bank for Construction of the USSR, the Statute on the Production Construction-Installation Association was affirmed (see: *Economic Gazette* 1979 No.49). Production associations (combines) also operate in other branches of the national economy.

independent enterprises are subordinate, acts in relation to them as a superior organ of economic management. If the structure of such an organization (trust, firm, combine) includes production units which are not independent enterprises, and if at the same time there are also enterprises which are subordinate to it, then such an organization (trust, firm, combine, etc.) forms with them a single juridical person and exercises in relationships with them those rights and obligations which are provided in the Statute on Enterprises as it does regarding its own production-economic activity; as regards those enterprises which are subordinate to it this organization acts in the capacity of an organ of economic management. In an analogous fashion, the Statute on Production Associations (Combines) in para. 6 provides for the possibility of the subordination to an association of independent enterprises, acting on the basis of the Statute on Enterprises. An association engaged in its own production-economic activity acts in relation to those enterprises in the capacity of a superior organization.

Everything which has been said about the juridical status of enterprises and associations also applies, in principle, to all branches of the national economy, irrespective of which concrete normative acts prevail in these branches.

The State Bank of the USSR, which is the single emission bank, the bank of the national economy, and the accounting center for the USSR (paras. 1 and 5, Charter of the State Bank of the USSR), is a juridical person. The branches, agencies, and cash disbursement points of the State Bank, even its republic and province offices, are not juridical persons, and the managers of these establishments of the State Bank act in the name of the State Bank of the USSR within the limits of the rights granted to them without special authority; i.e. they represent their organs. The All-Union Bank for the Financing of Capital Investment, *Stroibank* of the USSR, is organized in the same manner. The Foreign Trade Bank (*Vneshtorgbank* of the USSR) is also a juridical person.

In the capacity of uniform juridical persons with economic accountability are also organized the chief directorates of state insurance in the union republics, which have a far-flung system of management and insurance inspection, as well as the state workers' savings bank which is an all-union organization.

B. Institutions and other organizations operating on the state budget also are juridical persons if they have two features which express the necessary degree of property separateness of a juridical person: a) the institution has independent accounts from which it receives allocations to settle obligations; b) the manager of the institution enjoys the right of disposal of credits (allocations) from these accounts. If the allocation from accounts is not sufficient to cover a liability, then it is necessary to put before the proper superior instance the question as to the apportionment of the necessary sums from the accounts for the following year (Art. 33 Civil Code). Higher educational institutions, technical schools, hospitals, large museums, organs of state authority and management, etc., are examples of this type of juridical person.

Not every institution operating on the [state] budget has independent accounts and its managers do not always have the right to dispose of credits. Allocations necessary for his work may be provided for in the accounts of a superior organ and be disbursed at the discretion of its manager. Under these circumstances this organ acts as a juridical person, and the given institution is considered a part of its structure.

In regard to certain institutions operating on the [state] budget, the fact that they are juridical persons is directly provided either in their charters or in the proper normative acts.[6] Such an indication in the charter or other act may not, however, go counter to the requirements of Article 24 of the Civil Code: in the absence of features set forth in that article, an organization may not act as a juridical person. At

6. See para. 73 Statute on Higher Educational Institutions of the USSR, para. 64 Statute on Secondary Special Educational Institutions of the USSR (*Collection of Decrees of the USSR* 1969 No.4 items 24, 25).

the same time, if in the charter of the given institution or in the general statute on the institutions of the given type there are no instructions concerning its juridical personality, this question is decided without difficulty on the basis of general features given in the law (Art.24 Civil Code). Thus, republican and other ministries are juridical persons, regardless of the absence of instructions to this effect the General Statute on Ministries of the USSR.

C. The management of all-union and republican industrial associations is related to state organizations financed at the cost of other sources and having an independent account and independent balance, as are, in some instances, those republican and even all-union ministries heading an entire economic system, which are financed by deductions from economically accountable departments subordinate to them.

It is necessary to distinguish all-union and republican industrial associations from production associations (combines). In contrast to the latter they consist of a number of independent scientific-research, construction, technological, and other organizations and industrial enterprises, each of which operates on the basis of the Statute on Enterprises or another analogous special provision. These associations may also include production associations (combines). Every member of an industrial association is a juridical person, and the organizationally separate management of the association is also a juridical person. Industrial associations, which were created instead of central directorates of industrial ministries, occasionally embrace entire sub-branches of industry.

An all-union or republican industrial association as one entity does not act as a single juridical person, and it may not so act because it is a complex of separate juridical persons – enterprises and production enterprises (combines) of economic accountability, as well as the center of this complex, the management of the all-union or republican industrial association which is financed by them by means of deduction from them. All-union and republican industrial associations operate on the basis of the General Statute on All-Union and Republic Industrial Associations, confirmed by the Council of

Ministers of the USSR on 2 March 1973.[7] Analogous complexes also exist in other branches of the national economy.

Organizations not operating on the basis of economic accountability and financed by the so-called budget process – however, not from the state budget but rather from other sources, in part by allocations from other organizations – are also considered juridical persons if they possess the degree of separateness required by law, e.g. have an independent account and balance (para. 4, Art.24 Civil Code). The management of all-union and republican industrial organizations and ministries indicated above are considered juridical persons.

5. *Cooperative and social organizations as juridical persons.* Collective farms, inter-collective farm and other cooperative organizations and their associations, other social organizations, and even, under circumstances provided by legislation, enterprises and institutions of these organizations and their associations which have separate property and an independent balance, are juridical persons (para. 5, Art.24 Civil Code).

Cooperative organizations are distinguished from other social organizations by the fact that they are formed for the purpose of carrying out economic activity in the interests of their members: cooperative farms, fishery collectives, and gold-prospecting artels carry out production activity; consumers' societies and their unions carry out commercial and commercial-purchasing activity, and housing and vacation-home construction cooperatives carry out construction and exploit the housing fund, etc. The important (and sometimes the only) sources of the creation of the property of an organization consist of the shares and share accumulation of its members. The member who leaves a cooperative organization is returned his share accumulation.

The basis of the organizational unity of a cooperative organization as a juridical person is its *membership*. The members of the primary cooperative organizations are citizens and

7. *Collection of Decrees of the USSR* 1973 No.7 item 32.

members of cooperative unions and associations, e.g. organizations of the second rung and higher – juridical persons. Primary cooperative organizations and their associations are owners of property belonging to them (Art.99 Civil Code).

The members of a cooperative organization have the right to participate in the management of its affairs, and therefore its highest organ is the general assembly of members (sometimes an assembly of delegates selected by members). The executive organs (management board, presidium) are also elected and, as a rule, are collegial. The chairman of the management board is also usually assigned specific competence under the charter and, in this situation, acts as a juridical person. Cooperative democracy is apparent in the fact that the membership itself decides questions on the activity of the cooperative in accordance with the existing legislation.

Cooperative unions and associations of the second and subsequent rungs act in the capacity of individual juridical persons: a union and its members are organizationally separate, have separate property, are responsible for their own obligations, and each acts in his own name. Members of a cooperative or other social organization are not responsible for its obligations. However, such responsibility of members of a cooperative organization may be provided for by special legislation or by the charter of the cooperative organization, within the limits of a multiple amount of their shares (Art.36 Civil Code).

A collective farm is a cooperative organization of persons voluntarily forming an association in order to jointly carry out large-scale socialist agricultural production on the basis of the social means of production and collective labor. The general assembly of members of a collective farm receives a charter which must correspond to the Model Charter, adopted by the Third All-Union Congress of Collective Farm Workers and confirmed by decree of the Central Committee of the Communist Party of the Soviet Union and the Council of Ministers of the USSR of 28 November 1969, with changes and amendments, passed by edict of the Presidium of the Supreme Soviet of the USSR on 10 July 1980.

The content and range of the special legal aspects of a collective farm as a juridical person are defined in part V of the Model Charter. Besides its basic tasks of the rational conduct of agricultural production by means of its intensification and specialization taking into consideration natural and economic conditions, the collective farm may – for the purposes of a more complete and balanced utilization of labor resources and local sources of raw materials, and in order to increase the profitability of the social economy, while not being detrimental to agricultural production – form subsidiary enterprises and develop various industries, enter into contractual relationships with industrial enterprises and commercial organizations to create branches and shops within the collective farms for the production of various articles and merchandise through the efforts of the collective farm workers during periods when they are free from agricultural work. Collective farms have the right, by a decision of the general assembly, to form inter-collective farm and state-cooperative farm enterprises and organizations, to enter into unions and associations, to unite part of their financial resources with those of the local Soviets and other state and cooperative organizations for the construction of objects and to implement measures directed both towards developing collective farm production as well as towards the commonweal and improving the cultural and everyday services of collective farm workers.

A collective farm has the right to conclude various contracts which correspond with the aims of its activity, in part for the sake of selling its own agricultural produce and the production of its industries or purchasing machinery, cattle, or other objects necessary for the performance of its work and to render services; it opens an account in the State Bank, carries out accounting transactions, uses bank credits, etc.

The management of the affairs of a collective farm is carried out on a broad democratic basis. The highest organ of the collective farm as a juridical person is the general assembly of members (in very large collective farms, the assembly of delegates), and then accountable to it is the board of managers and the chairman of the collective farm (who may be granted specific individual competency by the charter).

Fishery collective farms form provincial fishery collective farm unions which carry out the organizational and economic guidance of the collective farms which are united in them. These unions are juridical persons.[8]

Consumers' cooperative societies carry out trade in items widely used in villages, prepare agricultural products and raw materials, and also carry out trade in surplus agricultural products purchased from collective farm workers. This type of cooperative forms a system in which the lowest link is the consumers' society enjoying the rights of a juridical person (para. 5, Model Charter of an Agricultural Consumers' Cooperative).[9] Agricultural consumers' cooperatives form district consumers' cooperative unions, and district consumer unions form provincial and republican unions of consumers' cooperatives; the entire system of consumers' cooperatives in the USSR is headed by the Central Consumers' Union. All these unions are juridical persons.

Organizations of consumer cooperatives may form enterprises of economic accountability, engaged in production, trade, or public catering. These enterprises, which are granted capital and circulating assets and have an independent balance, are juridical persons. Their legal position is basically analogous to the legal position of state enterprises, with this difference, however, that the property secured for them remains in the ownership of the cooperative organizations which constitute them.

Home and vacation-home construction cooperatives (*ZhSK* and *DSK*) are created for the purpose of constructing apartment houses or vacation homes to fill the demand of members of the cooperative for living space (vacation accommodation). The cooperative is considered a juridical person from the day of registration of its charter with the executive committee of the local Soviet.

8. See Model Charter of Fishery Collective Unions, confirmed by decree of the Council of Ministers of the RSFSR of 28 November 1960 (*Collection of Decrees of the USSR* 1960 No.40 item 202).

9. See *Collection of Charters of Consumers' Cooperative Organizations*, Moscow 1980.

In the union republics there exist Model Charters of the *ZhSK* and *DSK* which define their legal status and the relationship between the cooperative and its members. In the RSFSR there is the Model Charter of the *ZhSK*, which was confirmed by a decree of the Council of Ministers of the RSFSR on 2 October 1965,[10] and the Model Charter of the *DSK*, which was confirmed by a decree of the Council of Ministers of the RSFSR on 24 September 1958.[11]

There are *cooperatives for the construction and operation of collective parking garages for automobiles* of individual owners,[12] as well as cooperatives which have as their purpose the satisfaction of other needs of their members (for instance, cooperatives for the construction of creative studios for artists[13]). These cooperatives are juridical persons.

Horticultural societies exist on the basis of the Standard Charter of a horticultural society of workers and employees, which was confirmed by the Ministry of Communal Economy of the RSFSR and the Ministry of Agriculture of the RSFSR in agreement with the All-Union Central Trade Union Council and are juridical persons from the moment of their registration with the executive committee of the local Soviet.

The development of economic links between collective farms, consumer cooperatives, and state enterprises is leading to the ever increasing development of *inter-collective farm, collective-cooperative, and state-collective farm cooperative organizations* both of a production-economic as well as of a cultural-educational nature. Such inter-economic organizations, which have separate property and a separate balance, may act as juridical persons in the instances provided by law (para. 5, Art. 24 Civil Code). The legal position of inter-economic

10. *Collection of Decrees of the RSFSR* 1965 No.23 item 144.
11. *Collection of Decrees of the RSFSR* 1958 No.13 item 154.
12. The Model Charter of this type of cooperative was confirmed by a decree of the Council of Ministers of the RSFSR on 24 September 1960 (*Collection of Decrees of the RSFSR* 1960 No.33 item 160).
13. The Model Charter of this type of cooperative was confirmed by a decree of the Council of Ministers of the RSFSR on 4 December 1979 (*Collection of Decrees of the RSFSR* 1980 No.3 item 19).

enterprises and organizations in the rural areas is defined by the General Statute on the Inter-Economic Enterprise in the Rural Economy, which was confirmed by decree of the Council of Ministers of the USSR on 14 April 1977,[14] and the Statute on the Production Enterprise in the Rural Economy, confirmed by decree of the Council of Ministers of the USSR on 7 December 1978.[15]

Recently, particularly in connection with the decision of the May (1982) Plenum of the Central Committee of the CPSU, there has been an increase in the development of *regional agro-industrial enterprises* (RAIE), which include collective farms, state farms, inter-economic formations, and other agricultural enterprises, as well as enterprises and organizations which service them and which are linked with agro-economic production and the processing of its produce. According to the decree which was confirmed by the Plenum of the Central Committee of the CPSU and the Council of Ministers of the USSR, "On the Improvement of the Management of the Agricultural Economy and Other Branches of the Agro-Industrial Complex", enterprises and organizations which enter into an association retain their economic independence, their rights as a juridical person, and their institutional subordination; the associations basically carry out management and coordinating activities. Provincial, territorial, and republic (ASSR) agro-industrial associations are organized in an analogous fashion.[16]

Numerous trade unions and other public organizations which are formed for a variety of purposes – sport, defense, cultural, scientific and technical, professional, etc. – are also juridical persons. Some of them operate on the basis of special normative acts. For instance, there is a Model Statute

14. *Collection of Decrees of the USSR* 1977 No.13 item 80.
15. *Collection of Decrees of the USSR* 1979 No.3 item 15.
16. See: "The Food Program of the USSR for the Period to 1990 and Measures for its Realization", *Material of the May Plenum of the Central Committee of the CPSU 1982*, Moscow 1982, p.78, or *Collection of Decrees of the USSR* 1982 No.17 item 89.

on a Republic (ASSR), Territorial, Provincial, and District Regional (city) Volunteer Society of Horticulturalists.

Some social organizations form a system in which each territorial link is a juridical person alongside the central (republic and all-union) organs. In this manner are organized, for instance, the All-Russian Society for the Protection of Nature, the Union of Journalists of the USSR, and several others. In other instances the society or union as a whole is a juridical person, for instance, the All-Union Volunteer Society of the Army, Air Force and Navy (*DOSAAF SSSR*).

As is true of cooperative organizations, social organizations are characterized by the institution of membership, electoral voting and collegiality of the organs, and general assemblies (congresses) as the highest organs to which the executive organs are subordinate. The civil legal capacity of these organizations corresponds strictly with the aims for which a specific organization is formed.

Some societies have as one of their tasks the securing of the cultural, social, and professional interests and the protection of the rights of their members (All-Union Society of Inventors and Rationalizers, All-Russian Theatrical Society, etc.). The societies of creative professions perform an analogous function (Union of Writers of the USSR, Union of Architects of the USSR, Union of Artists of the USSR, etc.). These organizations form a Literary Foundation, an Artistic Foundation, etc., which are juridical persons and which are called upon to help satisfy the material needs of the corresponding creative profession. Under the auspices of these foundations, various enterprises and institutions may be established which may also be juridical persons if they meet the requirements of the law (Art.24 Civil Code).

The most important types of social organs are the professional trade unions which form a single system headed by the All-Union Central Trade Union Council. Committees of the professional trade unions, starting with the committee of the professional trade union of an enterprise and ending with the central committee of a professional trade union, as well as organs of inter-union associations (the All-Union Central

Trade Union Council and lower-level councils of professional trade unions) are juridical persons. This legally secures for all organizations of professional trade unions the possibility of entering into property relationships and helps them to execute various activities necessary to carry out the tasks confronting them. Professional trade unions may establish organizations which are juridical persons (clubs, palaces of labor, sanitariums, mutual benefit funds, etc.); the All-Union Central Trade Union Council may even establish enterprises which operate on the basis of Statutes relating to them which are confirmed by the Presidium of the All-Union Central Trade Union Council.

Chapter 6. The Soviet State as Subject of Civil Law

The state as a whole may act as a participant in civil legal relationships. Besides the Soviet Union there are also 15 union republics which are sovereign states. Therefore, Article 24 of the Civil Code provides that in circumstances envisaged by the legislation of the USSR and the union republics, state institutions may not act as juridical persons under their own name but rather in the name of the Soviet Union or a union republic. In these instances the subject of a concrete legal relationship is the state; it (the Soviet Union or a union republic) takes on itself the corresponding rights and obligations and property liability. This takes place, for instance, in the issue of state loans or lotteries. In these cases, no state legal persons of any kind bear any liability, as is directly stated in Article 33 of the Civil Code.

The participants of the state in legal relationships which are regulated by civil legislation do not turn it into the usual type of juridical person; the norms on juridical persons do not apply to it. The state as a sovereign entity itself defines the conditions, procedures, and limits of its participation in civil legal relationships, the character of its legal capacity, etc.; the general limits of the competence of the union republics is established by Articles 73, 76-81 of the Constitution of the USSR.

Of course, sometimes the content of the civil legal capacity of the state and the civil legal capacity of juridical persons (and even citizens) coincides. Thus, the state, like the citizen and like the juridical persons, may inherit both at law and by will (Arts.534, 552 Civil Code). However, the characteristic principle of special legal capacity of a juridical person is not applicable to the state; the state may not possess the right to personal property, etc. At the same time the legal capacity of the state includes the possibility of possessing rights which no one else may possess (only the Soviet Union may issue state bonds; only the union republics have the right of compulsory purchase of copyright).

The resources in the union and republican budgets, respectively, form the property basis for the participation of the state in civil legal relationships. In addition to this, this basis also includes other state property in its material form (and not only monetary resources) which is not allocated to state juridical persons. Furthermore, if the income from this property accrues to the federal budget or the expenses for its upkeep come from the federal budget, then the property itself must be considered as belonging to the Soviet Union. Where such income and expenses pass through a republican budget, the property belongs to that union republic.

In practice, the participation of the Soviet Union or a union republic in civil legal relationships is relatively rare. It takes place in internal turnover when the state issues domestic bonds taking on itself the obligation of repaying the holders of the bonds over a fixed period and in a pre-determined manner (the payment for earnings or the redemption of the bonds at nominal cost) with the sums given by them to the state. Relationships with the participation of the state are also formed when the state conducts various lotteries, with this difference, that in this instance the earnings do not go to the federal budget (as in the case of bonds) but into the budget of a union republic, which therefore must be acknowledged as the subject of the corresponding legal relationships. The state conducts these types of operations through the state workers' savings banks or other state organizations, acting as juridical persons, which in these circumstances act not under their own name but in the name of the state; therefore it is the state and not these organizations which are subject to any liability arising from these operations.

The state in the person of the Government of the USSR guarantees the safety of the monetary sums and valuables entrusted to savings banks and also is liable for obligations of the State Bank of the USSR for the deposits of citizens (Art.7 Charter of the State Bank of the USSR). This means that the USSR takes upon itself civil obligations to depositors to cover the debts of the savings banks in the event that they lack sufficient funds for that purpose in the same manner in which it takes on the obligations of the State Bank for the deposits of citizens.

The state participates in civil legal relationships in instances of the transfer to its ownership of ownerless property (Art. 143 Civil Code), of mismanaged houses and cultural treasures (Arts. 141 and 142 Civil Code), of treasure troves (Art. 148 Civil Code), and of unclaimed finds (Art. 144 Civil Code), as well as other instances set forth in law.

The state itself participates rarely in foreign commerce in civil legal relationships with foreign citizens or juridical persons, since the conduct of foreign trade transactions is basically carried out by special foreign trade associations which are subordinate to the Ministry of Foreign Trade of the USSR. These associations act as juridical persons and therefore themselves are liable for their operations with the state property allotted to them. The right to act independently in the foreign market is also granted to several other special organizations which are juridical persons. However, in some circumstances foreign trade transactions are conducted in the name of the Soviet Union by trade representatives of the USSR abroad, and all consequences of such transactions directly concern the state.

The participation of a sovereign state in these relationships endows them with certain particular features. In accordance with the norms of international law, a state and its property enjoy immunity abroad. This means that disputes in regard to these relationships may not be considered by a foreign court without a direct expression of consent on the part of the Government of the USSR. The property of the USSR located in another country may not be the object of execution, of alienation, or of arrest, and in general may not be the object of any enforcement proceeding. Related problems may be specially regulated by international treaties concluded by the USSR.

Chapter 7. Transactions

1. *Concept and significance of transactions.* In the sphere of civil law which regulates the varied and dynamic relationships of economic activity, civil rights and obligations arise most frequently as the result of purposeful and lawful actions of citizens and organizations. For instance, a citizen may draw up power of attorney or a will, give a promissory receipt and then liquidate the debt; a socialist organization may conclude a contract, execute it, and produce a bill for the services rendered, etc. Every one of these and analogous actions of citizens and organizations are accompanied by specific civil law consequences.

The lawful actions of citizens and organizations which are directed toward the establishment, change, or termination of civil rights and obligations are defined as transactions by Article 41 of the Civil Code. As is apparent from the above examples, transactions have wide application in everyday life.

Civil law transactions have a number of characteristics. These characteristics express their essence and, at the same time, make it possible to distinguish transactions from other legal facts which are outwardly similar to transactions but which have other functions and are regulated by rules other than the norms dealing with transactions.

In the first place, a transaction is a purposeful, i.e. *volitional* action of a citizen or organization. It is performed to attain a specific, legal result in which the citizen or organization is interested. For this the will of the given party must be expressed outwardly or else it is not perceptible to other parties. This outward expression of the will in the conclusion of transactions is characteristic of a transaction and makes it possible to distinguish it from legal events (for instance, natural calamities) and legal acts (for instance, the discovery of lost articles) which also lead to civil law consequences but have non-volitional character.

Secondly, a transaction is a *lawful* action which satisfies the requirements established by civil legislation. With this characteristic a transaction is distinguished from unlawful actions, which are described by the term "tort". Unlawful actions also lead to civil law consequences (in part to the obligation to compensate for the harm caused); however they are not transactions, they have special characteristics, and they are regulated by the rules on obligations resulting from the infliction of harm (Chapter 35).

Finally and *thirdly*, transactions are concluded by subjects of civil law and are directed at achieving exclusively civil law consequences. These characteristics make it possible to distinguish transactions from administrative, and in part planning, acts which are adopted by the organs of state management in the process of administering the national economy. For instance, the decisions of a ministry confirming a list of objects to be constructed (itemized lists) or issuing an order to a supply-marketing organization on the delivery of production are acts of planning not transactions. In the given example the acts issue from organs of management, which are subjects of administrative law, and involve not only civil law consequences (the obligation to conclude a contract), but also the administrative law obligation of the proper organizations to fulfill the decision of the competent state organ.

Civil law transactions are one of the basic legal forms used in the process of socialist economic activity and have a wide application. In executing transactions, citizens and organizations conclude and carry out contracts, settle accounts, and perform many other acts necessary to satisfy their economic, social, and cultural requirements. Insofar as civil legal transactions are executed within the sphere of the socialist economy, they manifest traits possessed by relationships of the socialist type. In part, transactions of organizations have a planning character and may not contradict the plan targets (for more details see Chapter 17).

2. *Types of transactions.* Civil law transactions are quite varied and may be divided into a number of types. The classification of transactions makes it possible to single out their legal characteristics and to understand better the purpose, sphere of application, and legal particularities of a given category.

First of all, it is necessary to distinguish *unilateral* from *bilateral* (multilateral) transactions. A unilateral transaction is one requiring the expressed will of only one party. This type of transaction includes the granting of powers of attorney, drawing up a will, accepting inheritance, announcing a competition, or performing a contract.

A *bilateral* transaction requires the expressed will of two parties, and a *multilateral* one requires the expressed will of three or more parties. Bilateral and multilateral transactions represent the agreement of several parties and are considered contracts. Contracts are the most widely used aspect of transactions; they will be examined in detail in Chapter 17.

Another classification of transactions is into the divisions of compensatory and gratuitous. Under a compensatory transaction, the participants on all sides perform a specific property consideration for the benefit of another side: the transfer of things or money, rendering a service or some other action. The contract of purchase and sale is a typical example of a compensatory transaction. The majority of civil law transactions are compensatory.

In a gratuitous transaction, the participants of only one side perform a property consideration. Examples of gratuitous transactions include contracts of gift, gratuitous use of property, or storage free of charge. Gratuitous transactions are less frequent than compensatory ones and are applied, basically, in the mutual relationships of citizens.

Depending on the moment that a transaction is considered to be performed, transactions are divided into those for which the agreement of the parties is sufficient (consensual transactions) and those which require, in addition to the agreement of the parties, the transfer of things or of money (real transactions). For instance, to conclude a contract of purchase and sale or a work contract, agreements among the parties are sufficient. To conclude an agreement of deposit for safe custody, in addition, requires the handing over of the article which is the object of the deposit. The following are also real transactions: contracts of loan, carriage of goods, or of gift.

A special group of transactions consists of the *conditional transactions*. These include those transactions which are concluded in such a way by the parties that the creation of rights and obligations depends on a circumstance which may or may not occur. This type of circumstance is known as a condition, and it may be a natural or some other event, for instance, reaping a good harvest, moving to another location, receiving specific property (vacation-home, inheritance), etc.

As an example, a conditional transaction is one under whose terms the seller is obliged to supply to the buyer an additional consignment of vegetables or fruit in the event of a good harvest. This type of transaction is said to be completed under delayed conditions, since the stipulated condition (a good harvest) delays the transaction from coming into effect. It is also possible to effect a transaction under condition of revocation, whose presence terminates the transaction. An example is a transaction which provides that in the event of a poor harvest (the presence of condition of revocation), there will be no sale of vegetables or fruits.

In order to protect the rights of the parties in a conditional transaction, Article 61 of the Civil Code provides that if the occurrence of a condition is hindered in bad faith by the party to whom the occurrence would be detrimental, then the condition is considered to have occurred and, conversely, if the party to whom the occurrence of a condition would be beneficial, in bad faith promotes the coming about of the condition, then the condition is considered not to have occurred.

It is necessary to distinguish the conditions of a transaction from the period of performance of a transaction. *The period of performance* is the period of time during which a transaction must be completed. A condition and a period of time have in common that their occurrence entails legal consequences. However, a condition may not be realized while the occurrence of the time period is inevitable. The period of time will be examined more closely in Chapter 18.

3. *The form of transactions.* The will of the participants in a transaction, reflecting their intention to conclude a transac-

tion, must receive specific external expression and consolidation. This is necessary so that the contents of the transaction become known to its participants and that the conditions of the transaction be clearly established so that it is possible without difficulty to decide the disputes of the parties should they arise in the future.

The method by which the will of the parties is expressed in the concluding of a contract is called the *form of the transaction.* There are two basic forms in civil law transaction: *oral and written.* The written form, in its turn, may be simple or notarial. As a general rule, for simple, everyday transactions the oral form is sufficient. For the conclusion of more complex and more important transactions, the law requires the written and sometimes the notarial form.

The simple, written form signifies that the conclusion of the transaction must be affirmed in written form by the parties effecting the transaction, or by their representatives (see Chapter 8 for representation). In this regard, the written form is recognized as not only comprising one document, signed by the parties, but also the exchange of letters, telegrams, a promissory note, or other written documents affirming the conclusion and conditions of the transaction. The notarial form requires that the contract which is being concluded be witnessed before a state notary or other official fulfilling that function.

According to Article 43 of the Civil Code, those transactions which are performed at the same time that they are concluded may be made orally unless the law provides otherwise. To this [group] belong those basic, every-day transactions which citizens conclude among themselves at a market, as well as in retail trade (without issuing a check). By virtue of Article 44 of the Civil Code, other transactions among citizens involving sums up to 100 rubles may also, as a general rule, be concluded orally.

According to Article 44 of the Civil Code, all transactions between organizations, between organizations and citizens (with the exception of the transactions mentioned above which are performed while they are being concluded), as well as transactions among citizens involving a sum of higher than 100 rubles, must be concluded in written form.

The notarial form is required by civil legislation only for some important and complex transactions: contracts of purchase and sale, the pledge of a house (Art.195, para. 1 Art.239 Civil Code), a gift of a sum of more than 500 rubles (Art.257 Civil Code), as well as a will (Art.540 Civil Code).

Sometimes, in addition to observing the written form (simple or notarial), it is necessary to register a transaction with a specific state organ. For instance, a contract involving the purchase and sale of a residential house must be registered with the proper executive committee of the local Soviet of People's Deputies.

Besides the oral and written form, civil legislation permits the conclusion of contracts by two other specific forms: first, by means of performing an action which signifies the conclusion of a transaction, and secondly, in the form of acquiescence [*molchanie*]. In these circumstances, the will of the participants in a transaction receives its expression through corresponding behavior.

Entering into an transaction by means of performing an action which signifies its completion takes place, for instance, when using a vending machine, or when a person receives a good or a service (storage in an automatic locker) upon putting in the proper amount of money. Another example is the sending of an article in response to an order received. In civil law, actions which testify to the intent of concluding a transaction are called implied [*konkliudentnye*].

According to Article 42 of the Civil Code, acquiescence indicates a will to conclude a contract only in circumstances provided in the legislation of the USSR and the RSFSR. In other words, there must be direct indication in legislation that acquiescence concludes a contract. These circumstances are quite limited. An example is the payment of a bill from the bank account of a socialist organization by means of acquiescence (tacit acceptance): the organization does not direct an objection to the bank against payment of the bill due, and the acquiescence indicates agreement to payment of the bill to be made through the bank.

4. *Conditions of the validity of transactions.* As has already been noted, transactions are lawful actions of citizens and organizations. Therefore, in order for a transaction to be valid and to give rise to results desired by its participants, it must conform to those requirements (conditions) which the law establishes for transactions.

Civil legislation provides that a transaction is valid if the following conditions are observed: a) the content of the transaction is legal; b) the participants to the transaction possess dispositive capacity; c) the expression of intent of the participants to a transaction corresponds to their actual will; d) the established form of the transaction is observed.

As a general rule, the violation of any one of these conditions renders a transaction invalid. There are two groups of invalid transactions in civil law: void (totally invalid) and voidable (relatively invalid).

The failure to observe the first two conditions is considered an especially serious violation of the norms of law, and as a rule causes a transaction to be considered void (totally invalid). This means that the transaction does not give rise to any of those legal consequences which a valid transaction of that type would engender. Besides this, the participants of a void transaction may be subjected to unfavorable property consequences as liability for violation the norms of law. The failure to observe the third condition, as a general rule, makes a transaction voidable (relatively invalid) rather than void. This means that such a transaction is valid and engenders the legal consequences for which it provides. However, the transaction may be disputed by interested parties or by the procurator in court or in *arbitrazh* which latter have the right to declare this transaction to be invalid either in whole or in part. In this instance the voidable transaction does not give rise to those legal consequences to which a valid transaction of that type would give rise, and the participants in the transaction may be subject to unfavorable property consequences which are a type of liability.

The legal consequences of violating the last of the enumerated conditions of a valid transaction – failure to observe the established form of a transaction – have special features and depend on which form of a transaction was violated. Therefore, it is necessary to treat with special care this condition of the validity of a transaction.

If the notarial form is prescribed for a transaction, then the failure to observe this form renders the transaction void (totally invalid). An analogous result occurs if the simple written form is violated where such a result is expressly provided for by law. For instance, failure to observe the written form renders void agreements on the liquidation of damages (Art.195 Civil Code), suretyship (Art.203 Civil Code), pledge (Art.195 Civil Code), and several other transactions. The law considers these transactions particularly important, and therefore it is always necessary to conclude them in written form.

In other circumstances, the failure to observe the simple written form leads to another result: the parties lose the right in a dispute to refer to the testimony of witnesses for confirmation of the transaction (Art.46 Civil Code). The transaction is considered valid and all other evidence for its confirmation is allowed with the exception of the testimony of witnesses.

This rule has important practical significance. In the event that one party refuses to carry out a transaction concluded orally for which the written form was prescribed, the other side will be in a difficult position since it does not have written proof and may not prove the fact of the conclusion of the transaction and its conditions. In introducing such a rule, civil legislation stimulates the participants in a civil activity to observe the established written form of transactions, and at the same time creates clarity with regard to the mutual rights and obligations of the parties.

In practice it happens that the requirements of law are violated not by the transaction as a whole, but only by some of its conditions. In such an instance, it would be unfair to consider the entire transaction invalid. Civil legislation takes note that such circumstances exist and in Article 60 of the Civil Code provides that the invalidity of a part of the transactions does not entail the invalidity of its other parts if it may be presumed that the transaction would have been concluded even without its invalid part.

5. *Types of invalid transactions and their consequences.* Deviation from the prescribed conditions of law necessary for the validity of transactions may manifest itself in various ways. Civil law therefore recognizes various types of invalid transactions, and establishes various legal consequences for them. Let us examine these consequences, beginning with void and progressing to voidable transactions.

Article 48 of the Civil Code establishes that a transaction which does not accord with the requirements of law is invalid. The term "law" is here interpreted in a broad sense, encompassing any legal norm. The consequences of an illegal transaction are as follows: each one of the parties is obliged to return to the other everything which was received under the transaction, and if it is not possible to return what was received, its value must be replaced by money. The mutual return of property under civil law is called by the term "bilateral restitution". Such are the general rules found in civil legislation on illegal transactions.

However, there are various types of illegal transactions. Among transactions which do not accord with the requirements of law, the Civil Code singles out those *transactions which were consciously concluded with an aim contrary to the interests of the socialist state and society*. These transactions are especially dangerous, and therefore their legal consequences are different. According to Article 49 of the Civil Code, if such a transaction is concluded intentionally by both parties, everything which was received and which will be due under the transaction is forfeited to the state; if only one party acted intentionally, everything given by the guilty party to the other is forfeited to the state and everything received by the guilty party under the transaction is returned to the other party.

For example, a youth sold a book bought by his father to a friend. It is clear that this type of transaction violates the requirements of law since the seller is not the owner of the book and could not dispose of it. This purchase and sale transaction is invalid, and under Article 48 of the Civil Code bilateral restitution is required: the book and the money received for it must each be returned to the proper party.

If an agreement for the sale of foreign currency valuables, which is a prohibited transaction (Art.137 Civil Code), was concluded and carried out between two citizens, the legal consequences are different. This transaction concerns not only its participants, but is contrary to the interests of the socialist state and society. Therefore, under Article 49 of the Civil Code, the transaction is invalid and everything received under it is forfeited to the state.

Another type of transaction which does not conform to the requirements of law and is invalid is a *transaction concluded by a juridical person in contradiction to the aim established by its charter, by the statute relating to it, or by the general statute on organizations of its kind.* If, for instance, a trading organization concludes a transaction providing that it carry out construction work, this transaction must be considered invalid. In accordance with Article 50 of the Civil Code, these types of transactions are dealt with by the rules of Articles 48 and 49 of the Civil Code, depending on their aims.

A transaction *which is concluded by a minor under the age of 15* is invalid, since the person does not have dispositive capacity.[1] The legal result of such a transaction, according to Article 51, is bilateral restitution. The party who has dispositive capacity is obliged, in addition, to repay to the party who does not have dispositive capacity the profits it received, as well as to reimburse for the loss or damage to the other's property if it knew or should have known that the other party did not have dispositive capacity. A transaction effected by a citizen who does not have dispositive capacity because of mental illness or mental retardation is also invalid and has the same legal consequences (Art.52 Civil Code).

1. This, it must be remembered, does not apply to petty, everyday and other transactions which a person under 15 years of age has the right to execute independently.

Finally, according to Article 53 of the Civil Code, *transactions which are effected only for appearances, without the intention of carrying out the juridical consequences* (pretended transactions), as well as *transactions which are concluded to conceal another transaction* (sham transactions) are also invalid. In the case of a sham transaction, those rules are applied which exist for the transaction which the parties actually had in mind.

For instance, a citizen, in order to escape an inventory of his property for compensation of harm caused, engages in a purchase and sale transaction with one of his relatives and does not receive any payment for the sale. This is considered a pretended transaction and under Article 53 of the Civil Code must be considered invalid and does not engender any legal consequences.

Let us now turn to an examination of voidable transactions, which become invalid if they are declared to be such by a decision of a court or *arbitrazh* in the course of a suit brought by interested parties or the procurator.

According to Article 54 of the Civil Code, a transaction effected by a minor between 15 and 18 years of age without the consent of his parents, adoptive parents, or guardians is considered invalid if a court proceeding is brought by the parents, adoptive parents, or guardian.[2] This situation results in the same legal consequences as do transactions concluded by minors under 15 years of age: bilateral restitution and reimbursement of losses brought about by the minor.

Under Article 58 of the Civil Code, a voidable transaction is one which is concluded under the influence of fraud, violence, threats, or collusion between an agent of one party with the other party, as well as a transaction which a citizen was forced to conclude as a result of the confluence of difficulties which are extremely unfavorable to his circumstances. These transactions may be disputed either in a suit brought by the victim, or in a suit brought by a state, cooperative, or public organization.

2. This rule does not apply to transactions which a person from 15 to 18 years of age has the right to execute independently (see Chapter 4).

In the event that such transactions are concluded, which sometimes happens in the interrelationships among citizens, the other party returns to the victim everything it received during the transaction, and if it is not possible to return everything, its value must be replaced in money. That which was received by the victim, or is due him, is forfeited to the state. This rule in civil law is called unilateral restitution. In addition, the other party must also reimburse the victim for any expense, losses, or damage to his property.

According to Article 57 of the Civil Code, a voidable transaction is also one which is effected under the influence of a mistake having substantive significance. The legal consequences of such a transaction are in the form of bilateral restitution. Besides this, the party guilty of the mistake is obliged to reimburse the other for any damage to its property caused by the mistake.

In order to apply Article 57 of the Civil Code properly, it is necessary to interpret correctly the term "mistake having substantive significance". This type of mistake is an incorrect conception of the circumstances of a specific transaction or lack of knowledge about them. A mistake having substantive significance may not be eliminated in general, or its elimination leads to significant expenses for the mistaken party. The reason for the mistake is not important. The question of whether the mistake is substantive or not is decided on the basis of concrete circumstances.

It is necessary to distinguish a mistake from fraud, whose presence leads to more serious legal consequences (see above). Fraud is characterized as intentional actions by parties in a transaction interested in its fulfillment. A mistake arises from lack of sufficient knowledge by the parties to a transaction.

For instance, a substantive mistake is lack of knowledge concerning the exchange of dwelling space designated for residential use for quarters for a mentally disturbed person. On the other hand, it is not a substantive mistake to lack knowledge of insignificant and easily eliminated defects in an exchanged premise. If the exchange of living space is linked to the presence of special amenities in the exchanged quarters, the presence of which was confirmed by the party giving it up for exchange, and these amenities do not exist, it is necessary to consider this transaction as concluded under fraud.

Chapter 8. Representation and Power of Attorney

1. *Concept and scope of the application of representation. In civil law, representation is the conclusion of transactions by one party, the representative of another party, who is being represented so that the juridical consequences to which these transactions are aimed arise directly for the represented party* (Art.62 Civil Code). The representative himself does not have any rights and obligations in the course of the concluding of the transactions: he acts for the represented party.

The concluding of transactions by representation is widely applied both in relationships involving citizens as well as in the activity of socialist organizations. It helps citizens and juridical persons to realize their legal capacity and to participate in civil activity. According to the law (para. 4, Art.62 Civil Code), a contract which must be effected personally, or contracts which are specifically [prohibited] by law [from being concluded by a representative] may not be made by representation. Included among such transactions, for instance, is the drawing up of wills; Article 540 of the Civil Code states directly that a will must be signed by the testator in his own hand. Also not allowed to [be concluded through representation] are juridical acts in the area of family law, for instance, the conclusion of marriage, adoption, etc.; these acts are not considered civil law transactions.

2. *Subjects of representation and demands which may be put to them.* A citizen acting as a representative, that is, one who actually exercises the dispositive capacity of the represented party, must himself possess civil dispositive capacity fully. A juridical person may act in the capacity of a representative if this corresponds to the aims of its activity and, therefore, enters into the content of its special legal capacity.

Any juridical person or citizen, whether capable or incapable, may be a representative. In regard to an incapable citizen, representation is carried out legally (see below, Part 4).

3. *Authority.* Representation without authority. *Authority is the right of the representative to effect juridical acts in the name of the represented party.* According to para. 1, Article 62 of the Civil Code, authority may be based directly on the law (for instance, the representation by parents in the name of their minor children under the age of 15), on the granting of a power of attorney, or on an administrative act (on this see Part 4). A representative must act in the interests of the represented party, and for this purpose the law establishes some general limitations on his power. Thus, in accordance with Article 62 of the Civil Code, a representative does not have the right to conclude a transaction for himself personally in the name of the represented party nor for another party for whom he is the representative at the same time. According to Article 33 of the Code on Marriage and the Family, a guardian acting in the capacity of legal representative of a person under his guardianship does not have the right to conclude transactions either for himself or for his spouse or close relatives in the name of the person under his guardianship. A guardian has the right to conclude transactions which exceed the limits of the ordinary only with the consent of the organs of guardianship and curatorship, while the conclusion of contracts of gift in the name of the person under curatorship is forbidden in general (on guardianship see Chapter 46).

Only those actions of the representative which are performed by him within the limits of his authority give rise to juridical consequences for the represented party. If the representative exceeds the limits of the authority granted to him, or if the transaction is effected by someone devoid of any authority, then such actions establish, change, or terminate the civil rights and obligations of the represented party only in the circumstances where these actions are subsequently approved by the represented party. Subsequent approval render the transaction valid from the moment it was concluded (Art.63 Civil Code). This situation is called representation without authority.

4. *Grounds for the origin of and types of representation.* The origin of authority may be linked with various juridical facts. It is necessary to distinguish between voluntary and legal representation.

In *voluntary representation*, the represented party chooses the representative himself and defines the range of his authority. Voluntary representation is usually based either on contract of agency (see Chapter 33) or on labor law if the representative is an employee of the represented party and the authority of the representative is based on the expression of the will of the represented party, as indicated by the issuance of a power of attorney or an administrative act emanating from the represented party. An example of the latter is the affirmation by a production association of the Statute on the Production Unit, by virtue of which its manager becomes the representative of the juridical person – the association within the limits established by this Statute. The same significance attaches to the issuance of a State Administration of Insurance certification to an agent, who is authorized to conclude contracts of insurance.

The presence of voluntary representation may also manifest itself from circumstances in which the representative operates: for instance, an employment situation where a person works as a sales person or cashier in a store, a clerk in a consumer-service work-shop, a cloak-room attendant, etc. (para. 2, Art. 62 Civil Code).

Legal representation was established to protect those persons, who do not have dispositive capacity who are not able to take care of themselves (the young, mentally disturbed, or mentally retarded). It does not depend on the will of the represented parties and arises from the circumstances of birth, adoption, and appointment (Arts. 14 and 15 Civil Code). There are also several other instances of legal representation.

5. *Power of attorney.* Power of attorney is the document given to a representative defining the range of authority of the latter for representation before third parties. Article 64 calls power of attorney *authority by letter*, from which it fol-

lows that the *written form* is obligatory for a power of attorney. A power of attorney may also be given in the name of a juridical person, but only, however, to conclude those transaction which do not contradict the charter (Statute) of the juridical person – the representative – or the general statute on organizations of that type (Art. 64 Civil Code). A power of attorney for concluding transactions which require the notarial form, as well as for some other conditions, must be certified notarially. Special rules regarding some aspects of the power of attorney are found in Article 65 of the Civil Code.

This is the usual type of power of attorney given to a citizen: "I, the undersigned, Ivan Ivanovich Ivanov, authorize Petr Petrovich Petrov (or: entrust P.P. Petrov) to receive the earned wages due to me for a certain period in such-and-such enterprise or institution (or: to receive correspondence; or, to sell some property belonging to me to a specific party for a certain sum; or, to any party and for a price at his discretion; or, not lower than a certain price; or, to conclude a certain transaction or number of transactions in my name). Place and date of the issuance of the power of attorney, and the period of its validity. Signature of the principal."

A power of attorney issued by a citizen for the receipt of earned wages, remuneration due to authors or inventors, pensions, grants and stipends, sums from State Savings Offices and for the receipt of correspondence, including both money and parcels, may also be authorized by an organization in which the principal works or studies, the manager of the house in which he lives, or the administration of a medical institution where he is undergoing treatment (para. 3, Art. 65 Civil Code).

In accordance with Article 66 of the Civil Code, a power of attorney in the name of a state organization is issued under the signature of its director with the seal of the organization; and in the name of a cooperative or social organization – under the signatures of the persons authorized to do so by its charter (Statute) with the seal of the organization. A power of attorney in the name of state, cooperative, or social organizations to receive or to issue money and other valuable property must also be signed by the chief (senior) bookkeeper of the organization.

According to Article 67 of the Civil Code, the period of a power of attorney may not exceed three years, and if the period of its validity is not indicated, then it has validity for one year from the day of its issue. An exception is made for powers of attorney authenticated by a state notary and intended for the conclusion of transactions abroad which do not contain instructions as to the period of their validity. These types of power of attorney remain in force until revoked by the person issuing them. Any power of attorney which does not include the date of its execution is invalid.

The person to whom power of attorney is issued must personally carry out those acts for which he has authority. But he may delegate authority for their performance to another person if he is authorized to do so by his power of attorney itself, or is forced to do so by circumstances in order to protect the interests of the principal (for instance, sudden illness of the agent). A power of attorney issued in the process of delegation must be notarially authenticated, and the period of its validity may not be greater than the period of the original power of attorney. The agent (representative) is obliged to inform his principal (the person authorizing the representation) of all essential information concerning the person to whom he is delegating authority; if he does not do so, he takes on the property liabilities of the delegate as if they were his own (Art. 68 Civil Code).

The person granting a power of attorney may revoke this authority at any time, and the person to whom a power of attorney was granted may renounce it (Art. 69 Civil Code). When a power of attorney is terminated, so is any delegation of this authority. A power of attorney is also terminated by death, or the termination of a juridical person (either as principal or as representative), as well as by recognizing the representative as incapable, partially incapable, or missing.

In instances when representation or authority is based on an administrative act, the document in which this act is expressed may replace a power of attorney. Thus, the director of a production unit of a production association (combine) may act as the representative of the latter, without being granted a power of attorney, on the basis of the Statute concerning this unit which was affirmed by the association.

Chapter 9. Limitation of Actions

1. *Concept and significance of limitation of actions.* As a rule, the defense of civil rights is realized through the means of filing a suit in court, *arbitrazh*, or other organ (Art. 6 Civil Code). Definite periods, called limitation of actions, are established for turning to these organs. In Article 78 of the Civil Code, limitation of actions is defined as "the general period for bringing an action to defend a right by the party whose right was violated."

Limitation of actions is essential to settle property relationships which have in fact been formed, and also because the determination of disputed circumstances may not always be accomplished after a long period of time with the necessary certitude. In the mutual relations of socialist organizations, limitation of actions helps to strengthen contractual, economic, and fiscal discipline. It alerts them to be concerned about recovering monetary sums and at the same time to exercise rights belonging to them.

Limitation of actions is not applied in all legal relationships. It is not used in claims concerning the defense of honor, the dignity of citizens, the right to a name, the violation of personal non-property rights of authors and inventors, or the defense of the right to state property by demanding property from social organizations and citizens. Limitation of actions does not extend to the claims of depositors on the release of deposits put into state workers' savings banks or into the State Bank of the USSR since these relationships are meant to last for the duration of the period that the money is safeguarded. In the law are also found other relationships which are excluded from the sphere of the effect of limitation of actions.

Considering the important role of limitation of actions in the protection of the rights of socialist organizations and citizens and the securing of state discipline, Article 82 of the Civil Code establishes that limitation of actions is applied by a court, *arbitrazh*, or arbitration tribunal whether or not the parties have requested it. The norms regulating limitation of actions are of an imperative character. This means that the periods of limitation of actions and the process of their calculation may not be altered by agreement of the parties. In the event of such agreements the court or *arbitrazh*, whether or not the parties so desire, applies the period of limitation of actions and the provisions regarding the period provided in law.

2. *Periods of limitation of action.* These are established by law. In regard thereto, it is necessary to distinguish: 1) general and 2) shortened or special periods of limitation of action. The latter are designated for claims specified in law. The general periods of limitation of actions are three years for relationships in which citizens participate, and one year for relationships among socialist organizations (Art. 78 Civil Code). The relatively short period of limitation of actions for mutual relationhips among socialist organizations has the aim of securing strict discipline in property relationships.

The special periods of limitation of actions are shorter than the general periods and have the purpose of alerting the party empowered by law to bring influence to bear more quickly on a faulty debtor.

In order to strengthen the educational influence of property liability in suits concerning the recovery of liquidated damages, penalties, or fines, a six-month period of limitation of actions has been established. This limitation is applied in suits which arise from the supply of goods of defective quality, as well as of incomplete products or those containing defects. The shortening of the period of limitations to various lengths of duration is applied in suits which arise from the carriage of goods, passengers, or baggage, relationships of communications organs with clients, and in several other situations.

3. *Commencement of the running of the period of limitation.* This has decisive significance for the correct calculation of periods of limitation. An erroneous determination of the beginning moment leads to a wrongful lengthening or, on the

contrary, shortening of the period of limitation. The law provides that the running of the period of limitation of action commences on the day that the right to file suit arises. And the right to file suit arises on the day when a party knows or should know of a violation of its right (Art.83 Civil Code).

In this manner, the commencement both of general as well as of shortened periods of the limitation of action are defined as that moment when the party who is entitled to do so may turn to the court, *arbitrazh*, or other organ for the enforced exercise of its right.

Usually a party knows or should know about a violation of its right at the moment of the violation. However, it happens that these moments do not coincide. For instance, a person who undergoes a trauma may find out about material harm caused to him not at the moment of violation of his right but when the loss of capacity to work or the extent of his losses becomes apparent. It is clear that while the person does not know of the violation of his right, he may not turn to organs of the state with claims for the defense of his right. However, if the plaintiff does not know about the violation of his right as a result of carelessness or lack of good management, it is proper to calculate the commencement of the running of the limitation of actions from that moment when he should have known about the violation of his right, and the running of the period of limitation of action begins on that day.

In order for the running of the period of limitation of action to commence, it is not necessary that the aggrieved person know the identity of the violator of his right. But the lack of knowledge of the identity of the violator or of the place where he is located may be considered a mitigating factor for allowing the period for bringing a suit to expire and a basis for a court or *arbitrazh* to restore this period.

In many legal relationships a necessary condition for filing suit is a preliminary declaration of claims by the aggrieved party to the debtor. This procedure was established by the Law on State *Arbitrazh* in the USSR, adopted by the Supreme Soviet of the USSR on 30 November 1979. Article 12 of this law provides that a dispute between socialist organizations may be transferred for decision to the State *Arbitrazh* only after the parties have taken measures to regulate the dispute directly through the established procedure. The periods for claims do not, as a rule, extend the periods of the limitation of action. According to an explanation of the State *Arbitrazh* of the USSR, the periods for the filing and review of claims are included in the periods of limitation of action established by law. The commencement of the running of the limitation of action coincides with the day on which the right to file a claim arises.

Another procedure for the calculation of the commencement of the running of the limitation of action is established for suits brought against transport organizations. Article 84 of the Civil Code establishes a period of two months for the limitation of action for such claims and provides that prior to filing a suit against a carrier in a case arising from carriage, it is necessary to make a claim against him. The running of the limtation of action in such instances begins from the day on which a reply is received or the expiration of the period established for receipt of a reply to the claim. This means that, in such instances, the period for bringing claims is not included in the period of the limitation of action. The commencement of the running of the period of the limitation of action is the day of receipt of the reply refusing the claim or the day on which the period for a reply to the claim expires. Under Article 84 of the Civil Code, the substitution of the parties to an obligation does not result in an alteration of the period of the limitation of action. The substitution of parties occurs in legal relationships when juridical persons are reorganized, in cases of inheritance, or assignment, and in the transfer of a debt. In such instances, the right to bring a claim under the period of limitation of action which still remains is transferred to the new creditor.

4. *Suspension and interruption of the period of the limitation of action.* Article 85 of the Civil Code lists those circumstances for the duration of which the limitation of action is halted, e.g. does not run. These circumstances are objective,

and their effect does not depend on the will of the participants in a disputed transaction. While they exist, they hinder the aggrieved party from bringing suit for the defense of a violated right. These circumstances include: a) force majeure; b) the establishment by the government of a postponement in the performance of obligations (moratorium); and c) membership of the plaintiff or defendant in the Armed Forces of the USSR during a mobilization.

A. *Force majeure* is an extraordinary and unavoidable event under given conditions. Included among force majeure are the destructive aspects of nature such as floods, avalanches, landslides, earthquakes, and some social events such as military actions, epidemics, or other circumstances which make it impossible to file a suit during that time.

An event is considered to be force majeure if it is extraordinary and cannot be predicted. If a party knows in good time of the possible occurrence of the given obstacles, references to force majeure have no basis. Force majeure is characterized by objective unavoidability, e.g. lack of necessary technical and other means with which it would be possible to prevent the consequences linked with that event. The reference in Article 85 to "given conditions" indicates the relativity of the concept of force majeure. Already in 1925 the Supreme Court of the RSFSR emphasized that "an obstacle becomes force majeure not on the basis of the inner qualities possessed by it but on the basis of the interrelationship of a number of conditions and concrete circumstances. That which in one place, in one situation, is easily avoidable may in another [place] be unavoidable."

B. *A moratorium* is a postponement in the execution of obligations under extreme circumstances which is established by the Council of Ministers of the USSR or the Council of Ministers of a union republic. In our country, a moratorium was declared, for instance, in the circumstances of World War II (the Great Patriotic War) in regard to the recovery of debts involving the financial liabilities of state and cooperative organizations which were at that time under the occupied, and later liberated, regions of the country. It is possible to declare a moratorium for a specific region of the country in connection with extreme circumstances.

C. *The presence of the plaintiff or defendant* in the Armed Forces of the USSR suspends the running of the limitation of action only for the length of the stay of the plaintiff or defendant in a unit of the army which is mobilized. Consequently, the recruitment of a citizen for service in the Soviet Army during peace-time does not constitute grounds for the suspension of the period of the limitation of action.

These three instances of the suspension of the running of the period of the limitation of action arc subject to obligatory application both for general as well as for special periods of limitation.

For suspension of the running of the limitation of action, it is necessary that the onset or the effect of the circumstances indicated in Article 85 was within the last six months of the period of the limitation of action, or within the limitation period if this period is less than six months. It is supposed that if the event arose and was terminated earlier, then the interested parties had sufficient time to defend their rights. From the day of the termination of the event which was the basis of the suspension of the limitation, the running of the period of the limitation of action continues for a further six months, or if it is a shorter special period of limitation, then it continues for the length of the shortened period of limitation.

The law may also make provisions for other grounds to suspend some periods of limitation. For instance, in accordance with Article 471 of the Civil Code, the running of the limitation period for suits for the compensation of harm to the health of a citizen is suspended during the time in which the proper organs decide the question of whether to grant a pension or relief to the interested party, since the amount of these sums directly influences the amount of the compensation for harm to the victim.

It is necessary to distinguish a suspension of the limitation from an *interruption* of the period of the limitation of action. An interruption signifies that the time which has run since before the event causing the interruption is not taken into account. After an interruption, the running of the limitation of action begins again for the entire period provided by law for a specific claim. Article 86 of the Civil Code establishes two grounds for the interruption of limitations: 1) the presentation of a claim or the 2) acknowledgment of a debt.

Filing a claim in the proper manner, e.g. in the proper court or *arbitrazh*, is the general basis for interrupting the limitation of action. Naturally, filing a claim suggests a decision of the dispute on its merits, and in those instances there is no possibility for interrupting the limitation period. But if the dispute is not decided and the case is not dismissed or if the period of limitation elapses while the case is being decided, the interruption of limitation presents the aggrieved person with the possibility of defending his rights. If the claim is not brought in the proper manner, for instance by violating the rules of jurisdiction, the interruption of the limitation period does not occur.

The second basis for interrupting the limitation of action is related only to disputes involving citizens. This is the performance by the liable party of an action testifying to his acknowledgment of his debt. Included among such actions are: a request to postpone the debt, partial payment, or various declarations to creditors announcing an intention to pay the debt. The actions of the debtor are also considered to be an acknowledgment of the debt in those instances when they are performed in relation to other parties if the creditor has knowledge of this. In all of these instances, the running of the limitation of actions is renewed every time to the full period from the moment the debtor performs the actions which were described.

5. *The restoration of the limitation of actions* by the court or *arbitrazh* is provided for by law in instances where it is acknowledged that there were mitigating circumstances for the lapse of the limitation of action (Art. 87 Civil Code). While the grounds for the suspension and interruption of the limitation of action are carefully defined by law, there is not even a listing of examples of grounds for the restoration of

the limitation of actions. This problem is decided by the court or *arbitrazh* in every instance by taking into account all concrete reasons which resulted in an untimely resort for the defense of a violated right. The question of the restoration of a lapsed period of the limitation of action is reviewed upon the application of the plaintiff or upon the initiative of the court or *arbitrazh* during the process of examining the case on its merits. In restoring the period of limitation, the reasons for considering the lapse to be mitigated are given in an order. Judicial practice acknowledges as valid reasons a lengthy business trip, a serious illness of the plaintiff, or lack of knowledge as to the address of the defendant. If the reasons for the lapse of the period of the limitation of action are carelessness or inattentiveness on the part of the plaintiff or of employees in a socialist organization, then the period of the limitation of action is not subject to restoration.

The restoration of the period of the limitation of action which lapsed due to mitigating circumstances does not in itself determine the outcome of a case. The verification and evaluation of the evidence of the claims filed lead either to the granting of relief or the dismissal of the suit if the claims of the plaintiff were unsubstantiated as to the merits.

6. *Consequences of the expiration of the period of the limitation of actions.* The expiration of the period of the limitation of action prior to the filing of a claim is a ground for the dismissal of the suit (para. 1, Art. 87 Civil Code). However, the claims for the defense of a violated right are taken under advisement by the court or *arbitrazh* whether or not the period of the limitation of action has expired (Art. 81 Civil Code). In these two norms of law, there are two concepts of the right to sue to be distinguished: 1) the right to sue in the material sense as the right of a person with the help of state organs to obtain relief as regarding the claim which has been brought and 2) the right to sue in the procedural sense as the right to resort to a court, *arbitrazh*, or other organ with a claim in order to defend a violated right.

The right to sue in the material sense is extinguished upon

the expiration of the period of the limitation of action. However, the right to sue in the procedural sense is not extinguished by the limitation of action. This means that the court or *arbitrazh* must take into consideration, on general grounds, suits filed after a lapse of the period of the limitation of action. The Plenum of the Supreme Court of the RSFSR has pointed out that the refusal by a court to recognize a suit, in part on the basis of the expiration of the period of the limitation of action, is an unlawful refusal to render justice. A suit which is filed is considered on its merits with an explanation of all the circumstances of the disputed legal relationship, including the reasons for the lapse of the period of the limitation of action and the presence of grounds for the suspension, interruption, or restoration of the limitation period.

It is possible that the debtor in a legal relationship fulfills his liability after the expiration of the limitation of action. In these instances, Article 89 of the Civil Code provides that the debtor does not have the right to demand request performance in return, even though at the moment of performance he does not know of the expiration of the limitation. From this it follows that the expiration of the period of the limitation of action does not terminate the legal relationship, but only results in the impossibility of the enforced realization of the liability of the debtor with the help of a state organ. This situation, as a rule, does not apply to the mutual relationships of socialist organizations.

When monetary sums are the subject of obligations in relationships involving state organizations, the debts for which the period of limitation has expired are written off from the balance of the creditor to be included in his losses by order of the manager of the organization, with a report of this going to a higher organization. However, this sum does not remain as profit of the debtor, but must be transferred by him to the state budget with a payment for forfeiture for a delay in its payment.[1] This procedure reflects the require-

1. See paras. 38 and 39, Statute on Bookkeeping Accounts and Balances, confirmed by decree of the Council of Ministers of the USSR, 29 June 1979. (*Collection of Decrees of the USSR* 1979 No.19 item 121).

ments of economic accountability, and is directed at preventing instances when the bookkeeping entry of a state organization does not reflect the true results of its economic activity.

Another rule is applied in relationships between cooperative and other social organizations which are the owners of their property. The amount of the creditor's indebtedness after the expiration of the period of the limitation of action is subject to being entered into the profits of the creditor-organization.

In regard to obligations concerning the transfer of other property or the performance of work, the performance of such liabilities is not allowed after the expiration of the limitation of action. The fate of the property which was subject to transfer between state organizations is decided in these instances by administrative act of a higher organization.

When the period of the limitation of action of a major claim expires, the period of the limitation of action of a supplementary claim is also considered to have expired. If it is impossible to attain execution of the primary obligation, then there are no grounds to submit claims for liabilities supplementary to it, for instance, payment of forfeits, interest, or other means to secure execution.

The Law of Property

Chapter 10. Rights of Ownership. General Concepts

1. *Concept of ownership and rights of ownership.* Ownership is an economic category. It is a social relationship, e.g. a relationship among people for the purpose of acquiring and owning things – the products of nature and of labor. Karl Marx pointed out that "every product is an acquisition by the individual of objects of nature within the limits of the specific social form and by means of it."[1] V.I. Lenin also wrote about property as acquisition.[2] It is an indispensable condition of production that property always accompanies the life of society; without acquisition, consumption is not possible.

In distinction from ownership, *the right of ownership is the totality of all legal norms which secure for individual persons, collectives, or society the condition of the possession (acquisition) of the means of production and the products of labor.* The right of ownership arises after the division of society into classes, when the state secures and defends property relationships by distributing the means of production and other [objects] of material value among the members of society. Every socio-economic formation contains its forms of ownership which define the economic system of society and its social relationships.

Under a slave-holding, feudal, and capitalistic system, private ownership is dominant although its forms significantly distinguish one [system] from another. All forms of private ownership are opposed by socialist ownership which is qualitatively different. In a socialist society, the means of production belong to the workers and are utilized in their interests for the systematic increase of the material and cultural level of the life of the people. The right of ownership secures and protects these property relationships.

1. K. Marx and F. Engels, *Works,* Vol. 12, p. 713.
2. See: V.I. Lenin, *Complete Collected Works,* Vol. 1, p. 178.

2. *Forms of ownership and types of ownership in the USSR.* According to Article 10 of the USSR Constitution, the basis of the economic system of the USSR consists of socialist ownership of the means of production in *the form of state (all-people's) and collective-farm/cooperative ownership.* These forms of ownership secure the ownership of the means of production for the people, and define the character of the national economy and the division of Soviet society into two friendly classes – workers and peasants.

The property of trade unions and other social organizations is also socialist property. It deals basically with the sphere of the consumption of things of material value.

Thus, socialist ownership in the USSR exists in three forms: a) *state ownership*, which is the property of all the Soviet people; b) *collective-farm/cooperative ownership*, which is the property of the individual collective-farms, other cooperative organizations, and their associations; c) *ownership of trade-unions and other social organizations.*

Under socialism there is also *personal property of citizens* based on earned income and meant to aid in the satisfaction of the material and cultural needs of citizens (Art. 13 USSR Constitution).

3. *Contents of the right of ownership.* The special feature of the right of ownership lies in the fact that this right is always linked with specific objects. But the right of ownership as a relationship is not a relationship between person and object, but a relationship between people which is based on objects. This is a relationship between the owner, who has certain rights, and all other members of society who are obliged not to violate his rights and not to infringe on property belonging to someone else.

The right of the owner consists of the totality of three powers: *possession, use, and disposition of the object* (Art. 92 Civil Code). These powers also characterize the content of the right of ownership. They are realized by the owner permanently, immediately, and at his discretion, but within the limits established by law. These limits depend on the form of the ownership and are different for different subjects of the right of ownership. In part, on the basis of Article 13 of the USSR Constitution, property which is in the personal possession of a citizen may not serve as a source of unearned income [or] be used to the detriment of the interests of society.

Possession is the actual ownership of an object, the possibility of physically acting on the object. The person who actually holds an article in his hands or in his household possesses it. The right of possession of property means that the possibility of actually owning it is secured by law.

The owner has the right of possession, but he is entitled to transfer the property into the possession of another person. For instance, under a contract of safekeeping an object is transferred into the possession of the safekeeper. A possessor who has an object on legal grounds is called a *lawful possessor*. The rights of lawful possession are secured by the state (Art. 157 Civil Code). The possession of an article may also be *illegal*, if the possessor does not have the corresponding right to possession. For instance, the possession of a stolen article, the appropriation of a lost article, or other unlawful retention of someone else's property is illegal. It is also illegal to obtain an article from a person who does not have the right to dispose of it, for instance, the purchase of stolen property.

Use means the possibility of extracting from the object its useful properties. By use also is meant the receiving of the fruits and profits from the object. The right of use means that the possibility of using the object and thus satisfying one's needs is secured by law. The right of use belongs to the owner, but may be transferred by him to another person. For instance, under a contract of property hire, the owner transfers the property for temporary use upon payment by the hirer (Art. 275 Civil Code).

As a rule, the right of use is linked to the right of possession, but these powers may also be realized independently. For instance, in transferring an object for delivery to an organ of transport or communication, or in depositing an article of clothing in a cloak-room, only the right of possession of the object is transferred, not the right to use it.

Disposition consists of the determination of the juridical fate of the object by means of alienation, destruction, etc. The right to dispose means that the possibility of determining and changing the juridical status of the object is secured by law. The right of disposition, as a rule, is linked with the right of ownership, since only the owner determines or changes the juridical states of the object. But under special legal instructions, another person may also effect disposition of property. A court executor sells the property of a debtor; a railroad transfers freight to another person if it is impossible to convey the freight to the consignee (Art. 76 Charter of the Railroads of the USSR).

4. *Objects of the Right of Ownership.* Those things which constitute the object of the right of ownership are always divided into groups depending on their legal conditions. The divisions are:

a) *things* which are the *exclusive property of the state* and may not be acquired or belong to the right of ownership of a collective-farm/cooperative, other social organizations, or individual citizens. The earth, its minerals, waters, and forests are objects which are the exclusive property of the state. They are allotted to organizations and citizens for use only in the manner established by law.

b) *things* which may be *acquired* only by *special permission.* These include: weapons, aircraft, powerful poisons, precious metals in the form of coins, ingots, or raw condition, foreign currency, and others (Art. 137 Civil Code). The non-observance of the special procedure established for acquiring these objects carries with it the invalidity of the transaction and the withdrawal of such items without compensation to the ownership of the state (Art. 49 Civil Code).

c) *things* which may be *acquired for ownership* by anyone. However, special rules exist for acquiring certain objects for ownership. For instance, the law establishes a maximum number of livestock which may be under the personal ownership of a citizen (Art. 112 Civil Code). Certain conditions are established for the contract of purchase-sale of a dwelling (Arts. 106, 238 Civil Code).

d) *things individually determined* and things *determined by generic traits*. The former have individual traits and may be distinguished within a general mass of homogeneous objects. The latter are determined only by number, weight, or measure. The right of ownership exists only in relation to individually-determined objects, since it is impossible to be the owner of potatoes, apples, or grain in general. The right of ownership is aimed at a specific object. But after the isolating a definite part of such things, these objects become individualized.

The contract of property-hire may be concluded only in relation to an individually-determined object since, at the expiration of the agreed term, the hirer must return precisely that object which was taken for use. On the other hand, the object of a contract of loan is a thing which is defined only by a generic trait. Upon the expiration of the term of the contract, the borrower does not return the objects received by him but rather the same number of things of that type and quantity.

e) *divisible and indivisible things*. The former includes objects which may be divided without harming their economic function, for instance, a quantity for milk, grain, or cement. On the other hand, a book, table, piano, or similar objects are impossible to divide. This classification has significance in the division of common property, the return of property, and in inheritance.

Divisible things can be divided and their various parts given to various people; indivisible [things] are either given to one of a general group of owners who must pay monetary compensation to the others, or they are sold and the money received for them is divided among the owners.

In some instances, it is possible to divide property without harming its economic purpose, but the division affects the material or artistic value of the property (for instance, a library, collection, service, and so on). In case of dispute, judicial practice considers such things as indivisible and not subject to division.

The law also provides for the division of things into the *principal thing and the accessory* (Art. 139 Civil Code). *By accessory* is meant a thing whose purpose is to serve the principal object and which is linked to it by general economic design. An accessory is always a separate object, for instance, a key for a lock, a frame for a picture, a case for a violin, etc. The question of whether to acknowledge a thing as an accessory is often decided by law, a standard, or by contract. As a rule, the accessory shares the fate of the main object. However, the law or a contract may stipulate otherwise. For instance, the parties may agree that in the purchase of a violin the case remains with the seller.

Money and securities are special objects of the right of ownership.

Money is the common equivalent in civil turn-over; with its help the value (price) of the other things, with the exception of objects which are the exclusive property of the state, is fixed. As a rule, money lacks individually-determined traits. It is characterized only by the sum and is not defined by the number of notes but by the monetary units which are contained in them. It is the legal means of payment. This means that no one has the right to refuse to accept money in clearing a payment. On the territory of the USSR, monetary obligations must be carried out, and payment effected, in Soviet currency.

Money may be individualized by recording the number of a specific bank note, for instance, in an investigative report. Then it acquires the traits of an individually-determined thing. It is also possible to individualize money when it is used for purposes other than as a means of payment, for instance, in collections.

A security is a document by which a property rights is concluded and which must be presented in realizing that right. Bonds for state loans, lottery tickets, and letters of credit are widely used [securities]. Checks and other securities are used in the payment relations of socialist organizations. Without possession of a security and the ability to present it, it is impossible to secure the right expressed in it, for instance, to receive winnings or a monetary sum from a lottery even if there are witnesses or documents testifying to the right of ownership of the security. The receipt upon receiving a certain amount of money or a thing is not a security. It may be substituted by other proof provided for by law.

5. *Methods of acquiring the right of ownership*. The law links the acquisition and the termination of the right of ownership to various circumstances (grounds). These methods are commonly divided into *primary* and *derivative*.

In *primary* methods, the right of ownership to a certain object arises for the first time because its origin is not linked to the right of another person to this property. In *derivative* methods, there is legal succession, e.g. the conditionality of the rights of a new owner by the rights of the former owner.

The primary method for the origin of the right of ownership of state property is nationalization, e.g. the enforced confiscation, based on law, of property to the ownership of the state. During the first years after the October Revolution, all basic means of production belonging to the capitalists and landowners passed to the Soviet government without compensation through the process of nationalization.

At the present time, the basic primary method for the origin of state and collective farm-cooperative ownership is the economic activity of social organizations, as a result of which new objects of production and dwellings are created, and industrial and agricultural products and other merchandise are produced. The decision of the XXVI Congress of the CPSU and of the May 1982 Plenum of the Central Committee of the CPSU, which adopted the Food-Supply Program of the USSR for the period to 1990, provides for a significant increase in goods to satisfy the vitally important needs of Soviet citizens, requiring in the 1980s the orderly transition to the most beneficial intensive factors of economic growth and of a sharp increase in labor productivity.

The primary methods of acquisition of the right of ownership include: *the mismanagement of property, recognition of property as ownerless, unclaimed found objects, treasure trove, requisition, and confiscation*. On the basis of these grounds, the right of ownership arises chiefly for the state. In accordance with Article 141 of the Civil Code, it is possible by decision of a court to transfer the right of ownership of a dwelling to the state in case of its mismanagement. By *mismanagement* is meant such treatment of the property that it falls into ruin. A dwelling is taken away from the owner only in instances of culpable mismanagement of the house by the owner. The executive committee of the local Soviet of People's Deputies at first gives a deadline for the repair of the house to the owner, and if the citizen does not carry out the necessary repairs without good reason, the court may confiscate the house without compensation for the fund of the local Soviet. Ownerless property may also pass to the ownership of the state by decision of a court (Art. 143 Civil Code). In contrast to property that is mismanaged, which does not have an owner although he may not take care of his property, ownerless property has no owner at all or he is unknown. Such property is discovered by financial organs. One year after such property has been registered, the people's court declares the property ownerless and transfers it to the ownership of the state. In rural areas, property which had previously belonged to a collective farm household passes to the ownership of the collective farm on whose property it is located. The decision of the court is rendered on the basis of an application made by the collective farm one year from the day this property was registered by the executive committee of the local Soviet.

Unclaimed found property passes to the ownership of the state. Found property is an object which is lost by the owner or possessor and found by another person; it should be returned to the owner (Art. 144 Civil Code). If the owner is not known to the finder, the latter must give the object to the police [militsiia] or to the executive committee of the village Soviet of People's Deputies. An object found in an institution, enterprise, or on a transport vehicle is given to the management of the particular organization. As a rule, found articles are safeguarded to be returned to the owner for six months, after which time the right of the owner to the lost article terminates and the right of the state to the article arises. The person finding the article does not receive a reward, but has the right to request imbursement for his expenses from the owner (Art. 146 Civil Code).

Special rules are established with regard to *untended* and *stray livestock* (Art. 147 Civil Code). The person detaining such livestock must return them to the owner or give information about the detention to the police or to the executive committee of the village Soviet of People's Deputies so that a search may be made for the owner. The livestock is given for temporary use to the nearest state or collective farm. If the owner is not found within a defined period of time, he loses the right of ownership and the livestock passes without compensation to the ownership of the collective farm or to the state (if the livestock is kept on a state farm).

A treasure trove is money or other valuable objects hidden underground or in some other fashion whose owner cannot be found or who lost the right to [the objects] under law (Art. 148 Civil Code). A treasure trove passes into the ownership of the state. The persons finding a treasure trove are obliged to pass it to a financial organ and have the right to a reward of 25% of the worth of the valuables handed over. Appropriation of a found treasure trove is punishable by criminal law (Art. 97 Criminal Code). If the search for such valuables falls within the scope of a person's official duties, he does not have the right to receive the stipulated reward.

Requisition is the enforced alienation of property for the interests of the state or society with payment to the owner for the value of the property. Requisition is permitted only in instances and by procedure established by law (Art. 149 Civil Code). Requisition is occasionally carried out under extreme circumstances on the basis of special acts of the organs of government and management, for instance during natural calamities or a state of war. Payment for the requisitioned property is made at state prices by the organ to which the property is transferred. Specific instances of requisition are indicated in the Civil Code. Article 142 of the Civil Code provides that in case of the mismanagement of property having significant historic, artistic, or other value for society, the owner must first be given a warning to cease the improper treatment. If this request is not fulfilled, the cultural valuables may be seized by a decision of a court and placed under the ownership of the state; the citizen receives compensation for the value of the property. The seizure of such property takes place on the basis of a suit brought by a museum, archive, writers' union, or other state or social organization. The seizure of valuable metals and diamonds is also requisition, as provided in Article 150 of the Civil Code.

Confiscation is the seizure by the state of property without compensation from persons who have committed a criminal act. Confiscation is used as a means of punishment for certain crimes: embezzlement of state or social property, speculation, defrauding buyers and clients, engaging in a forbidden trade, receiving bribes, and in other cases.

Derivative is the term used for the origin of ownership when the rights of a new owner are conditioned by the rights of the previous owner. By this procedure ownership arises on the basis of agreement of purchase-sale, contract, exchange, gift. In this situation, the specific moment of the transfer of ownership has significant meaning.

Article 135 of the Civil Code provides that the right of ownership of the acquirer of property by contract arises at the moment of the transfer of the thing unless otherwise stipulated by agreement of the parties. The parties may, for instance, state in the contract that the transfer of the right of ownership occurs at the moment of the conclusion of the contract or with the payment of a price for the thing independently from when the object is transferred. A transfer is the actual yielding of a thing to the possession of the acquirer and also: a) yielding a thing to the post office to be sent, or b) yielding a thing to a transport organization to be delivered to the acquirer.

It follows, therefore, that the right of ownership passes to the acquirer from the moment the thing is yielded to the post office or to the carrier. From that moment, as a rule, he bears the risk of the accidental destruction or damage of the thing (Art. 138 Civil Code). If a fire or natural calamity occurs on the journey and the thing is destroyed or damaged, the buyer to whom the right of ownership was transferred at the moment that the thing was sent must pay the full amount to the seller. The rule concerning the moment of transfer of risk due to the accidental destruction of a thing has dispositive character and may be changed by agreement. If the transfer or the acquisition of the thing is delayed, the party responsible for the delay bears the risk of the accidental destruction or damage of the thing.

Derivative methods also include the inheritance of property by law or by will, the transfer of a thing on the basis of a decision of a superior organ of management, and on several other grounds.

Chapter 11. The Right of State Socialist Ownership

1. *Concept and meaning of the right of state socialist ownership.* The USSR Constitution defines state ownership as "the common heritage of the entire Soviet people, the basic form of socialist ownership (Art. 11). More than 90 % of the country's productive assets belong to the state. The most valuable types of property are under the ownership of the state. The distinctive feature of the right of state ownership in the USSR is the combination of the highest degree of state power with the competence of an owner. The Soviet state itself defines its competence for managing state property and the procedure and legal forms by which it will act as manager.

State property is the main source of public wealth, the welfare of the nation, and the assurance of the security of the country. Most of the social rights of citizens are realized on the basis of the utilization of state property. The requirements of the USSR Constitution to safeguard and strengthen socialist property aid in realizing the rights of citizens. These requirements not only assume a careful attitude on the part of citizens toward the national welfare but also positive acts directed toward increasing it, e.g. raising the effectiveness of labor, improving the quality of production, expedient use of funds, economizing in materials and energy supplies, or reducing the cost of capital construction. A decree of the Central Committee of the CPSU and the Council of Minister of 12 June 1979, "On the Improvement of the Planning and Strengthening of the Activity of the Economic Mechanism and the Effectiveness of Production and the Quality of Production"[1] establishes the basis for further perfection of the management of state property by means of improving planning and the organization of industrial and building production.

The subject of the right of state socialist ownership – the Soviet state. Article 21 of the Principles states: "The state is the sole owner of all state property." The unified fund of state property belong to the Soviet Union, e.g. to all the

1. *Collection of Decrees of the USSR* 1979 No. 18 item 118.

republics united in the Soviet state. Irrespective of which state organ uses the property, or under what jurisdiction or administration it is located, it is included in the unified fund of state property. The transfer of this property from the jurisdiction of one organization to another does not lead to a change in ownership – it remains the Soviet state.

2. *Objects of the right of state ownership.* The USSR Constitution defines the objects of the right of state socialist ownership as follows: "The following are under the exclusive ownership of the state: the land, its minerals, waters, and forests. The state owns the basic means of production in industry, construction, and agriculture, the means of transportation and communication, and the banks, the property of commercial, communal and other enterprises organized by the state, the basic municipal housing fund, as well as other property necessary to perform the tasks of government" (Art. 11). It is clear from this list that the primary and decisive means of production are under state ownership.

Among the objects of the right of state ownership, the earth, its minerals, waters, and forests are singled out. These objects have special significance for the national economy and are the exclusive property of the state, e.g. may belong only to the state under the right of ownership. The basic natural resources are subject to a special legal regime. They do not have monetary value, are withdrawn from commercial use, and are offered to organizations and citizens only for use. Transactions of purchase-sale, pledge, succession, gift, exchange or assignment of parcels of land, which overtly or covertly violate the right of state ownership of land are invalid. The transfer of the right of the use water and forests, and other transactions which overtly or covertly violate the right of state ownership of the waters, minerals, and forests are also invalid.

3. *The content of the right of state socialist ownership.* The state, like every owner, possesses the powers of *possession, use, and disposition of property*. The state realizes these

powers as owner of all state property and as bearer of supreme state power in the country, which fact influences the legal regulation of state ownership. The Soviet state uses the power of an owner in following the aim of all possible development and protection of the nation's property and the securing of public interests. At the same time, the state as bearer of supreme political power defines the legal regulation of property relations in the country in the interests of the all-round development and protection of state property.

The Soviet state realizes its right of ownership chiefly through organizations especially created to manage state property – enterprises, institutions, or organs of management. To each of these is secured part of the state property.

State juridical persons share the rights of possession, use, and disposition in regard to this property. This totality of powers forms a specific right, which is called the *right of operative management*. Article 26-1 of the Principles directs that property secured for state, state/collective-farm, or other state/cooperative organizations is under the operative management of these organizations which realize their rights of possession, use, and disposition of property within the limits established by law in agreement with the aims of their activity, the targets of the plan, and the purpose of the property.

The right of operative management serves as a juridical means and a necessary condition of the participation of state organizations in economic activity. The right of operative management expresses the principle of democratic centralism, combining the centralized planned leadership of the state with the economic separateness and independence of state organizations.

By its content, the right of operative management is narrower than the content of the right of ownership. If the owner realizes the right of possession, use, and disposition of the property within the limits established by law (Art. 92 Civil Code), then a state organization realizes its powers in the operative management of property not only within the limits established by law but also in accordance with the aims of the activity, plan tasks, and with the purpose of the property (Art. 94 Civil Code). This means that the powers of an organization, which realizes the right of operative management, are limited not only by law but also by the will of the owner of the property, since the former is always subordinate to the latter and obliged to act in accordance with the instructions thereof.

In the process of realizing the right of operative management, a state organization may transfer to another party the right of operative management or the specific powers of possession or use of property by virtue of contract, administrative act, or on the basis of another legal ground. In these cases, the purpose of the property may be changed, and the character of the transferred and received power may not coincide. For instance, in acquiring goods (equipment) for the organs of material-technical supply, the buyer – a state organization – will realize its rights of possession and their use, while the rights of the seller to the property are contained in the right of disposal and possession. In the transfer of property from a state organization to a cooperative, a social organization, or a citizen, the owner changes: the right of operative management ceases and the buyer acquires the right of ownership to this property.

State organs of management and social-cultural state organizations also realize the right of operative management of property. To ministries which manage industrial enterprises and to other organizations which are juridical persons are secured buildings, equipment, means of transportation, financial resources to pay for wages, for payment of business trip expenses of employees, and other property. The organs of management do not realize the right of operative management in regard to property secured to subordinate enterprises and organizations which are recognized as juridical persons, since the right of operative management of specific property may belong to only one organization. In regard to property which is secured to subordinate organizations which are juridical persons, the organs of management carry out administrative-legal dispositive actions.

4. *Legal regime of the property of state economic organizations.* The property of state organizations of eocnomic accountability forms various property funds depending on its economic and designated purpose. Each fund has its specific purpose, and in their totality they secure the property base of the productive-economic activity of the enterprise. The legal regulation of the fund is determined by the procedure of the acquisition of property by the organization, the range of its rights in regard to the property, and the limits of dispositivity of the property on the part of superior organs and the procedure of levying execution on the property by creditors.

The composition of the property of state organizations of economic accountability is divided into capital and *circulating funds* and *special financial funds.*

The division into capital and circulating funds stems from the economic division of the means of production into the means of labor and the objects of labor. The former are used gradually in the process of labor and partially transfer their value to the finished product. The latter are used to the fullest extent in every production cycle, and their entire value consists in the expenditures for the production goods. The legal concepts of capital and circulating funds must be based on juridical criteria and therefore differ from economic criteria.

In accordance with the Statute on Bookkeeping Accounts and Balances, which was confirmed by a decree of the Council of Ministers of the USSR on 29 June 1979,[2] capital production funds of industry include: buildings, structures, equipment, transport equipment, agricultural machines and equipment, construction equipment, mature work and producing livestock, and other objects which are used for more than one year and whose value is higher than the 100-ruble limit established by the ministries and institutions. A special instrument or special clothing or shoes are not included in the fixed assets irrespective of value or length of service. Circulating production

2. *Collection of Decrees of the USSR* 1979 No. 19 item 121.

funds include: supplies, raw materials, and financial assets, as well as objects of minor value or those which are quickly exhausted, etc.

It is characteristic of the legal regime of capital funds that the organization which holds these assets on its account, on the whole, realizes their production use. As a rule, it is the organs of management which dispose of these assets. The rights of state organizations are especially limited in regard to buildings and structures which are secured to them. In accordance with the Statute on the Procedure for the Transfer of Enterprises, Associations, Organizations, Institutions, Buildings, and Structures, which was confirmed by decree of the Council of Ministers of the USSR on 16 October 1979,[3] the transfer of buildings and structures from one state organization to another takes place gratuitously by decision of the central organs of management of the USSR and the Councils of Ministers of the union republics, together with all funds, assets, and allocations necessary for their activity, and the transfer of impartially completed buildings and structures is carried out with the plans of capital construction and financing, funds for materials and equipment, and with documentation on the estimates for the project. The transfer of buildings and structures from a state organization to a cooperative or other social organization is also carried out by decision of the organs of management, but with compensation, e.g. for payment, since in this instance the right of ownership is passed from the state to a social organization. The Statute mentioned above also provides a method for the transfer of enterprises, production associations, associations, and institutions from the subordination or administration of certain organs of state management to the subordination of others. In this case, the transfer without compensation between state organizations takes place together with the property secured to the transferred organization, since the owner of the property was and remains the state. In the transfer of enterprises, associations, organizations, and institutions under the

3. *Collection of Decrees of the USSR* 1979 No. 26 item 172.

management of state organs to cooperative or other social organizations, or from cooperative or other social organizations to state organizations, all the property of the transferred organizations is paid for, unless the legislation of the USSR provides for another procedure.

If at an enterprise or other organization, material valuables (equipment, fuel, etc.) are created which are above the norm or which [constitute] a surplus, the realization of this property is also carried out by special procedure.[4] Keeping in mind the economic and rational use of this property, it has been established that enterprises and organizations exercising operative management over surplus material valuables may dispose of them independently only in the instance of a refusal to redistribute this property on the part of superior organs of management or the organizations of material-technical supply to whom the realization of these material valuables is entrusted. The realization of above-normal and surplus material valuables is carried out for payment. However, legislation also provides for instances of the independent and gratuitous transfer by enterprises, associations, and other social organizations of the property of schools, children's institutes, educational institutions, medical, and other departments and organizations.

Capital funds are acquired from special allocations for capital investment. Capital funds may not, as a rule, be alienated to citizens. To secure the effective utilization of capital funds, a document [*pasport*] is compiled for production associations and enterprises in which are kept the facts on the existence and use of productive capacity. As a rule, organizations give to the state budget from their profits a payment for capital assets in the range of 6% of their value.

Circulating assets in the process of production and the realization of products comprise a circle of activity. Sums received for the realization of production or the performance of services are directed anew toward the acquisition of materials,

4. See the decree of the Council of Ministers of the USSR of 31 July 1981, "On the Method of Realizing Above-Normal and Unused Material Valuables," in *Collection of Decrees of the USSR* 1981 No. 22 item 126.

fuel, semi-finished products, and the restoration of other expenditures for the manufacture of products. Therefore, the rights of organizations in regard to circulating funds are wider than in regard to capital funds. Organizations may independently dispose of circulating assets secured to them in accordance with their designated purpose and plan tasks, and bear property liability for the state of these assets as a result of their management.

The general range of circulating assets of organizations is defined in the financial plan based on plan tasks and in accordance with economically based norms of expenditures and reserves of goods and material valuables. There are provisions for a number of stimuli for the rational utilization of circulating assets: payment of interest on the value of circulating assets into the state budget, an increase of interest for bank credit to supplement shortages in circulating assets, lower deductions from profits for incentives when there is an insufficient reserve of circulating assets, etc. Surplus circulating assets may be requisitioned by a superior organ only by procedure of redistribution in the annual account or by changing the norms in relation to changing production plans (para. 42, Statute on the Production Association; para. 12, Statute on the Enterprise).

Special funds are financial means which are used to secure specific production requirements of an organization (an amortization fund, a fund to develop new technology, to further production for export, etc.), and means to encourage the activity of the enterprise. The latter includes the funds for economic stimulation, namely: the material incentive fund, the socio-cultural fund, the fund for the construction of dwellings, and the fund for the development of production. These funds are based primarily on profit and stimulate the general results of labor, above all the improvement of quality indicators and the fulfillment of obligations for the delivery of production in accordance with concluded contracts. Special funds also include stimulating funds to encourage the attainment of special indices (fund for awarding a bonus for the creation and introduction of new technology, fund for a bonus for economizing on fuel, electricity, etc.).

Stimulating funds are disbursed in accordance with the estimate and the provisions on the fund, which are affirmed by the administration jointly with the trade-union organization and with the active participation of the labor collective.

5. *Legal regime of the property of organizations on the state budget.* The legal regime of the capital funds (buildings, equipment and others) of institutions on the state budget and other organizations which are financed by the state budget coincide with the legal regime established for the fixed assets of organizations of economic accountability. Besides capital funds, the composition of the property of organizations on the state budget includes financial assets and materials (office supplies, medicines, and others). The financial assets are given for one year only and are disbursed strictly according to the estimate taking into account the designed purpose of these assets. The remainder is returned to the budget.

Besides the assets which are apportioned from the state budget, some organizations on the state budget may have special means, which they receive from rendering services for compensation, from the profits of subsidiary enterprises, etc. These assets are spent according to special estimates and remain with the organization on the state budget even after the end of the year. Organizations on the state budget may also have funds for the material stimulation of the collective and centralized funds and reserves for financial assistance to subordinate enterprises and other economic organizations. These assets are spent in accordance with special regulations.

6. *Levying execution against the property of state organizations.* Every state organization, as a juridical person, is responsible for the obligations of the property which belongs (is secured) to it. The state is not responsible for the obligations of the organizations, and the organizations are not responsible for the obligations of the state (Art. 33 Civil Code).

Soviet law does not provide for the possibility of liquidating a state organization by bankruptcy, e.g. the enforced selling off of all the property secured to an organization in order to liquidate its debts. It is not possible to levy execution in general on many objects comprising the property of a state organization, e.g. they may not be forcibly realized in order to liquidate the claims of creditors. These include: buildings, structures, equipment, and other property included within the capital assets (Art. 98 Civil Code).

As a rule, claims are levied against financial assets by the method of warrant of execution directed to a banking institution to enforce deductions from the account of the debtor. If an organization does not have financial assets on which execution may be levied, the execution may be directed against the circulating assets with the exceptions provided in legislation of civil procedure (Art. 412 RSFSR Code of Civil Procedure), which has the aim of safeguarding the normal activity of the state organization. All completed production may be forcibly sold, and the sums gained applied to the liquidation of debts.

Levying execution against the property of institutions and other organizations subsisting on the state budget is allowed only in regard to financial assets.

Chapter 12. The Right of Collective-Farm/Cooperative Socialist Property. The Right of Ownership of Trade Unions and Other Social Organizations

1. *Concept and meaning of the right of collective-farm/cooperative socialist ownership.* The USSR Constitution defines collective-farm/cooperative ownership as a form of socialist ownership which, with state (all-people's) ownership, constitutes the basis of the economic system of the USSR. The socialist nature of collective-farm/cooperative ownership is expressed in the fact that, since it is a form of social ownership, it excludes the exploitation of man by man, and is based on socialist social means of production and collective labor. State means of production play a basic role in the development of collective-farm/cooperative ownership, and are the basis for the formation and activity of collective-farm/cooperative organizations. Article 12 of the Constitution states that "the state aids the development of collective-farm/cooperative ownership and its drawing-together with state ownership."

The process of the drawing-together of the two forms of socialist ownership is expressed in the fact that collective-farm/cooperative organizations, like state organizations, form units of a single socialist system of the national economy and act on the basis of common state plans for economic and social development. Such a drawing-together is shown in the intensification of agricultural production, the establishment of inter-economic and agro-industrial enterprises and associations with the participation both of state and cooperative organizations, which actively carry out the Food-Supply Program of the USSR. However, in spite of the uniformity of collective-farm/cooperative and state ownership and the significant drawing-together, collective-farm/cooperative ownership has its special features.

The individual collective farms and other cooperative organs with their associations are the subject of the right of collective-farm/cooperative ownership. State (all-people's) ownership has a higher level of collectivization than does cooperative ownership. The latter is characterized by a muliplicity of subjects. In contrast to state organizations, each collective farm, fishery collective, village general store, district and regional (territorial) consumers' union, the Central Union, dwelling-house building, dacha-building, and other cooperatives are subjects of the right of collective-farm/cooperative ownership. The members of a cooperative do not have right of ownership to the common property, which belongs to the cooperative organization. Their rights are defined by their membership in a specific organization.

In recent years, there has been wide development in the agricultural economy of inter-economic organizations and production associations, which were created on the basis of the specialization, concentration, and cooperation of production. Property is allotted to these organizations under operative management. They realize the right of possession, use, and disposal of the property secured to them, which belongs to the collective farms, to other cooperative organizations, and to the state.

The object of the right of ownership of collective farms, other cooperative organizations, and their associations is the means of production and other property which is necessary to carry out statutory functions (Art. 12 USSR Constitution). Article 100 of the Civil Code specifies that the property of collective farms and other cooperative organizations and their associations consists of buildings, structures, tractors, combines, other machinery, means of transportation, and other property considered to be a part of capital assets, as well as the products produced by these organizations. The property of cooperative organizations includes monetary assets, seed and fodder stock, and other property which corresponds to the activity of these organizations. From this list it is obvious that the range of objects of the right of collective-farm/cooperative ownership is also the range of objects of state ownership. The land, which is the exclusive property of the state, is secured to the collective farms for gratuitous, indefinite use. The collective farms, like other users of the land, are obliged to use the land efficiently, to treat it with care, and to increase its productiveness (Art. 12 USSR Constitution).

The objects of collective-farm/cooperative ownership may not be the same for all cooperative organizations. Each type of cooperative has its own objects of the right of ownership, which are set forth in the charters (statutes) of the cooperatives and in other normative acts. Fishery collectives have in their possession a fishing fleet; a consumer cooperative has commercial and administrative buildings, equipment, bakeries, and other objects; a dwelling-house building cooperative has dwellings; etc. The range of objects of the right of ownership of a cooperative organization always corresponds to its purpose and special legal capacity.

The basic source for the origin of the right of ownership of collective-farm/cooperative organizations is their productive and other economic activity, as a result of which property in kind is created and monetary income arises. The state gives multifaceted material aid to collective-farm/cooperative organizations on a systematic basis. This aid is reflected in the gratuitous transfer of various property to the cooperative organizations, in advancing them credit on favorable terms, in the waiving of obligations, etc.

Another source for the origin of the right of ownership of collective-farm/cooperative organizations is the collectivity of the persons who are members of the cooperative in the form of monetary dues. This is the primary source for the basis of the origin of the right of cooperative ownership in some types of cooperatives, for instance in dwelling-house building, dacha-building, and garage-building cooperative. For collective farms this source does not, practically speaking, have any significance at the present time. Collective-farm/cooperative ownership may arise also on the grounds of civil law transactions: various contracts on acquiring property for ownership, succession, etc.

The property of cooperative organizations is subdivided into funds. Depending on the designated purpose and the legal regime, distinctions are made between *capital* and *circulating, productive* and *non-productive, special, centralized, share,* and other funds. The procedure for the formation, replenishment, and disposal of the funds is defined in the charter of the cooperative organizations or in special statutes on the funds.

The method of the formation and the legal regime concerning the funds are divided into *indivisible* and *divisible*. The *indivisible* sector includes the means of production and other property which is not subject to disbursement among the member of the cooperative and is used strictly for a planned purpose. On collective farms, indivisible means include property which is used in the production process (production funds), as well as property which is designated for the satisfaction of cultural-social requirements (non-productive funds). These funds are replenished every year from profits and several other sources. Indivisible funds on collective farms also include seed funds, insurance funds, funds for the sale of products of the collective farm to the state, funds to pay for taxes and other payments to the state, and other special funds.[1]

Divisible funds include share funds, payments for labor, material stimulation, and social security. These means are also the property of the cooperative, but in the case of a member's leaving the cooperative his share is returned to him in monetary form; the remaining divisible funds are designated for distribution among the members of the cooperative in proportion to the quantity and quality of work invested in social production.

In those types of cooperative which have such systems, for instance, in consumer cooperatives, there are centralized funds. These funds are created by deductions from the profits of lower-level organizations, are secured to the center of the system, and are spent in the interests of the system in accordance with the intended purpose of the fund.

Every cooperative farm or other cooperative organization which is the owner of its property realizes in connection

1. The legal regime of collective farm property is studied in detail in the course "Collective Farm Law."

therewith the rights of possession, use, and disposition. Article 99 of the Civil Code emphasizes that the right of disposition of property, constituting the property of cooperative organizations, belongs solely to the owners. The same rule is written into the corresponding charters of cooperative organs, and in part in Article 13 of the Model Charter of a Collective Farm. Neither the lower organs of a given cooperative system nor state organizations have the right to dispose of the property of cooperative organizations. In many normative acts, attention is drawn to the impermissibility of the arbitrary disposition of the property of cooperative organizations or their production and monetary means.

Possession, use, and disposition of the property of cooperative organs is realized on democratic grounds by agreement of the members of organizations, expressed, as a rule, in decisions of general assemblies and other organs of the cooperative. Such decisions are carried out in accordance with the charters (statutes) of the cooperatives and the plan tasks of competent state organs and cooperative centers, and take into account the designated purpose of the property. Paragraph 11 of the Model Charter of a Cooperative Farm states that the management and all the members of the cooperatives farm are obliged to secure the effective use and to preserve collective farm property.

In the process of exploiting property, cooperative organizations enter into contractual agreements with other organizations and citizens, realize the distribution of production and income, effect the sale and lease of property, and write off things of material value. Property relationships among various units of a single cooperative unit are compensatory in nature. Deducting means from lower units for the use of superior units of a given cooperative system is allowed only in instances provided in the charter (statute) of the corresponding organization or by normative acts.

2. *The right of ownership of trade unions and other social organizations constitutes a third form of socialist property in the USSR.* In Article 10, the Constitution affirms for the first time at the highest level the [status] of the property of trade

unions and other social organizations in the capacity of an independent form of socialist property. The right of ownership of trade unions and other social organizations secures the material base of the activity of these organizations, aiding the development of political activity and initiative, and the satisfaction of the varied interests of citizens. The property of trade unions and other, numerous social organizations (sport, defense, creative unions, and others) is not a means of production, since these organizations do not engage in production activity, but constitute mainly non-production funds.

The organization in its entirety is the subject of the right of ownership of social organizations. Individual subdivisions of the system, as well as clubs, sanitoriums, printing houses, and other organizations which are part of the composition of the social organizations, are not subjects of the right of ownership even if they utilize the rights of juridical persons.

For instance, the right of trade union ownership belongs to the entire system of trade unions in the USSR, and individual trade union committees, sanitoriums, rest houses, clubs, Young Pioneer camps, and other trade union organizations are not subjects of the right to ownership. As independent subjects of a right, e.g. juridical persons, they have the right of operative management of the property allotted to them. The All-Union Central Trade Union Council *(VTsSPS)*, as the highest organ of the system, oversees all of the trade unions' property, and the individual trade-union organizations manage that portion of the property which lies within the limits allotted to them.

The objects of the right of ownership of trade unions and other public organizations are given in Article 103 of the Civil Code. They include, in part: buildings, structures, equipment, and other property of enterprises; sanitoriums, rest homes, palaces of culture, clubs, stadiums, and Young Pioneer camps; cultural-educational funds and other property which corresponds to the aims of these organizations. The specific range of objects of each public organization is defined in its charter (statute) and depends on the aims of its social and cultural activity.

The basic source for the origin of the right of ownership of trade unions and other social organizations is membership dues, which are consolidated in a specific society, as well as state aid. For instance, on the basis of the decree of the Council of Ministers of the USSR of 10 March 1960,[2] sanitoriums, rest homes, boarding homes and other institutions with the property belonging to them were transferred free of charge to trade unions in order to improve the organization of leisure and sanitorium/health-resort services to workers and to increase the role of trade unions in this important function of the trade unions. In the Statute on the Production Combine (para. 47) and the Statute on the Enterprise (para. 17), provisions are made on their obligation to provide the trade union committee with premises, equipment, means of transportation and communication, [and] to transfer cultural/social and sporting inventory to trade-union organizations. Economic organizations assign financial means for mass-cultural and sports activity to trade-union organizations on a systematic basis. These means are included in the assets of the property of the trade-union organizations from the moment that they are entered into the current account.

The property of trade-union and other social organizations may also originate as a result of civil law transactions, publishing activity, paid public performances, and other nonproduction activity of these organizations.

The content of the right of ownership of trade-union and other social organizations, like that of other owners, comprises the rights of possession, use, and disposal, which are realized in agreement with the charters (statutes) of these organizations. In Article 102 of the Civil Code it is emphasized that the right of disposal of the property comprising the property of trade-union and other social organizations belongs exclusively to the owners.

The disposition of the property is effected on democratic grounds on the basis of the estimate, affirmed by the general assembly of the members of a specific organization or by another of its organs which is designated in the charter

2. *Collection of Decrees of the USSR* 1960 No. 8 item 52.

(statute). The legal capacity to dispose of capital assets is realized, as a rule, by the central organs of a specific system which has the right to disperse the property among the organizations of the system. The transfer of property from one unit to another within the limits of one system takes place without compensation.

3. *Executing liability on collective-farm/cooperative property and on the property of trade-union and other social organizations.* Each of the named organizations, [and] their superior units, are juridical persons, bear independent responsibility for their liabilities, and are not responsible for the debts of other organizations (Art. 32 Civil Code).

A significant portion of the property of collective-farm/cooperative and other social organizations is exempt from execution. According to Article 101 of the Civil Code, it is not permitted to levy execution against the buildings, structures, tractors, combines, other machines, means of transportation, and other property related to the capital assets of enterprises, their socio-cultural institutions, as well as their seed and fodder reserves for the debts of collective-farm/cooperative organizations. Execution is levied against the financial means and other property within the limits provided in Article 412 of the Code of Civil Procedure, e.g. with the circulating assets which are necessary for the continuation of the normal activity of these organizations remaining at the disposal of the collective-farm/cooperative organizations. Finshed products may be sold forcibly and the sums received [for them] used to liquidate debts.

Execution against the property of trade-union and other social organizations is also levied in the first instance against their monetary means (Art. 104 Civil Code). If there are insufficient monetary means to liquidate liabilities, execution may be levied against property belonging to them, with the exception of buildings, structures, equipment, and other property considered part of the capital assets of enterprises, sanitoriums, rest homes, palaces of culture, clubs, stadiums, and Young Pioneer camps, as well as against cultural and educational funds and several other objects.

Chapter 13. The Right of Personal Ownership

1. *Concept and meaning of personal ownership.* Personal property provides citizens with the possibility of satisfying their various material and spiritual needs. All, or almost all, of the objects which surround us at home (clothing, furniture, a television, radio, refrigerator, or bicycle and much more), and for some, the house itself constitute personal property. The money belonging to a citizen by right of personal ownership is spent on food, various purchases, services from personal service shops, transportation fees, housing rental, etc.

The right of personal property is *the right of a citizen, granted under force of law, to possess, use, and dispose of property with the aim of satisfying personal needs.*

Public funds meet almost one-fourth of the requirements of citizens of the Soviet Union, who do not have to spend personal means on them. The USSR Constitution guarantees to citizens a free education, the use of libraries, free text-books for school, and medical aid. Citizens of the USSR are granted free, or for partial payment, trips to sanitariums and rest homes; pensions and benifits are paid from public funds. The Party and the state devote much attention to this source for raising of material well-being. The payments and benefits received by the populace from public funds are increasing steadily.

Besides this, the Program of the CPSU emphasizes that "the transition to communist distribution will be accomplished when the principle of distribution according to work is completely resolved, that is, when there is an abundance of material and cultural benefits and work is transformed into the first and foremost requirement for all classes of citizens." Therefore, until such time, personal property remains the main source for the satisfaction of the needs of citizens. Personal property possesses a number of important features:

a) Personal property has a *work character.* The USSR Constitution provides that earned income constitutes the basis of personal property. Here is meant, in the first place, earned income linked to the participation of citizens in the production of society. It is possible to evaluate the specific proportions of this earned income from the following facts: 74.2% of the entire income of the family of an industrial worker comes from such income, and for the family of a collective-farm worker the figure is 54.5%

Among the number of other sources of personal property one ought to mention the earnings of citizens from individual work activity in the sphere of trades and crafts, and agriculture, as well as other types of activity which are based solely on the individual labor of citizens and the members of their family. The possibility of realizing such activity is provided in Article 17 of the USSR Constitution which, at the same time, emphasizes that "the state regulates individual labor activity, ensuring its utilization in the interests of society." This is shown by the fact that legislation does not permit activity in those aspects of trades and crafts which are harmful to social interests (for instance, the preparation of some types of weaponry, ammunition, and explosive materials, medications and objects of medical technology, as well as the circulating of phonographic records and magnetic recordings). In order to engage in a specific activity, it is necessary beforehand to receive a special document of registration from the executive committee of the local Soviet of People's Deputies. Certain trades are not subject to registration. For instance, citizens who perform domestic services for others (sawing and chopping wood, washing windows, etc.) are exempt from registration.

Subsidiary husbandry is also one of the sources of personal property. This has particularly important significance for the family of collective farm workers: more than one-fourth of their entire income is derived from subsidiary husbandry.

There also exist other sources for the acquisition of property by citizens: inheritance, winnings from a lottery ticket or state-loan bond, interest from deposits in a savings bank, and gifts. It is not hard to realize that these sources of income also, if not directly, are indirectly linked to the personal labor of the owner himself or of other persons.

b) *Personal property derived from socialist property.* As was mentioned previously, earnings which are connected with participation in the production of society constitute the basis of personal income. Therefore, it is clear that the level of material well-being of the population of our country depends on the further growth of the socialist economy. As far as the individual citizen is concerned, the availability of property which belongs to him by right of personal ownership is directly linked to the quality and quantity of his work in a plant or factory, or a collective or state farm, etc.

c) Personal property designated first of all and for the main part *for the satisfaction of the personal requirements of citizens.* This means that property under the personal ownership of citizens may not be used to derive unearned income to the detriments of the interests of society.

2. *Subjects of the right of personal ownership.* The ability to possess property under personal ownership is the most important element of the legal capacity of citizens. Like all other rights listed in Article 10 of the Civil Code, the right of personal ownership may belong to a citizen irrespective of age or the state of his health, that is, irrespective of his dispositive capacity. A two-year old child may become the owner of a house bequeathed to him by his grandfather.

The right of personal ownership is related to the category of such rights which may belong only to citizens. The bearer of such a right may be either a single citizen or several citizens (co-owners). The membership of a collective farm may also be the subject of personal property (see Chapter 14).

3. *Objects of the right of personal property.* A listing of the property which may belong under the personal ownership of citizens is found in the USSR Constitution (Art. 13), The Principles (Art. 25), and the civil codes of the union republics. Under the personal ownership of citizens may be found articles of household utility, of personal concumption and comfort, and of the subsidiary household, a dwelling, and savings from labor. As a general rule, the law does not estab-

lish a maximum for the listed articles. Thus, there is no limit to the amount of money which may be in a citizen's account in a savings bank. This also applies to furniture, clothing, and other articles. It is only for certain articles, as an exception, that it is necessary to establish definite limits in order to retain their utilitarian character. Such definite limits are established, in part, for dwellings. Legislation in general regulates a wide range of questions connected with the right of personal ownership of a dwelling. This is explained by the fact that a dwelling retains its significance in its capacity as an important source for satisfying the demands of citizens for shelter: at the present time, one-fourth of the housing stock of towns is under personal ownership.

Limiting norms are also established for the amount of livestock which may be under personal ownership. A family in which there are no members of a collective farm may have under its personal ownership one cow, one sow with a litter within two months or two fattening pigs, three sheep and goats older than one year, not inlcuding litters; if there are no cows or pigs, not more than five sheep and goats older than one year, not counting litter, are allowed. In addition to this, taking into consideration local conditions and national particularities, the executive committees of district and city Soviets of People's Deputies have the right to allow a citizen to keep under personal ownership one work animal (horse, mule, donkey, camel, ox, or buffalo). There is no limit on how many domestic fowl or beehives may be kept. It must be borne in mind that citizens who are able to work must concern themselves above all with the socialist usefulness of labor. Therefore, in regard to those who do not fulfill this requirement, the executive committee of a local Soviet may apply a sanction: to deprive them of their right to keep any type of livestock, as well as fowl or beehives under personal ownership.[1]

1. See Edict of the Presidium of the Supreme Soviet of the RSFSR, 13 Nov. 1964 *Gazette of the Supreme Soviet of the RSFSR* 1964 No. 51 item 887; 1970 No. 13 item 221; 1982 No. 32 item 1193.

The quantity and type of livestock which the family of a collective-farm worker (collective-farm household) may own is defined in the Model Charter of the Collective Farm (para. 43).[2] In certain districts, taking into account national particularities and local conditions, an increase in the amount of livestock under the personal ownership of the family of a collective-farm worker or changing one type of livestock for another is permitted by decisions of the Council of Ministers of the union republics. Within the limits of established norms, each collective farm defines the quantity and type of livestock which the family of a collective-farm worker on a specific collective farm may own.

More and more Soviet citizens are becoming the owners of various means of transportation; every year a million or more automobiles and more than one million motorcycles and motorscooters are sold to them. The quantity of automotive means which may be under the personal ownership of a citizen is not limited by legislation. Rather, the right of personal ownership of means of transportation is also regulated by several special norms. In the interest of public safety there are provisions for their obligatory registration and technical inspection. Special rules are also established for the sale to citizens of automobiles and heavy motorcycles with sidecars. The sale may be carried out only on a commission basis through stores of state and commission trade.

4. *The right of personal ownership of a dwelling.* Like all other objects of the right of personal ownership, a dwelling must be used to satisfy the personal interest of those to whom it belongs. Therefore, it is provided by law that every citizen has the right to own one dwelling (or part of a dwelling). Spouses and minor children living at the same place also have the right to own only one house (or part thereof). Thus, such a family does not have a right to own two houses or even two parts of different houses. A citizen who is a member of a housing construction cooperative also may not own a

2. *Collection of Decrees of the USSR* 1969 No. 26 item 150.

house or part of one.

In practice there occur instances when the owner of a dwelling is given or inherits another house by will, or spouses entering into marriage each have a house by right of personal ownership, that is, a citizen or his family becomes owners of several houses on legal grounds. In this instance, the law requires that the citizen choose only one house. He is obliged to sell, give, or in some other fashion alienate the other house. The citizen owning a apartment in a housing construction cooperative who becomes the owner of a house must act in the same fashion. In this instance the owner must either retain the house and leave the membership of the housing construction cooperative or, remaining a member of the cooperative, sell (give or in some other fashion alienate) the house. The owner must fulfill these requirements in the course of one year. If the owner does not use the opportunity available to him, the house is subject to enforced sale by decision of the executive committee of a local Soviet of People's Deputies. The money received from the sale (with a deduction for the expenses involved in the sale) is given to the former owner. If the enforced sale of the house does not occur, the house is transferred without compensation to the ownership of the state.

The law also establishes the maximum dimensions for the dwelling or the part (parts) which may belong to the ownership of a citizen. The house may not exceed 60 square meters of floor space. However, the executive committee of a local Soviet may permit the building, acquiring, or securing of a house of larger dimensions to be owned by those who have a large family or the right to supplementary floor space. In this instance, the floor space of the house (part of a house) may not exceed the established norm for floor space (in the RSFSR it is 9 square meters per person), and for those who use the right to supplementary floor space (persons suffering from certain illnesses, members of some creative unions, citizens having academic degrees, and others), the maximum dimensions of the house may not exceed the basic norm and the supplementary [norm] (for various categories or persons the norm is established at 10 or 20 square meters).

If it is shown that a house (part of a house) under the ownership of citizen exceeds the limits established by law, the same results occur as in instances of acquiring a second house.

The construction of an individual dwelling may take place only with the permission of a local Soviet on the basis of an approved design and with the observance of established construction norms and rules. If it is discovered that the house or part of a house is constructed without permission, or without an approved design, or with a significant deviation from the design, or with obvious violations of basic construction norms and rules (more floors in the house, a significant increase in the area of the house, carrying out construction which destroys the design, structure, and organizations of public services in a populated area), then the consequences provided in Article 109 of the Civil Code take place: the improperly erected house is considered to be illegally built and the owner does not have the right to dispose of it (he may not sell, trade, lease it, etc.). By decision of the executive committee of a local Soviet of People's Deputies, such a house must be demolished by the citizen himself or at his expense or it may be confiscated by decision of a court and placed without compensation into the fund of the local Soviet. The court may deprive the citizen of the right to live in such a house; however, if the citizen has no place to live, the executive committee of the local Soviet, to whom the confiscated house is given, must provide him with other living quarters.

In connection with various state and social requirements there may arise the necessity of confiscating from a citizen the plot of land allocated to him, and consequently, demolishing his house. In this instance, according to established norms, the owner of the house and members of his family, as well as other citizens permanently living in the house, must be provided with apartments in houses in the state or public housing fund or, if they so desire, the possibility of entering into membership in a housing construction cooperative ahead of those who may already be on a waiting list is secured [for them]. Furthermore, the owner of the house has the choice of being paid the value of the demolished house and of other buildings and structures, or being given the right to use the materials from the demolition of the house at his discretion. If the owner so desires, the dwelling and buildings subject to demolition may be moved and established in a different place.

The dimensions of the plots of land which are granted to citizens for personal dwelling construction are strictly limited. Thus, a citizen building a dwelling in a city may be allotted 300 to 600 square meters, and if outside a city, from 700 to 1200 square meters. The citizen has the obligation to use the allotted plots of land in a rational manner. Besides this, the plots of land, like property belonging to citizens by right of personal ownership, may not be used to the detriment of the interest of society to derive unearned income.

5. *Content of the right of personal ownership*. The content of the right of personal ownership contains these powers of the owner: *possession, use,* and *disposition* which he may realize within the limits established by law. As a rule, a personal owner himself *possesses* and *uses* the property belonging to him. However, he is allowed to transfer his property to the possession and use of other parties even with the purpose of receiving income from this. Thus, the owner of a house or *dacha* has the right to give living quarters for use for payment which must not exceed the maximum established by law. The maximum rates are established by decrees of the Council of Ministers of the union republics. In the RSFSR this payment for living quarters is established at the limit of not more than 16 kopecks per square meter per month.

The USSR Constitution establishes that "property under the personal ownership or use of citizens must not serve for the derivation of unearned income." Unearned income is that which is received as the result of acts by the citizen which violate the law. For instance, a citizen gives an apartment or *dacha* for temporary use for payment which exceeds the established maximum rate, or uses an automobile belonging to him to transfer passengers for payment. Extremely severe sanctions are established for the violation of these requirements. According to Article 111 of the Civil Code, a house, *dacha*, or other property which is systematically used by the owner to derive unearned income is subject to confiscation without compensation in court proceedings in a suit brought by the executive committee of a local Soviet. The confiscated house (*dacha*) is added to the housing fund of the local Soviet.

The USSR Constitution prohibits citizens from using their property "to the detriment of the interests of society." An owner acts in such a fashion, for instance, if he takes improper care of a house belonging to him, allowing it to fall into ruin, or if he takes improper care of property which has significant historical, artistic, or other value for society (the picture of a famous painter, an ancient icon, or old manuscript). The consequences of the improper care of such property by the owner have already been discussed in Chapter 10.

The owner has the right to *dispose* of the property belonging to him at his discretion. He may sell, give, or trade it, give it for use, etc. In exceptional circumstances, the law may establish defined limits on the right of disposal of some property. Thus, Article 238 of the Civil Code provides that an owner does not have the right to sell more than one house (or part of a house) in the course of three years. The right of disposal of a dwelling is also limited where the owner wants to sell his share of common property to an outsider (see Chapter 14 on the right of preferential sale). The owner who realizes the right of disposition of property without observing the requirements of the law incurs unfavorable consequences: the transaction by which the article was sold, exchanged, or given may be recognized as invalid with the return of the parties to their original position, or the article (or the money received for it) is turned over to the state, etc. (see Chapter 7).

Chapter 14. The Right of Common Ownership

1. *Concept and meaning of the right of common ownership.* The right of common ownership arises when one and the same property belongs to two or more parties. Such property may be one object (for instance, a dwelling or automobile), as well as an aggregate of objects (for instance, an inter-collective-farm enterprise whose property comprises buildings, equipment, raw materials, finished products, etc.).

The participants in common ownership are called *co-owners.* They may all be subjects of the right of ownership: the state, cooperative or other social organizations, or citizens. However, the law establishes that property may not belong to a citizen and the state, or to a citizen and a cooperative or other social organization under the right of common ownership (Art. 116 Civil Code). If such a situation nevertheless arises on grounds allowed by law (for instance, the confiscation of a co-owner's share of a house, or where a share in the common property of a citizen is transferred to the state by will), then the right of common ownership of the state and a citizen or a cooperative or other social organization and a citizen must be terminated within one year.

The methods for such termination are set forth in Article 123 of the Civil Code: the property is either distributed in kind (if, of course, this is physically feasible), the state (or social organization) sells its share to the citizen, or, conversely, the citizen sells his share to the state (or social organization), or, finally, the entire property is sold and the money received for it divided among the co-owners. The co-owners themselves choose one of these methods of termination. If they cannot reach agreement, the issue is decided by the courts.

Common property originates from various circumstances: inheritance (property consisting of a dwelling and an automobile is transferred by inheritance to the children and wife of the deceased), joint purchase of a particular object (for instance, a garage for two vehicles), and so on. In relationships between the state, cooperative, and other social organizations, the most widely used source of common property is the amalgamation of property in order to attain various economic aims. It is for this purpose that inter-economic enterprises (organizations) are established by relevant state and cooperative organizations. The XXVI Congress of the CPSU set the task: "To develop state-cooperative, inter-collective-farm, inter-state farm and collective-farm/state farm associations and agro-industrial organizations."[1]

Under these conditions, organizations, that is, juridical persons, are established. They possess, use, and dispose of property secured to them by right of operative management, while the property itself belongs, by right of common ownership, to the state, collective farms, other cooperative and social organizations, that is, to the participants in the creation of this inter-economic enterprise (organization).[2]

The law distinguishes between two types of common ownership which are essentially different from each other: *shared* and *joint* ownership.

2. *The right of common shared property*. Common ownership is shared if a specific share belongs to each of its participants by right of common ownership. These shares may be equal (the share of two children in the property received by inheritance) or unequal (the wife is willed two-thirds and the daughter one-third of the property). Having a share under right of ownership does not mean that the owner possesses a specific portion of the object. On the contrary, each owner has the right to a share of the entire object and at the same time to each of its parts. This situation is linked to an important consequence. For instance, two co-owners (brother and sister) own equal shares in a building complex consisting of a house and its additions. Furthermore, the brother lived in the house, and the sister lived in the additon. If the addition is burned down in a fire, the sister has the same right as her brother to use the house, since her right to common

1. *Materials of the XXVI Congress of the CPSU*, Moscow 1981 p. 201.
2. See: General Statute on the Inter-Economic Enterprise (Organization) in the Agricultural Economy (*Collection of Decrees of the USSR* 1977 No. 13 item 80).

ownership of the house is preserved.

The participants in a shared ownership together possess, use, and dispose of the property belonging to them. All questions arising from this situation (the manner of using the object, giving it for use to another party, carrying out repairs, etc.) must be decided by agreement of all participants. If no agreement is reached, any one of the participants may bring a suit. The decision of the court is binding on all participants to the common property (Art. 117 Civil Code).

Moreover, each of the co-owners has the right to decide independently how to alienate his share: he may sell, exchange, or give it away, etc. In all of these instances, as well as in inheritance of a share of common ownership, all the rights and liabilities of the former owner are transferred to the new owner.

In instances of the sale of a share by the owner, the law considers it necessary to secure the interests of his co-owners. For this purpose, the right of "preemptive purchase" is recognized for them. Article 120 of the Civil Code charges the co-owner with the obligation of notifying the other participants to the common property in writing of his intention to sell his share and of the conditions of such a sale (above all, the price). Each of the participants in common shared property has the right to demand that the share be sold to him under the conditions given in the notification. A participant in common property has the right to sell his share to an outsider only where the other co-owners decline to realize the right of preemptive purchase or do not make known their desire to use their preemptive right within the established time (one month in the sale of a share of a house, ten days in the sale of a share of some other property). If a participant in common property concludes a contract of purchase-sale with an outsider in violation of these conditions, not informing his co-owners of the sale, any one of the co-owners has the right to bring suit demanding that the rights and obligations of the buyer in such a transaction be transferred to him.

The co-owners have the right at any time to leave the membership of participants in common property, e.g. to demand an allotment of their share in kind from the common property. The co-owners define the method of allotment jointly, and if they cannot agree, the court helps them. The court is required to extract the share, and only if this is impossible to do without causing harm to the economic purpose of the property is the co-owner given a determined sum of money as compensation (Art. 121 Civil Code).

In agriculture, the property of inter-economic enterprises (organizations) is also common share property. Its means are formed from the shared contributions of the industries which participate in the creation of such an enterprise and deductions from profits gained as a result of their own production-economic activity, as well as from other sources. The participants in an inter-economic enterprise may produce share contributions in the form of money as well as objects of material value. In the latter instance, production buildings, structures, equipment, means of transportation, and other property which is given as a shared contribution are included in the balance of the inter-economic enterprise. The method of determining the extent of the shared contribution of the participants is established by decision of the assembly of authorized representatives of these economic entities.

3. *Right of common joint ownership.* Joint ownership is also called non-share ownership. The share of each of the co-owners is not defined, and the necessity for establishing the extent of the share belonging to each of the participants arises only in case of the termination of common joint ownership. In contrast to shared ownership, joint ownership originates only under circumstances provided in law. Among citizens, joint ownership arises under two circumstances: the common joint ownership of spouses and the common joint ownership of a collective-farm household. The first of these two types of joint ownership is regulated by family rather than by civil legislation and therefore is examined in Part II of this text (Chapter 43).

The property of a production combine in agriculture is also common joint property if its structure includes state farms, collective farms, other state, cooperative, and other social organizations.[3]

4. *Ownership of the collective-farm household.* A collective-farm household is a family-labor combine of citizens whose basic source of property is the income received from participating by personal labor in collective-farm production, and whose supplementary income derives from subsidiary farming on a personal plot. As a rule, a collective-farm household consists of the members of one family: husband, wife, children (irrespective of age), parents, etc. Not only collective-farm members, but also workers, officials, pensioners, and dependents permanently living on a collective farm may be member of a collective-farm household.

A garden plot of land is allotted for the use of a collective-farm household. The subsidiary economy located on this plot constitutes the property of the collective-farm household: a dwelling, production cattle, fowl, and light agricultural equipment in accordance with the charter of the cooperative farm. The common property of the collective-farm household also includes the earned income given to the members of the household for their participation in the communal economy or other income given to them, as well as articles of household utility and personal use acquired through common means. Besides common property, the members of a household also have personal property belonging to them. This is the personal earned income and savings of a member of the household, as well as property acquired by him through personal means or received through inheritance or as a gift and not transferred to the ownership of the household (Art. 113 Civil Code).

The possession, use, and disposal of the property of a collective-farm household is realized upon agreement of all its members. Usually the head of the collective-farm household acts in its name, but other members may also do so. If a dispute arises among the members of a collective-farm household

3. See: Statute on the Production Combine in Agriculture (*Collection of Decrees of the USSR* 1979 No. 3 item 15).

in regard to the possession, use, and disposal of the common property, this dispute is decided by a court where a suit is brought by any member of the household who is at least 16 years old (Art. 127 Civil Code).

As has already been mentioned, insofar as the property of a collective-farm household is related to common joint property, e.g. non-share property, the proportionate share of an individual member is determined only in instances when it is a matter of apportionment (leaving the household without creating a new household, for instance, when one of the members of a household leaves for permanent residence in a city) or a matter of division (for instance, a daughter marries and forms a new collective-farm household with her husband), or when execution is levied against the common property for the personal debts of a member of the household.

Article 129 of the Civil Code provides that all members of a household, including those under age and those unable to work, have the right to an equal share of the common property. This means if the household consists of husband, wife, children of three and seven years of age, as well as of the father of the head of the household, an 80-year old pensioner, the share of each member of the household is one-fifth of the common property. As an exception, it is permitted to reduce the share of those who were members of the household for a short period or who participated insufficiently through their labor or assets in the running of the household. However, this exception does not apply to those under age or to other members of the household who are unable to work.

Members who are able to work may lose their right to a share of the common property. They are those who have not participated by their labor or assets in the general running of the household for a period of three years. However, this rule does not apply in instances when the member of a household is unable to participate due to mitigating circumstances (being called to military service for a certain period, studying at an educational institution, long illness, etc.).

The legal regime of the common *property of an individual peasant household* is the same as that of a collective-farm household. The difference lies only in the range of objects which may belong to the ownership of the household. An individual peasant household may have supplementary working cattle (with the permission of the proper organs of authority) and agricultural equipment necessary to work the plot of land granted to the use of household without employing other labor.

Chapter 15. Protection of the Right of Ownership

1. *Meaning and methods of protecting the right of ownership*. The USSR Constitution provides for the duty of all citizens of the USSR to safeguard and strengthen socialist property which is the basis of the economic system of the USSR. The Constitution states that "it is the duty of a citizen of the USSR to fight theft and waste of state and social property, to treat the wealth of the people with care" (Art. 61). In the Constitution it is also established that the state safeguards the personal property of citizens (Art. 13).

The protection of the right of ownership has its aim to secure socialist legality in property relationships, to help in the development and strengthening of ownership relationships, and in case of violations, to reestablish the interests and rights of the owner by placing unfavorable property consequences on the guilty parties.

Socialist and personal property are protected by many branches of Soviet law, including the norms of administrative, land, financial, criminal law, and others. In civil law the protection of the right of ownership is realized in various ways depending on the type of violation: bringing suits for the recovery of property and the elimination of the violation of the rights of the owner which are not linked with deprivation of possession of the object; bringing suits for compensation for losses caused by non-fulfillment of contractual obligations; compensation for harm caused without a contract, and so on.

However, Chapter 14 of the Civil Code, entitled "Protection of the right of ownership," provides for only two suits for the owner: 1) a suit for the recovery by the owner of property from another's unlawful possession (*revindicatio*) and 2) a suit to eliminate violation of the rights of an owner which are not linked with deprivation of possession (*actio negatoria*). The special features of these suits consist in the fact that they protect the absolute rights of the owner from violations of the right of possession, use, and disposition of an object belonging to him and are applied only in instances when the disputed article actually exists. The rules of Chapter 14 on the protection of the right of ownership are accordingly also applied to protect the right of operative management of socialist organizations.

2. *The suit of an owner to recover property from the unlawful possession of another party* is called *revindicatio* (from the Latin word *vindico*, I demand) [or *replevin*]. It is applied in instances when the owner has lost possession of an object and another person possesses this object unlawfully, for instance, by appropriateing a lost article, theft, purchase of a stolen article, etc.

The owner has the right in all instances to demand his object from the unlawfull possessor if the latter knows or should have known that his possession is unlawful (possessor in bad faith). The unlawful possessor must return all profits received during the time of unlawful possession along with the object. For instance, when a cow is returned, the calf born to it must also be transferred to the owner and there must be reimbursement for the value of the yielded milk minus deductions for necessary expenditures on the property. By necessary are meant expenses for the maintenance of the property, its repair, and other expenditures.

But if the unlawful possessor of another's article did not know and was not obliged to know of the unlawfulness of the possession (acquirer in good faith), then it would be unfair to return the object to the owner under all circumstances. It is possible that the acquirer in good faith bought the object in a commission store, where it was placed by the party who appropriated a lost article, or the acquirer in good faith obtained a stolen article under circumstances which did not cause suspicion regarding the good faith of the seller. In such instances Article 152 of the Civil Code establishes the rule that articles bought for compensation by an acquirer in good faith may be reclaimed by the owner from the latter if: a) the objects were lost by the owner or the party to whom the objects were given for possession by the owner; b) the objects were stolen from one or the other; or c) the objects were taken from their possession against their will.

If the owner transferred his object in good faith to a party who does not justify his trust, recovery of the object from the acquirer in good faith is not permitted. In such instances it is presumed that the owner was imprudent. For instance, the owner left his article for safekeeping with an acquaintance who sold it. If the buyer is an acquirer in good faith, he is not obliged to return it to the owner. The latter may claim compensation for losses from the acquaintance with whom he left the article, but he has no right to recover the article itself from the acquirer in good faith. If in an analogous situation the article is stolen from the acquaintance, the owner has the right under a vindication suit to recover it from the acquirer in good faith.

In instances of recovery of the article, the acquirer in good faith is obliged to reimburse the owner for all income which he derived or ought to have derived from the object from the time that he learned or ought to have learned that his possession was unlawful or received notice about the suit of the owner for return of his property. In turn, the possessor has the right to demand compensation from the owner for any necessary expenditures in connection with the property from the time that the owner was entitled to income from the property (Art. 155 Civil Code).

An acquirer in good faith who received the article gratuitously, e.g. without any sort of expense on his part, for instance, through contract of gift or by inheritance, must return the article to the owner under all circumstances. The primary protection of the rights of the owner in this instance is based on the fact that the possessor in good faith does not suffer losses in returning the article.

There are instances when the possessor carries out improvements raising the value of the object during the time of unlawful possession. A possessor in good faith has the right to keep the improvements carried out by him if they can be removed without harming the object; otherwise, he has the right to be reimbursed for his expenses, but in an amount no greater than the amount of the increased value of the object.

There are exceptions to these general rules; according to para. 2, Article 152 of the Civil Code, it is not possible to recover an object from an acquirer in good faith if it was sold in accordance with the procedure established for carrying out judicial decisions. In these instances, no consideration is given to how the article left the possession of the owner. However, the owner has the right to obtain the value of the article from the party for whose benefit the illegal distraint of the property took place.

Special rules are established for the recovery of property of socialist organizations. Taking into account the role of socialist property in the economy of the country, the law established that unlike citizens, state organizations, collective farms, and other cooperative and social organizations have the right to recover property from any unlawful possessor irregardless of how it was alienated (Art. 153 Civil Code). This also establishes the primary defense of socialist property and the right to its unlimited vindication.

The rule of the unlimited vindication of the property of socialist organizations is applied irrespective of who is the defendant in an action to recover the property – a socialist organization or a citizen, and what type of property is being claimed. Moreover, according to Article 90 of the Civil Code, the period of limitations does not extend to claims of state organizations for the recovery of state property from the unlawful possession of collective farms, other cooperative and socialist organizations, and citizens.

There is a special rule for recovering money and securities for a claimant. Article 154 of the Civil Code establishes that these objects may not be recovered from an acquirer in good faith. This rule is in effect regardless of how these objects left the possession of the owner and whether the plaintiff is a socialist organization or a citizen. The purpose of the rule is to secure the stability of economic activity, since money and securities are means of circulation. This rule is applied in relation to an acquirer in good faith. If the money or securities are being kept by a citizen who stole or appropriated them, they must be taken from him and returned to the owner.

3. *A suit of an owner to eliminate violations which are not linked with deprivation of possession (actio negatoria)* is brought to eliminate violations of the right to use and dispose of an object, for instance, when as the result of the unlawful actions of another person the owner is deprived of the possibility of using a road, residential or subsidiary premises, or is unable to dispose of his property freely.

The difference between an *actio negatoria* and *revindicatio* is that in an *actio negatoria* the owner possesses the object but is unable to use or dispose of it in a normal manner. In examining an actual case, the USSR Supreme Court has recognized that on the basis of an *actio negatoria* a citizen who had a right to personal ownership of a dwelling could demand the eviction of an outside party living in her house without her consent. The contents of an *actio negatoria* consist of the fact that it obliges the defendant to conclude certain actions or forbids him to realize certain actions which violate the rights of the owner. The actions which violate the right of the owner must be against the law. The digging of a hole beside the house of the owner causes him inconvenience, yet the owner must accommodate himself to such actions if they are carried out on legal grounds. But if the hole is not filled at the end of the job, the owner has the right to demand that the violation of his right to use his property is eliminated. In order to bring an *actio negatoria*, it is necessary that the violations of the rights of the owner exist on the day on which the suit is brought and during the examination of the case. If they are terminated, there are no grounds for an *actio negatoria*, but a claim for compensation for losses may be presented on the basis of Article 444 of the Civil Code.

An *actio negatoria* is also employed when the owner is deprived of the possibility of disposing of his object as the result of the illegal action of a sheriff, who includes the objects of the owner as part of the property of another person in distraint or who imposes an arrest on the property. The owner asks the court to recognize his right of ownership to the disputed property and to free the latter from distraint or arrest as those were unlawfully carried out. This type of suit is an *actio negatoria* only if the object remains in the possession of the owner. If it is taken from him, *revindicatio* must be brought.

4. *Protection of the rights of an owner who is not the possessor.* According to Article 157 of the Civil Code, the right to recover an object from another's unlawful possession and the right to eliminate violations of use and disposition of an object belong not only to the owner but also to the legal possessor of another's property – leaser, carrier, safekeeper, agent, and other parties. For example, these parties may bring suit to free property from arrest on the grounds that the property belongs not to the debtor but to another person from whom they received it for possession and use. The placing of property under arrest and the related possibility of its enforced alienation threaten not only the rights of the owner but also the rights of the possessor. Therefore, they are granted the possibility of trying to protect their rights independently of the owner. The granting of the claim restores the right of ownership of the lawful possessor of another's property and not of the owner. The lawful possessor may also bring claims against the owner if the latter hinders the possessor in realizing the possession and use of the property.

The protection of the rights of the possessor of another's property is also established to serve the interests of the owner. The latter must be assured that the party to whom he transferred his object will protect it from unlawful encroachment by third parties. The possessor's right, established by law, to present a claim to third parties for the return of property or the elimination of violations to its use and disposition does not exclude the right of the owner to present claims to these same third parties on the basis of *revindicatio* or *actio negatoria*.

The Law of Obligations

Chapter 16. The Concept of Obligation. Grounds for the Origin of an Obligation

1. *Concept and significance of obligation. An obligation is a legal relationship in which one party (the debtor) is obliged to perform a specific action for another party or to refrain from a specific action, and the creditor has the right to demand that the debtor perform his obligation (Art. 158 Civil Code).*

Obligation is the most prevalent and at the same time the most varied type of civil legal relationship. When a transport organization transfers goods and passengers, an industrial enterprise supplies raw materials and goods to a plant or consumer goods to a trade organization, an institute carries out scientific-research or design work at the direction of an enterprise, a construction organization builds a dwelling house or a production unit, a citizen acquires merchandise in a store or orders a coat from a work-shop, or a seamstress sews a dress – in these and many other situations, organizations and citizens enter into obligation. Obligation has great significance in securing the normal functioning of any production enterprise, as well as in satisfying the daily material and cultural needs of citizens.

The norms which regulate obligations constitute an important section of civil law – *the law of obligations.* It encompasses a significant portion of civil legal relationships. It is sufficient to point out that Part III of the Civil Code, "The Law of Obligations," contains more than one-half of all the articles in the code. The part consists of a general section (entitled "General Principles") and a specific section (entitled "Particular Types of Obligations"). The latter part is devoted basically to various contracts. It also contains three types of the so-called "non-contractual obligations." These are obligations which arise as a result of inflicting harm, or rescuing socialist property, as well as of acquiring and retaining property without cause.

The contents of an obligation contains the rights of the creditor and the obligations of the debtor. The rights and obligations of the parties in an obligation usually have as their subject some specific property (a seller is obliged to turn over a purchased article to a buyer, a buyer – to pay money for it, a loan center gives out articles for temporary use, etc.). Besides this, there are also obligations of non-property character (an example is the obligation of an author to correct the proofs of his work).

As a general rule, an obligation is directed at the performance of specific actions: handing over objects, carrying out work, rendering services, paying money, etc. Only in rare instances does the obligation of a debtor manifest itself in the necessity of refraining from certain actions; for instance, unless otherwise arranged by contract, a custodian takes on himself the obligation not to use the object given to him for safekeeping.

2. *Parties to obligations*. There are always two parties in an obligation. The *creditor* is the party which has the right to demand the performance of actions (or restraint from action), and the *debtor* is the party which must carry out these actions (or must refrain from carrying them out).

Although there are only two sides in an obligation, more parties may be involved in it. This happens when two or more citizens or juridical persons act in the capacity of creditor or debtor at the same time. Let us take as an example the obligation binding the family of a renter to the owner of a dwelling house. If this family consists of seven people, and the house belongs to five owners, it is not difficult to establish that on one side seven, and on the other side five, persons will enter into obligation. This type of situation is called *plurality of parties in obligation*. In regard to plurality of debtors, it is important to determine whether each one of them must perform the obligation in its entirety or only a specified part, and in regard to plurality of creditors, does each one separately have a right to demand performance of an obligation in its entirety or only a part thereof? Depending on the answer to these questions, the law distinguishes two types of obligations in regard to plurality of persons: first, a *shared* obligation, and secondly, a *joint and several obligation.*

Unless otherwise provided by law or contract, an obligation involving a plurality of parties is a shared obligation. This means that each one of the creditors has the right to demand performance, and each one of the debtors must perform a strictly defined share of the obligation. If there are no other provisions in the law or contract in regard to this, the shares must be equal. A share obligation with unequal share arises, for example, in the relationships of industrial enterprises with consumer cooperative organizations which they supply with merchandise. In this type of obligation, the regional consumer union and several dozen district consumer unions all act in the capacity of buyer at the same time. The contract secures the share of the regional consumer union and each of the district consumer unions (the quantity and type of merchandise which must be supplied to them).

A special feature of any share obligation consists of the fact that no single debtor is responsible for the other and no single creditor has the right to receive more than his share from the debtors. After performing his share of the obligation, a creditor simply leaves the legal relationship. Analogous consequences arise also for a creditor whose obligations are performed to the extent of his share.

If a share obligation is the rule, a joint and several obligation is the exception. A joint and several obligation may arise only in instances when directly provided by law or contract. Specifically, the law recognizes as a joint and several obligation one which is concerned with an indivisible subject (Art. 180 Civil Code), and also the obligation of persons who jointly caused harm to compensate for it (Art. 455 Civil Code).

In an obligation involving joint and several creditors, each one of them has a right to make a claim to the debtor for the entire amount. Thus, if a citizen lives in a house belonging to two owners, both of them may demand payment from the renter which is owed for the apartment. It is understood that the renter has performed his obligations fully once he has paid the money to one of the creditor-owners. In regard to the relationship between creditors, Article 184 of the Civil Code provides that the creditor who has received performance from the debtor must reimburse the other creditors the share allotted to them. In the above example, the owner who has received the rent payment is obliged to pay one-half of the amount to the co-owner.

If joint and several debtors participate in an obligation, a creditor has the right to demand performance of the obligation from all debtors jointly or from any one of them separately, either to the fullest extent or for a specific part (Art. 181 Civil Code).

Let us take an example: three people have committed a robbery in a store, causing losses of 12,000 rubles. One of them is the owner of a house, but the others do no have any property sufficient to cover the loss. Since according to the Civil Code, as has already been noted, those who jointly caused the loss are considered joint debtors, the creditor-store has the right to demand the entire amount owed to it from the owner of the house in bringing a claim against the house. If the sale of the house does not bring the necessary sum, the store has the right to demand the remainder from any one of the three participants in the theft to the full extent or a specific share.

However, it is not allowed that only one of the debtors, the one selected by the creditor, suffer. Therefore, Article 183 of the Civil Code allows the debtor who performs an obligation to file a counter claim against the remaining debtors. However, in relation to him, they are now considered share and not joint and several debtors, and specifically, equal share debtors (of course, with the deduction of his own share). In this example, the one who reimbursed the store for the loss of the 12,000 rubles may demand that each of the other two debtors repay him 4000 rubles each.

Such a claim (which is called "a claim for exoneration" [*regressnoe*]) only in joint and several obligations, but also in other instances when one party has to pay for another. For instance, when two insured motor vehicles collide, their owners receive collision insurance from the State Insurance Authority [*Gosstrakh*]. And in its turn, the State Insurance Authority may demand the paid sum from the owner of the motor vehicle responsible for the collision by a claim for exoneration. Or, to take another example: if a buyer discovers substantive defects in purchased furniture, the store is obliged to return its cost. In instances where the defects are those of production, the store may, by a claim for exoneration, demand the money paid for it from its supplier, a furniture combine.

From the moment of the origin of the obligation to its discharge, the same parties usually act in the capacity of debtor and creditor. However, for various reasons, a creditor or debtor may leave the obligation if another party replaces him. Then the so-called change of *parties in obligation* occurs.

Three questions arise in relation to a change of parties in obligation: is the change always possible, by what procedure is it carried out, and what consequences does it have? The answer depends on whether the debtor or the creditor is replaced.

The law calls the substitution of the creditor assignment of claims. According to Article 211 of the Civil Code, such an assignment is in principle possible. This article sets forth only a limited set of circumstances prohibiting assignment of claims. It is impossible to assign a claim if this is directly forbidden by the contract or expressly by law. Finally, Article 211 of the Civil Code does not allow the assignment of two types of claims: those linked to the personality of the creditor (for instance, an author may not transfer his right to receive royalties for his work to someone else), and those directed at compensating for harm which led to injury or caused death.

The assignment of claims is based on a contract concluded by the original and new creditors. The consent of the debtor is not required: the law proceeds from the assumption that it makes no difference to a debtor to whom he must pay money, perform work, etc. The original creditor has only one obligation: to inform the debtor of the change. If he does not do this, the debtor has the right to perform his obligation not to the new, but to the original creditor (Art. 213 Civil Code).

All the rights of the original creditor are transferred in their entirety to the new creditor; therefore, if someone guarantees to perform a debtor's obligation for the original creditor, the guarantee remains in force for the new creditor. The original creditor is responsible to the new creditor only in regard to validity of the assigned claims (for instance, that it has not lost juridical force as a result of the lapse of a time-limit, etc.). If the debtor violates his obligation, the original creditor is not responsible for this to the new creditor. Such a responsibility may ensue only under the condition that this is provided in the contract between the original and new creditor (Art. 212 Civil Code).

In contrast to the assignment of claims, a *transfer of a debt* may substantially harm the interests of the creditor: it makes a great deal of difference to him who has the obligation to perform a certain action. Therefore, Article 215 of the Civil Code permits the transfer of a debt only with the permission of the creditor. Thus, the transfer of a debt requires the expression of the will of three parties: the original debtor, the new debtor, and the creditor. As in assignment of claims, a debt is transferred to a new debtor in the same amount as it was for the original debtor. Therefore, if an owner received a loan from a bank for capital improvements to his house and if a penalty is imposed because of overdue repayment of the money, the obligation to repay both the basic amount of money and the penalty for tardiness to the bank are transferred to the buyer if the house is sold.

3. *Grounds for the origin of obligations.* The grounds for the origin of the obligations are called *legal facts*, and the law links them with the origin of civil legal relationships concerning obligations. Legal facts include the following: administrative (including plan) acts, transactions (including contracts), tort, unjust acquisition and retention of property,

and other actions of citizens and organizations. In some instances the obligation is created by one of these legal facts, and in others, by their specified totality (legal component).

Administrative (plan) acts express the will of the state organ issuing them. Some administrative acts can create obligations involving citizens and organizations (for instance, by virtue of an order an obligation arises of an organization managing dwellings to conclude a contract of hire of a structure with the party indicated in the order). Plan acts give rise to obligations only in mutual relationships between socialist organizations. For example: the obligation of a transport organization to supply the means of transportation, and of the shipper to use them, arises on the basis of an approved plan for the transport of goods.

The most prevalent basis for the origin of obligations between socialist organizations in the transfer of objects of production, rendering services and work, is a contract which is concluded in accordance with a plan act. The expediency of using a complex basis (the plan act plus contract) may be explained by the fact that the plan act defines only the general conditions of the mutual relationship of the organizations. They are later defined and detailed by the parties in concluding the contract. The content of the contract which is concluded on the basis of the plan act must agree with the plan task (Art. 159 Civil Code).

Under these circumstances, the plan and the contract are equally necessary. If the parties do not conclude a contract between themselves, then, even if there is a plan act, no rights and obligations are created between them. In such instances, an enterprise dispatching products may not demand payment from their recipient, nor may the buyer demand that the supplier dispatch products in agreement with a plan act.

An administrative act and a contract are sometimes necessary for the creation of obligations involving citizens. For instance, the obligation linking an organization which manages dwellings with the renter of a dwelling in a house belonging by right of ownership to the state, a cooperative, or other social organization stems from two legal facts: an order issued by the executive committee of the local Soviet (administrative act), and a contract concluded between the manager of the house and the party named in this order. And in this instance, if there is no order, the contract is considered invalid and a very unfavorable consequence ensues for the party: to vacate the dwelling without being granted another one.

Obligations between socialist organizations and citizens, as well as between citizens, originate for the most part directly from contracts. Obligations between socialist organizations may also originate directly from contracts. For instance, it is not necessary that a special plan act exist in order to conclude a contract for the supply of types of products which are not regulated by the plan; it is not necessary to have a special plan act to perform design work if the value does not exceed the established maximum.

Unilateral transactions give rise to obligations less frequently. An example would be a public promise or reward. The party making the announcement must pay the reward to the person who performs the actions stipulated in the announcement (finds a dog or discovers documents). One type of public promise of reward is an announcement by a socialist organization to hold a contest for the best execution of a certain task (a story, picture, structural design, etc.).

The law imposes on juridical persons and citizens *the obligation to compensate for any harm caused by them*. The basis for the origin of an obligation in such an instance is the tortious act carried out by a party (running someone down with a motor vehicle, releasing polluted sewage, failing to take administrative measures for safety in carrying out construction work, etc.).

Obligations may arise from the *unjust acquisition* or the *unjust retention of property at another's expense*. One party may acquire property at the expense of another only if grounds are established by law (for instance, as a result of inheritance, gift, purchase and sale, etc.). The party which acquires property unjustly (a postal package delivered by mistake to someone bearing the same surname as the intended recipient), or who retains it unjustly (products were dispatched to an automobile plant but a motorcycle plant paid for them), must return the property acquired (retained) unjustly to the party at whose expense it was acquired and retained. In the above example, the actual recipient must give the package to the person who is entitled to get it, and the automobile plant must give the sum paid for the products to the motorcycle plant.

Obligations may arise as a result of a discovery, an invention, a rationalization proposal, or the creation of works of literature, science, or art.

In accordance with Articles 158 and 4 of the Civil Code, obligations may also arise "as the result of other actions by citizens and organizations." An example is the obligation arising from the rescuing of socialist property. If a citizen who took part in rescuing socialist property (extinguishing a fire in a plant, preventing theft in a store, etc.) is injured, the organization (plant, store) whose property the victim rescued must compensate him. This obligation is substantially different from the obligation for causing harm. However, in this instance the socialist organization must compensate for the injury, although it did not commit any unlawful acts.

Chapter 17. Contracts. General Principles

1. *Meaning and significance of contract. A contract is an agreement between two or more parties directed toward the creation, change, or termination of obligations.* Contract is the most prevalent basis for the origin of obligations linking socialist organizations, socialist organizations and citizens, as well as citizens.

In the mutual relationships between socialist organizations, contracts are usually concluded and performed in agreement with the plan tasks approved for their participants. Such contracts (called economic contracts) have the aim of coordinating the acitivity of the parties: of enterprises – suppliers and buyers, of customers and contractors, of shippers and organizations, and so forth. After concluding a contract, the parties may by agreement define the best means of fulfilling each of their plans and tasks. As a result of the contract, the obligations of the enterprises to the state arising from the approved production plan are at the same time obligations of each of them toward the other party. This makes possible the effective use of the system of property sanctions to strengthen plan and contractual discipline.

The decisions of the Party Congresses and the joint decrees of the CC CPSU and the Council of Ministers of the USSR contain important instructions in regard to the more effective utilization of contracts in the nation's economy. The XXVI Congress of the Party called for: "The improvement of economic ties between industrial, agricultural and commercial enterprises and organizations. The elevation of the role of the economic contracts in working out the production plans and in the realization of consumer goods."[1] The decree of the CC CPSU of 12 July 1979, "On the Further Improvement of the Economic Mechanism and the Tasks of Party and State Organizations" emphasizes the necessity "to unflinchingly strengthen state, production, and labor discipline, and raise the sense of responsibility of the work force for fulfilling the

1. *Materials of the XXVI Congress of the CPSU*, p. 180.

established plans and contractual obligations."[2]

In our country much attention is also directed to contracts in which citizens are the parties. The great concern which the Soviet state has for raising the material well-being of the population is reflected in the development of diverse contractual links involving citizens. The network of consumer stores and enterprises is spreading continuously, and the commodities and services granted to them are increasing in number and more varied in their scope.

The state spends enormous sums every year on the construction of dwellings and on the development of all types of transportation. Of course, among the contracts involving citizens, the most important are those concerning citizens and socialist organizations.

The Soviet state takes effective measures for the development of contractual relationships among citizens. The decisions of the XXVI Congress of the Party emphasize that it is necessary "to develop the network and to improve the work of collective-farm markets. To help the population to deliver and to dispose of a surplus in the agricultural production carried out on the subsidiary household plots, and in garden, orchard, and other cooperatives."[3]

2. *Contents of a contract.* Any contract consists of specific conditions (points) which secure the rights and obligations of citizens. In their entirety, these conditions constitute the *contents of the contract.*

The conditions of a contract are of three types: *essential, incidental, and usual.* Essential conditions include conditions which are necessary and sufficient to consider a contract concluded. Therefore, if one of the essential conditions is missing from a contract, then there is no contract. And, conversely, if the contract contains all the necessary conditions it is considered concluded even if no other conditions besides the essential ones appear in it.

2. *Perfection of the Economic Mechanism. Collection of Documents,* Moscow 1980, p. 4.
3. *Materials of the XXVI Congress of the CPSU,* p. 180.

Insofar as the origin of a contract is linked to the reaching of an agreement by the parties in regard to its essential conditions, it is important to define their range. Instructions regarding this are found in Article 160 of the Civil Code. This establishes the following types of conditions among the essential: a) one which is regarded as such by law, and b) any other conditions which one of the parties claims must be agreed upon.

An enumeration of essential conditions is usually contained in norms concerned with the contracts of socialist organizations. Such an enumeration for a procurement contract is contained in Article 268 of the Civil Code (quantity, quality, time period, procedure and conditions of delivery and the place of delivery of agricultural products, etc.); [an enumeration] for a contract of supply is found in the Statute on the Delivery of Production-Technical Products and in the Statute on the Delivery of Consumer Goods (both of these acts name five essential conditions: name, quantity, nomenclature (assortment), quality, and price of the products).

In any contract at least those conditions which define its subject are considered essential – quantity of services rendered or property transferred (of what type and how much). The conditions concerning the time periods are also usually considered essential with the exception of those which are performed at the moment of the concluding of the contract (for instance, the purchase of goods in a store). Finally, in contracts for consideration, the conditions concerning the value of property or services are considered essential.

Usual conditions are those which reproduce the dispositive norm of the law, and those which provide rules not secured by law are called incidental. Let us take as example Article 425 of the Civil Code. This indicates that "the custodian does not have a right to use the property delivered to him for safekeeping unless otherwise provided by contract." The condition prohibiting the custodian from using the property is considered usual. On the other hand, the condition under which the custodian is granted the right to use the delivered property belongs to the incidental category. This example makes it clear that incidental conditions widen the scope of the contents of the contract. This is not true of usual conditions. Whether or not a specific norm is repeated in the contract or not, it nevertheless is in force in the contract concluded by the parties.

3. *The conclusion of contracts.* Insofar as a contract represents the agreement of two or more parties, it is in order for it to come into being necessary that all of them express their will and that the contents reflects the agreement of the will. If one party proposes that another buy a motorcycle from him for 500 rubles but the other party is prepared to pay only 300 rubles, there is no contract. The parties usually express their will to conclude a contract by the words "I will buy," "I will sell," "I would like a loan," "I will take it for safe-keeping," etc.

Regardless of the form in which the will of the parties is expressed, the conclusion of contracts falls into two successive stages: first, *the proposal* (offer), and secondly, *the acceptance of the proposal* (acceptance).

An offer is a legal fact which has an important consequence: the party which made it is bound by it, and if the party to whom this offer is directed accepts it, the contract is considered concluded. Various types of offers are possible: *oral or written, with instructions for a time period for reply, or without instructions.* If an offer to conclude a contract is made orally and does not fix a time period for reply, the contract is considered concluded only if the party which received the offer immediately informs the party making the offer of its acceptance (Art. 163 Civil Code). It is in this way that goods are acquired in a store or market, tickets at a box-office, airline office, etc.

A written contract which does not contain a time period for replying is considered accepted if a positive reply is received within a period normally considered necessary for this (Art. 163 Civil Code). For instance, if a proposal is sent from Moscow to Kiev the length of time necessary for mail to get from Moscow to Kiev and back is considered, as is the time necessary to prepare the reply.

Finally, a third variant is possible: an oral or written offer with instructions for a period in which to reply. In this instance, the contract is generally considered concluded if there is communication concerning acceptance of the proposal within the limits of the fixed period.

In order for the offer to give rise to the enumerated consequences, it must comply with two necessary features. First, the offer must encompass all the essential conditions of the contract. The appeal to "buy my tape-recorder" is not a proposal since this does not indicate what kind of tape-recorder it is nor what its price is. Secondly, an offer must be directed toward a specific party (a specific organization or citizen). A notice in the newspaper to the effect that "Sawmill No. 3 always has lumber for sale" is not considered an offer. Therefore, if an enterprise sends a pick-up truck for lumber based on such a notice it takes the risk that the combine will refuse to conclude a contract and will incur no obligation for this.

The rule concerning the specific addressee of an offer does have an exception. This is when it is possible to conclude from the offer itself that although it is addressed to everyone, the party making the offer is prepared to conclude a contract with any and every party responding. An example of this type of proposal addressed to everyone is a display of goods in the window of a store. Thus, if a citizen expresses a desire to acquire an object displayed in the window, his action might be considered an acceptance of the offer. Therefore, after he pays the price of the object it must be given to the buyer. If such articles are not yet available in the sale room, the buyer has the right to demand that the article bought by him be taken from the window.

Acceptance of an offer must also comply with certain requirements.

First, as has already been noted, an agreement to conclude a contract must be expressed in good time: if the offer does not contain a period for a reply, this must be expressed immediately; if such a period is given, not later than this period. And what happens if the reply to an offer is sent quickly but, because of circumstances in no way involving the sender, is delayed past the established term? Such a reply is considered late only in the circumstance that the party making the offer at once informs the other party of the receipt of the belated reply (Art. 164 Civil Code). If this does not occur, even if the reply is overdue, the contract is concluded.

Secondly, the acceptance of an offer must be unconditional (such as, "yes, agreed"). If the reply to an offer contains an agreement to conclude a contract but under different conditions (for instance, at a lower price, higher quality, shorter period of time, etc.), this reply is considered a counter offer. At the same time the parties exchange positions. Now the party which acted first with the offer has to either agree or not to agree to the conditions proposed to it.

Generally, a contract is considered concluded from the moment when the party which made the offer receives information as to its unconditional acceptance. However, if by law or by agreement reached by the parties it is necessary to put the contract into a specific form (for instance, notarial), it is considered concluded only after it takes such a form.

A contract in written form is usually a single document signed by the parties. But there are exceptions to this rule. If the law or the contract provides that the written form is necessary, the contract may be concluded through the exchange of letters, telegrams, telex messages, etc. It is necessary only that each of these documents is properly signed by the proper parties.

4. *The conclusion of contracts between socialist organizations.* The procedure for the conclusion of these contracts has a number of special features. This mainly concerns contracts based on a plan act which is binding for both sides. If one of the organizations designated in the plan act refuses to conclude the contract, the other has the right to initiate a dispute directed toward forcing it to conclude the contract. The dispute is settled either by court (if one of the parties

is a collective farm, inter/collective-farm, or state/collective-farm organization) or by *arbitrazh* (if both parties are socialist organizations other than collective farms, state/collective-farms, or inter/collective-farm organizations).

In such instances, the party which went to court or *arbitrazh* may present a draft contract. And if the other side does not present substantiating objections against the necessity of concluding the contract or against specific conditions of the presented draft, the court (or *arbitrazh*) affirms the draft and instructs that party to sign the contract within a period specified in the decision. If the contract is not signed by the end of that period, it is considered valid.

Special normative acts which are devoted to specific types of contracts establish which of the parties must present the proposal. For a contract of supply this is usually the supplier and only exceptionally the buyer; for a work contract for capital construction, it is always the contractor. If there are no such instructions, the initiative for the conclusion of contracts must come from the party which will give over property or render services. Special rules usually fix a time limit for taking such initiative (for instance, the supplier of products for production-technical purposes must submit a draft plan not later than twenty days after the receipt of a plan act, and a construction contractor, not later than two months after the affirmation of the State Plan for the Economic and Social Development of the USSR).

The party to whom the draft contract is sent must sign it and return one of the two received copies within an established period (if special rules do not provide otherwise, within ten days). The party which does not agree with specific conditions of the draft contract which is sent to it, or which suggests that it is necessary to supplement it, must present a protocol of disagreement. In it it presents its own version of specific conditions which are different from the proposed [version], or of the supplemental conditions attached to it. At the same time, it must stipulate "with protocol of disagreement" on the signed contract.

The contract together with the protocol of disagreement is

returned to the party which drew up the draft. And now the latter decides whether or not it will accept the conditions in the protocol of disagreement. If these conditions are accepted, they are put into the text of the contract. Conversely, the party which receives the contract with the protocol of disagreement may initiate a pre-contractual dispute in court (or *arbitrazh*). If the fixed period of time has elapsed, it is not possible to go to court (or *arbitrazh*). It is assumed that the party agrees with the proposal contained in the protocol of disagreement.

In the course of examining a pre-contractual dispute, the court (or *arbitrazh*) hears the arguments of the parties and tries to reconcile them on legal grounds. If it does not succeed in this, the court (or *arbitrazh*) itself creates a version of the disputed condition in its decision and this is included in the concluded contract. As a general rule, the court (or arbitrazh) has the right to examine a pre-contractual dispute only in instances when the contract is concluded on the basis of plan acts binding on both parties. All other pre-contractual disputes may be examined only if two conditions are present: a) explicit instructions for this are given in law (for example, para. 28 of the Statute on the Supply of Production-Technical Goods provides for the possibility of giving disagreements on conditions of a contract concluded by the parties to *arbitrazh* for a decision) where the parties have reached agreement on the quantity, nomenclature (assortment), and time periods for supplying the products; or b) the parties agree to give their disputes over to decision by *arbitrazh*.

In other circumstances, it is not possible to turn to the court or *arbitrazh* to settle pre-contract disputes. Therefore, if an organization which makes an offer does not agree with the counter offer of the other party, the contract may not be considered concluded. In instances when special instructions on this are given in law, contracts between socialist organizations may be concluded by the acceptance and performance of an order or by the acceptance and performance of a warrant for goods (Art. 160 Civil Code).

Such a procedure is used mainly in the supply of goods. It is possible, for instance, to conclude a contract by the acceptance and performance of an order in instances when enterprises – manufacturer and consumer – have long-standing economic links. The order of the buyer is considered accepted by the supplier if he does not express objection within thirty days.

It is possible to conclude a contract with the aid of an acceptance for performance of a warrant for goods only when the warrant contains all the essential conditions of the contract of supply (quantity, nomenclature or assortment, quality of goods, etc.) and in the opinion of the supplier and the buyer there is nothing else on which to agree. If both parties, on receipt of such a detailed, comprehensive contract, remain silent (neither one voices objections against the order and does not demand agreement on additional conditions within a period of ten days), the warrant takes on the force of contract. In this way, an acceptance for performance of a warrant is a contract of supply.

5. *Types of contracts.* Contracts used in civil law are varied and may be divided into the following types:

a) *Plan and non-plan contracts. Plan* contracts are those which are concluded by the designated parties in performing a plan act. These acts may differ in contents; some oblige both organizations mentioned in them to conclude the contract, and others – only one of them. Let us compare two plan acts: a monthly plan to transport goods by railroad and limit of credit. The first act creates an obligation to conclude the contract for both parties: the railroad and the shipper. Therefore, if either one of them abstains from concluding the contract to transport the goods (the railroad refuses to supply the means of transport or the shipper does not supply the freight for shipment), it is subject to liability established by law. Limit of credit gives rise to other consequences. It imposes on a bank the obligation to give credit to a specific organization to the extent mentioned in the limit. As far as the organization is concerned, it has the right not to receive any loan from the bank at all or to conclude a contract to receive an amount less than that mentioned in the limit.

Non-plan contracts are those concluded at the discretion of the parties. These are mainly contracts of citizens with a store, transport organization, or consumer service enterprise. In all these instances, as well as in the conclusion of a contract between one citizen with another, no plan acts are necessary. As has already been noted, non-plan contracts may also be concluded by socialist organizations. This happens when the economic relationships embodied in the form of a contract are not directly based on the plan.

b) *Unilateral and bilateral contracts. Unilateral contracts* are those by which one side has only rights and the other only obligations. For instance, a contract for loan leads to the obligation of the borrower to return the received article (money). This is matched by the right of the lender to demand return of the articles (money). The lender has no obligation in respect to the borrower.

A bilateral contract is one in which each of the participants has both rights and obligations at the same time. A large number of contracts are of this type. Examples are purchase and sale and lease. The contract of purchase and sale creates the obligation of a seller to transfer an object and demand payment for it, and the obligation of a buyer to pay the fixed price and the right to demand transfer of the bought object. A lease gives rise to the obligation of the leasor to give an object for use of the leasee and the right to ask for the fixed payment, and for the leasee the right to demand the object be given to him for use and the obligation to pay the fixed price.

c) *Contracts in favor of the parties and contracts in favor of third parties*. In concluding a contract, parties usually act in their own interests; accordingly, the rights based on the contract are theirs (of the buyer and seller, the contractor and client, etc.). However, there are instances when one of the parties acts in the interests of a third party in entering into a contract, and the right created by the contract devolves to the latter. An example is depositing money into a savings bank in another's name. If a grandfather makes a deposit in the name of the grandson, the latter has the right to dispose of the deposit. Only if the grandson renounces his right may the grandfather dispose of it. Such a contract is known as a *contract in favor of a third party* (Art. 167 Civil Code).

The Civil Code lists nineteen types of contracts (purchase and sale, exchange, lease, work contract, transport, etc.). Each of them has its own specific features. Citizens and organizations select the proper contract depending on need. However, the list of contracts contained in the law is not exhaustive. Parties may conclude a contract which is not provided in law. This contract is considered valid, and consequently, may create rights and obligations between its parties if, of course, it satisfies the requirements expected of contracts in general, [and] of their particular type, e.g. of specific contracts. In general, when a contract which is not provided for in the Civil Code is concluded, specific elements of well-known contracts which are regulated by law are utilized. This results in *mixed contracts*, which are a combination of purchase and sale, custody, lease, etc.

Chapter 18. Performance of Obligations

1. *Significance and principles of performance of obligation. Performance of obligation* is the execution of an act provided therein (transfer of an object, payment of money, building of a house, etc.). And in those rare instances when the obligation is directed toward restraint of actions, its performance is expressed in the non-execution of such actions (an author does not give his manuscript to another publisher, a custodian does not use the article given to him, etc.).

Legislation in force regulates in detail who must perform an obligation, for whom, how, and when. Besides this, the norms of the Civil Code secure the following general principles with which the performance of obligation in every specific circumstance must agree: 1) *proper performance;* 2) *specific performance*; 3) *comradely cooperation and mutual aid*; and 4) *economy*.

Article 168 of the Civil Code provides that an obligation must be performed *properly*. This means that it is necessary to perform the action provided in the obligation (to refrain from the action provided in the obligation) in strict conformity with the instructions of the law, the plan act, or contract, and in their absence, to the usual requirements. Proper performance is a complex concept. It includes a number of requirements since the obligation must be performed not only by the proper party, but also in the proper manner in regard to the subject, place, time, and method of performance. Detailed instructions concerning these demands are found in the Civil Code, and also in acts regulating specific types of obligations (retail purchase sale, supply, house lease, rental, etc.).

The requirement of proper performance undoubtedly includes the impermissibility of unilateral refusal of performance of an obligation or the unilateral alteration of the conditions of a contract (Art. 169 Civil Code). Unilateral refusal and unilateral alteration are possible only when there are explicit instructions in law. This is most frequently connected with the unlawful behavior of the other side. Thus, in the instance of the purchase of an article of defective quality, the buyer has a right to demand a proportionate reduction of the price (resulting in the alteration of the corresponding condition of the contract) or recission of the contract (Art. 246 Civil Code). Here is another example: an organization which has announced a contest for the best performance of some type of work has the right to alter its conditions during the first half of the period announced for the submission of the work (Art. 440 Civil Code).

The principle of specific performance is expressed in the necessity for a debtor to perform an obligation in kind: to carry out precisely that action which constitutes the subject of the obligation (to transfer an object, execute a specific job, perform a suitable service). To substitute payment of money for performance in kind is not allowed. For example, a factory is obliged to supply cloth worth 10,000 rubles to a commercial organization every month. In January, the delivery did not take place. The buyer has the right to seek liquidated damages from the supplier for this. However, due to the principle of specific performance, the factory is obliged in addition to fulfill the January delivery in February. This means, that in February it must ship cloth worth not 10,000 but 20,000 rubles.

Soviet civil law does not grant to the debtor an opportunity "of buying himself out of performance" by compensating for losses. This is linked to the fact that the aim of socialist society is not to derive profits, but the maximum satisfaction of the material and cultural needs of society, and the obligations themselves are directly or indirectly directed to the fulfillment of the plan tasks. The fixed goal may be reached only under the conditions that the obligations which are concluded by socialist organizations are performed in kind: a supplier delivers the appropriate amount of products to a buyer, a contractor builds and transfers to a customer the object ordered in the contract, a railroad delivers goods to the designated station, etc. If this were not so, a creditor would not be able to execute his own production plan. No amount of money can replace undelivered raw material or equipment, an unbuilt ship, or undelivered goods. In exactly the same way a citizen who ordered a coat in an atelier or who bought a ticket for a trip is interested above all in the actual performance of the second part of obligations: that the coat or suit be sewn, that the trip take place.

The principle of the specific performance of obligations is secured in a general fashion in Articles 217 and 218 of the Civil Code. The first of these applies to obligations which have as their subject the delivery of an individually defined article. If the debtor in breaching the contract does not deliver the article, the creditor is granted the right to demand that it be taken from the debtor and delivered to him. The same demand may be made by a buyer to whom the seller did not give a purchased article, a leasee to whom the lessor did not give an article for use, etc.

The principle of specific performance encompasses not only the obligation to transfer an object but also the obligation to perform defined work. In the latter instance, its effect is evident in the fact that if the obligation is breached by the debtor, the creditor has the right to choose either to perform the work at his expense (for instance, if a defective object is delivered, the creditor may give it to a third party to be repaired and then to demand from the debtor the sum which he paid for this) or to demand reimbursement for his losses (Art. 218 Civil Code).

The *principle of comradely cooperation and mutual aid*, which stems directly from the socialist nature of obligations, is secured in Article 168 of the Civil Code. It provides that each side must render to the other side all possible assistance in the performance by it of its obligations. Thus, although this goes beyond the limits of the usual contract of supply, if long-standing economic ties exist between enterprises-producers and users, the obligation may be imposed upon the buyer (user) to help the producer in the task of raising the quality of the product turned out, specifically to give information on the results of using its products; and the obligation on the producer to advise the buyer (user) during the process that the received goods are being used.

Article 168 of the Civil Code also secures the principle of *economy*. Its requirement is expressed by the necessity for each of the parties to perform their obligations in the manner most economical to the socialist national economy. The action of this principle may be illustrated by the example of the relationships involved in capital construction and supply. Thus, a contractor with the agreement of the client is granted the right to make changes in a project under his supervision is order to lower the cost of construction below the estimate. In obligations linking a railroad with a shipper (recipient) of goods, the parties may not allow the wasteful use of the means of transportation, and particularly of cross-hauls or excessively long and repetitive hauls.

2. *Subjects of performance*. As a general rule, the debtor himself performs an obligation. However, he is granted the opportunity to transfer either performance in whole or in part to a third party. Correspondingly, the creditor is placed under obligation to accept performance not only from the debtor but also from a third party. This rule also has exceptions. For instance, when a publisher commissions a story, a novel, or a poem from a specific author or a philharmonic society concludes a contract with a specific person to perform in a concert, the interests of the publisher and philharmonic society may undoubtedly suffer if instead of well-known authors and actors, someone else carries out the activity envisaged in the obligation. Therefore, the law provides that in such instances the debtor must carry out the obligation himself. Besides this general norm, the law prohibits the transfer of certain specific types of obligations to third parties. Thus, Article 398 of the Civil Code places the obligation of performing a task entrusted to him by the principal on the agent personally (to receive money for him, to sell an object belonging to him, to buy something for him, etc.).

Article 171 of the Civil Code permits placing the obligation of a debtor onto a third party if one of three conditions is present:

First, if this is provided by established rules. For instance, in sending parcels from Kuibyshev to Khabarovsk, the sender concludes a contract with the Kuibyshev Communications Enterprise, but according to the USSR Communications Charter, the communications enterprise in the place of receipt also participates in the performance of the contract.

Secondly, if a third party is related to the debtor through administrative subordination. For instance, when a production enterprise concludes a contract with a regional department of the State Committee of Agricultural Technology and the obligations of the latter (to accept and pay for products) must be performed by district and inter-district departments which are subordinate to it.

Thirdly, if the third party is related to the debtor by a relevant contract. For instance, a shoe factory concludes a contract of supply with a distributor in the shoe business, and the distributor, in its turn, concludes a contract with a department store. At the instruction of the distributor, the shoe factory ships the goods directly to the department store; at the same time, the obligation of the distributor to the department store is performed by its contractual party, the factory.

As a general rule, a debtor is himself obliged to perform for the creditor. However, a creditor has a right to demand that the debtor perform the obligation for a third party (for instance, to deposit apartment rent into a savings bank). If such a demand results in additional expenses for the debtor, he must be reimbursed by the creditor.

The law particularly singles out circumstances under which a debtor does not have the opportunity to perform a monetary obligation only because the creditor is not present at a specific location or refuses to accept the performance. In order to prevent the occurrence of unfavorable consequences established for an unpunctual debtor (payment of liquidated damages, compensation for losses, and, if it concerns a person living in the house of a personal owner – eviction, etc.), the debtor is granted the right to bring the money which is due and deposit it in a notary's office. The creditor may later receive the deposited sum. But this need not concern the debtor, since the bringing of the monetary sum or security for deposit in the notary's office is considered performance of obligation (Art. 185 Civil Code).

3. *Period for performance.* The most important condition of obligation is the *period.* Some obligations contain exact instructions regarding the time in which an action must be performed, others do not provide a period for performance in general, and finally, a third group defines the time as the moment of demand (for instance, the obligations to safeguard hand luggage in a railroad station, landing pier, port).

In obligations with unspecified periods, as in obligations with periods defined at the moment of demand, the creditor may demand performance and the debtor is obliged to carry out performance at any time (Art. 172 Civil Code). As a general rule, in such instances the debtor is granted seven days from the time the creditor demands performance. However, the law or contract may provide that the obligation be performed immediately. The necessity for immediate performance sometimes arises from the nature of the obligation (for example, when a citizen demands his deposited money from a savings bank). If the obligation has a period, then belated performance is improper performance resulting in various unfavorable consequences for the debtor.

And how is the behavior of a debtor who fulfills his obligation prematurely to be judged? Sometimes such actions are considered permissible, and sometimes not. In a number of circumstances, premature performance is provided for in the law. Thus, Article 290 of the Civil Code grants to a citizen the opportunity of returning a hired object at any time. Besides this, in relationships involving citizens, a general rule is in force: the debtor has a right to perform an obligation prior to expiration of the period if it is not otherwise provided by law, the contract, the nature of the obligation (Art. 173 Civil Code). A renter has the right to pay his apartment rent before the fixed time, a contractor to carry out the work, etc. But an airplane must not take off or a ship leave earlier than the scheduled time, a theater performance may not start earlier than the time listed on the poster, etc.

In relationships between socialist organizations, there are different rules than in those involving citizens: a debtor in principle does not have a right to prematurely perform an obligation stipulated by law. Plan obligations are based on assumptions of rhythmical performance. The transfer of articles, rendering of services, or performance of work before the fixed time frequently is as inconvenient to a creditor as their performance after the fixed time. Thus, the interests of a construction organization suffer substantially if a supplier delivers equipment long before the building in which it is to be placed is ready. It is not hard to imagine the difficult position of a department store if a factory delivers the entire year's supply of fishing equipment, boats, and summer sport inventory in January. In order to prevent this, Article 173 of the Civil Code permits premature performance of obligations involving socialist organizations only in instances when this is provided by law, the contract, or with the consent of the customer.

The Civil Code establishes specific rules for the determination and calculation of period. Periods must be determined by law, the plan act, or contract in terms of a calendar date, the expiration of the period, calculated in years, months, weeks, days, or even in hours (for instance, for the delivery of baked goods to a store), or with instructions concerning events which must transpire. The period begins on the day following the calendar date or the events which fix its beginning. A period calculated in years terminates in the appropriate month and day of the last year of the period; one calculated in months, on the appropriate day of the last day of the month of the period; in weeks, on the appropriate day of the last week of the period. When the last day of the period falls on a non-working day, the period is considered to have ended on the first subsequent working day. Finally, it must be borne in mind that an act may be performed on the last day of the period, in the 24 hours of that day. However, if an action must be performed in an organization, the period expires within the hour when, according to the rules governing its operation, work is terminated.

4. *Place of performance.* The place of performance answers the question of exactly where a debtor may perform his obligation to the creditor (delivering an article, giving over a construction object, etc.). It may be provided by law, or by contract (in relationships between socialist organizations – by a plan act); frequently it may be defined by the nature of the obligation. It is usually established by normative procedure or by agreement of the parties whether the supplier must transfer a buyer's product to his own storehouse or to ship them by a specific mode of transport to the buyer's address. From the nature of a rental contract for living quarters, for example, it is apparent that it must be performed at the place of the location of the quarters, and from the contract concluded with a store for the delivery of items intended as gifts it is not hard to realize that the place of its performance is the residence of the recipient indicated in the order.

In the instance of the lack of precisely defined instruction on the place of the performance in the law, contract, or plan act and the impossibility of arriving at a decision on this from the nature of the obligation, Article 174 of the Civil Code supplies an additional rule. It defines the place of performance depending on the type of obligation. An obligation to transfer a building must be concluded at the place where the building is located, and monetary obligations (excluding monetary obligations of state, cooperative, and other social organizations for which special rules are established) – at the place of residence of the creditor at the moment the obligation arose; however, if the creditor has changed his residence before this time and has informed the debtor thereof, it is necessary to perform the obligation at the creditor's new place of residence (any additional expenses incurred for this must be borne by the creditor). Finally, for all other obligations the place of performance is the debtor's place of residence (if the debtor is a juridical person, the place of its situs).

Guided by the rule given above, it is not difficult to establish that in instances when the creditor finds it necessary to transfer money to the creditor by mail or telegram in order to perform his obligation, the expenses for this are borne by him. Conversely, if the performance of obligation entails the shipping of articles as baggage or in the form of postal parcels, then insofar as the obligations is not of a monetary character, it must be performed at the place where the debtor is located, and therefore the expenses incurred by the debtor are reimbursed by the creditor.

5. *Subject and means of performance.* The actions which a debtor must perform are the subject of performance. If the performance of such actions is connected with the transfer of an article, then the article itself is called the subject of performance. If money constitutes the subject of performance, it must always be expressed and paid in Soviet currency, except in circumstances explicitly provided by law (Art. 175 Civil Code). As a general rule, it is forbidden to add interest to any monetary or other obligation. Exceptions are made only: for the operation of special credit institutions (a savings bank pays interest for deposits, and when a bank gives credit to socialist organizations and citizens it has a right to demand from then not only the amount of the principal but also the fixed interest [Art. 176 Civil Code]); and for foreign trade obligations as well as several other circumstances provided by law (for instance, in the sale of merchandise on credit).

Usually there is only one specified subject in an obligation: in a purchase made in a store, the article selected; in the conclusion of a rental contract, the apartment or room indicated in the lease, etc. However, an obligation may indicate that its subject may be one of several articles. In a contract with a cafeteria, a collective farm not knowing what the crop of vegetables will be assumes the obligation to sell "in the fall either 10 tons of carrots or 20 tons of potatoes." This type of obligation, which is constructed on the principle of "either/or" is called an *alternative obligation*. Article 178 of the Civil Code indicates that unless otherwise provided by law, the contract, or the nature of the obligation, the right of choice among several subjects belongs to the debtor (in the example given above, to the collective farm).

The method of performance is of interest mainly when a debtor is obliged to transfer some type of collective article (a specific amount of money, a complete set of equipment, etc.). In this situation, the question may arise: does he have a right to perform the obligation in part or must he perform it in its entirety?

In a number of instances the law, plan act, or contract provides for the necessity of the creditor's accepting obligations in part. Sometimes the possibility of performance in part arises from the nature of the obligation (for instance, the purchase of goods by installments). In other instances, the creditor has a right not to accept performance of obligations in part (Art. 170 Civil Code). In relationships between socialist organizations, a stricter norm is sometimes applied prohibiting the creditor from accepting performance in part. Thus, a buyer who is presented with an unassembled product (an automobile without a regular or spare tire or without necessary accessories) has not only a right but also an obligation to refuse to pay for the product until it is completed or exchanged for one which is complete.

Chapter 19. Securing Performance of Obligations

1. *Concept and significance.* As a general rule, debtors – organizations as well as citizens – perform their obligations properly. However, there are exceptions. In such instances a creditor may claim losses caused by a debtor who has breached his obligations. The threat of having to reimburse for losses forces the debtor to take action to perform his obligation. However, often this basic means of securing performance of obligations – compensation for losses – is, for one reason or another, insufficient to satisfy the violated interests of the creditor. In this situation, the parties may avail themselves of supplementary legal means as part of their mutual relationships.

The Civil Code (Art. 186) lists five means of securing performance of obligation: liquidated damages, pledge, suretyship, guarantee, and earnest. Of these, liquidated damages, pledge, and suretyship may be used under any circumstances, earnest only in obligations involving citizens, and guarantee only in relationships between socialist organizations.

The significance of these means of securing obligation consists in the fact that a primary obligation (transfer of an article, performance of work, or of services) is linked to an auxiliary obligation. This becomes a factor when a debtor violates the primary obligation. The auxiliary obligation is firmly linked to the primary one, it arises only when a primary obligation does, and terminates at the same time as does the latter.

All five measures of securing obligation are directed to the same goal – to secure the performance of obligations. However, each one carries out its function of securing in a unique manner.

2. *Liquidated damages. Liquidated damages is an amount of money determined by law or by contract which a debtor must pay to the creditor in the instance of failure to perform or improper performance of an obligation* (Art. 187 Civil Code).

The term "liquidated damages" [*neustoika*] is based on the verb *ne ustoial*. Someone has not upheld his bargain, has violated the obligation, and therefore must pay.

Compensation for losses also means that a faulty debtor must pay a specified sum to the creditor. However, there are distinct advantages in liquidation of damages. It is necessary to show the cause and extent of the losses, and this is not always possible – or at least not always simple. Furthermore, there are instances when a debtor causes no losses to the creditor in violating a contract, but nevertheless violates the interests of the state, society, or other persons.

Here is an example of this: A store receives shoes of defective quality and marks them down. If it settles accounts with the manufacturer on the basis of this reduced price at which the shoes were sold to citizens, the store will suffer no losses. And this means that the action of the factory producing shoddy products remains unpunished. Another situation may arise under liquidation of damages. It is formed on the principle "if you transgress – you pay", regardless of the consequences this may cause for the creditor. And yet another important aspect of liquidation of damages: it is very simple to calculate its extent. Because of this, liquidation of damages is widely used.

Liquidation of damages is a very flexible sanction. There are various types of liquidation of damages, and the lawmaker and sometimes the parties themselves may choose precisely that type which most suits the specific issues of their mutual relationships. Above all it is necessary to point out that liquidation of damages exists either in the form of a *penalty* [*penia*] or a *fine* [*shtraf*]. A penalty is established for overdue performance. It is exacted interruptedly in an increasing sum (for example, 0.1% of the unpaid sum for each day of delay).

In distinction to penalty, a fine is a one-time payment of liquidated damages. Sometimes it is a set sum. A railroad which breaches an obligation provided for in the plan to supply means of transport must pay a fine of 50 rubles for each railroad car not supplied for the shipment of goods to a shipper. However, a fine is usually expressed in percentages. Thus, a supplier of defective goods is fined 20% of the cost of the shoddy merchandise.

In relationships between socialist organizations, a penalty and fine may be combined. An organization which is late in executing design work must pay the client a penalty over a period of 30 days (.01% per day) after which the extra charge of the penalty terminates and a fine is payable – 1% of the cost of the unfinished work.

It often happens that when a creditor demands liquidated damages, he discovers that the breach of the obligation by the debtor has caused him specific losses. This gives rise to the question: does the creditor have a right to claim compensation for losses in addition to liquidated damages? It depends on what type of liquidated damages is applied in this situation. In regard to the possibility of combining liquidated damages with compensation for losses, the law distinguishes four types of liquidated damages: *calculated, fine, exclusionary*, and *alternative*. Article 189 of the Civil Code makes calculated liquidated damages the rule and the other three the exceptions, thus making their application possible only if there are specific instruction for this in the law or in the contract.

Under *calculated liquidated damages*, the debtor retains the obligation to compensate for losses but only to the extent that these are not covered by liquidated damages (fine, penalty). For instance, if a buyer demands a fine of 12,000 rubles from a plant for failing to supply raw materials, and later discovers that the lack of raw material caused him to incur losses of 16,000 rubles, he may demand only an additional 4,000 rubles (16,000 less 12,000) from the plant. Calculated liquidated damages are specifically provided for the breach of various obligations by third parties in two basic economic contracts – the contract of supply and the work contract for capital construction.

In *fine* liquidated damages, the creditor has the right to demand compensation for the full amount of losses in addition to the amount of liquidated damages. This most severe type of liquidated damages is applied in one of the most flagrant violations of contractual discipline: the supply of defective or incomplete goods.

The essence of *exclusionary* liquidated damages is that the creditor has no right either to compensation of losses which exceed the liquidated damages or in addition to liquidated damages. Exclusionary liquidated damages exist in relationships between clients and transport organizations, banks, the post office, telegraph office, etc.

Finally, in *alternative* liquidated damages, the creditor has the right to choose between liquidated damages and compensation for losses. If the creditor chooses liquidated damages, he loses the right to compensation for losses. In contrast to the other types of liquidated damages, alternative liquidated damages have not received practical application.

A distinction is made between legal and contractual liquidated damages depending on their origin.

The *extent of legal liquidated damages* and the circumstances under which they are demanded are provided in rules which are obligatory for the parties – (in transport charters and codes, in the Rules on Work Contracts for Capital Construction, the Statutes on Delivery, etc.). Thus, in relationships between socialist organizations, the law establishes a penalty of 0.04% of the unpaid sum per day for late payment. The penalty for late payment for communal services or for late rental payment by citizens is 0.1% of the unpaid sum per day. The terms of a contract which are directed to avoiding payment of liquidated damages or to decrease their extent are invalid.

The parties themselves establish contractual liquidated damages. An agreement concerning liquidated damages must be concluded in written form. The violation of this requirement results in the invalidity of the agreement (Art. 188 Civil Code).

Legal liquidated damages are applied mainly in relationships involving socialist organizations, while contractual liquidated damages are applied primarily in relationships between citizens. But in exceptional situations, contractual liquidated damages may also be applied by socialist organizations. Thus, by agreeing between themselves, they may provide for higher liquidated damages than the law provides for a specific violation. They also have a right to provide for liquidated damages in cases of failure to perform or of improper performance of such obligations for which legislation in general makes no provisions.

As a rule, when obligations are breached, the court and *arbitrazh* must award liquidated damages in the amount provided for by law or by contract. However, Article 190 of the Civil Code allows for the possibility of reducing liquidated damages. The court has a right to do so when it recognizes that the liquidated damages are too high in comparison with the creditor's losses. Various factors may be taken into account: the extent of performance of the obligation by the debtor, the property situation of the citizens involved in the obligation, or the property or any other interests of the creditors which must be considered. Liquidated damages may be reduced by *arbitrazh* only in exceptional circumstances after taking into account those interests of the debtor and creditor which deserve attention. Here is an example: A buyer notices that a complex and expensive machine which has been delivered to him is lacking a part and brings a suit demanding that the manufacturer pay a fine for supplying an incomplete product. Such a fine is 20% of the value of the defective product, which in this instance would be about 10,000 rubles. However, *arbitrazh* found it possible to lower the liquidated damages to 1000 rubles after taking into account the insignificant cost of the missing part (1 ruble, 22 kopecks).

The agencies of *arbitrazh* have a right not only to reduce but also to increase liquidated damages. If one party flagrantly breaches the terms of a contract, arbitrazh may award a fine and penalty which is up to 50% higher than the rate (for instance, 750 instead of 500 rubles). However, it may do this only if the amount (which would be 250 rubles in the example given above) is entered into the income of the budget of the union republic. Among the possible grounds for increasing the rate of liquidated damages the State *Arbitrazh* attached to the USSR Council of Ministers lists repeated breach, length of the delay in the performance of an obligation, or causing substantive harm to a creditor or the national economy, etc.

3. *Pledge. Pledge* is a means of securing obligation by *which a creditor has a right to receive satisfaction on the value of the pledged property in priority to other creditors in case of the failure of the debtor to perform an obligation* (Art. 192 Civil Code).

When no pledge is given, the creditor must avoid a situation where a debtor who violates an obligation does not possess sufficient property to satisfy the creditor's claims (creditors who have priority by law may be particularly vulnerable here). This danger is avoided by pledge.

Pledge is useful mainly in securing the claims of a creditor in a contract of hire. It is applied in relationships between a bank and a socialist organization. Socialist organizations are lent money by pledging goods of material value. When a citizen is extended credit to build a house, the subject of the pledge is the house which is built. Besides banks, pawnshops also use pledge. They lend money to citizens on a pledge of various articles (clothes, watches, furniture, etc.).

A pledge may be established by *law* or by *contract* (Art. 195 Civil Code). However, only the latter form of pledge is used in practice – pledge based on contract. A contract of pledge is made between two parties, the one giving the pledge (the pledgor), and the one receiving it (the pledgee). Such a contract must be in writing, and, in addition to this, a contract of pledge for a dwelling-house, must be notarized and registered with the executive committee of a local Soviet. When property is pledged to a pawnbroker, a pawnticket is issued. A violation in regard to the above-mentioned requirements in regard to its form invalidates a contract of pledge.

Either an owner of property or a state organization which possesses a right of operational management over the property may act as pledgor. Only property having monetary value and which may by law be used to satisfy claims may be used as pledge. The earth, its mineral wealth, forests and waters, the basic means of socialist organizations, and grain and fodder reserves, as well as other circulating funds which are necessary for normal activity may not be used for pledge. Special rules are provided for pawnbrokers. They may accept only articles of personal consumption and use as pledge.

Unless otherwise provided in law or contract, pledges property (with the exception of structures) must be given over to the pledgee. However, in actuality this occurs only in relationships involving pawnbrokers. In all other instances, the owner of the pledge property keeps it and may possess, use, and dispose of it. In legal terms, pledge "follows the object". This means that if the right of ownership or of operational management of the pledged property is transferred to another party, the pledge remains in effect for the latter.

The law gives the pledgees some rights which match those of the owners. If the object which is given as pledge to a pledgee leaves his possession, he has the same right to demand it from another's unlawful possession, and on the same grounds, as does the owner. Furthermore, he may bring suit in *vindicatio* even when the property does not leave his own possession but that of the debtor.

When a debtor breaches an obligation, a court (or *arbitrazh*) judgment is levied in execution against the pledged article. Banks sell the property pledged to them by socialist organizations through their own means. A pawnbroker does not need a court judgment in order to sell his articles; after the expiration of the established period (after the period for which the object was given on loan), the debtor is granted an additional month of grace. The pawnbroker gives the pledged article to a trade organization to be sold.

When the amount received from the sale of a pledged article is insufficient to satisfy the claims of a creditor for the full amount, he has a right to bring a claim for the unpaid portion against the remaining property of the debtor. The only difference is that the creditor has no privilege in relation to other creditors as regards property which was not pledged.

4. *Suretyship.* In a contract of suretyship, *the surety binds himself to the creditor of another party to answer for the performance of an obligation by the latter in full or in part* (Art. 203 Civil Code). Except in foreign trade, suretyship is used very infrequently; it is used in particular when a citizen is given suretyship for a library by the enterprise for which he works. Suretyship must be in writing; the violation of this requirement leads to the invalidity of the contract. The securing function of suretyship lies in the fact that it provides a supplementary debtor in the person of the surety in addition to the main debtor, furthermore, the debtor and the surety answer to creditors as joint and several debtors unless otherwise provided by law (Art. 204 Civil Code).

In surety, a debtor and a suretyship are linked to a creditor by an obligation; there is no obligation between the debtor and the surety. The surety takes on the role of creditor only if he, rather than the debtor, performs the obligation. When this occurs, he is granted all the rights which belonged to the previous creditor (including, for example, the right to demand liquidated damages, the right arising from a pledge, etc.).

A creditor may bring a suit against the debtor, against the surety, or against both at the same time. If the suit is brought against the surety only, he is obliged to bring the debtor into the suit. The violation of the requirement leads to unfavorable consequences for the surety: when he later presents a claim in subrogation against the debtor, the latter has a right to contrapose any objection he has against the creditor (for instance, he may allege that the creditor violated a counter obligation or that the obligation was actually already performed, or even though not performed, was no longer in effect, etc.).

Specific obligations are also imposed on a debtor who has performed his obligation. He must inform the surety of this fact. If the debtor does not do so, and the surety, not knowing that the obligation has already been performed, performs it again, the debtor is subject to all the unfavorable consequences of this twofold performance.

A creditor has a right to bring a claim against a surety no later than three months from the day that an obligation must be performed. If no date is given for the obligation or it is defined as the moment when it is requested, a creditor may bring a claim against the surety during the course of one year from the moment of the conclusion of the contract of suretyship (Art. 208 Civil Code).

5. *Guarantee.* By a contract of guarantee a socialist organization – *the guarantor – binds itself to a bank to answer for the performance of an obligation linked to the obtaining of loans by a subsidiary enterprise (organization).* Guarantees are usually used when enterprises and economic organizations are operating poorly (not fulfilling the production plans approved by them, not securing their own circulating assets, etc.). A bank gives loans to such enterprises and organizations only if a guarantee is given by a superior organ of economic management (the management of an industrial association, the management or chief management of a ministry, etc.).

Guarantee is similar to surety. Because of this, the Civil Code provides the most important rules regulating surety for guarantee as well (specifically, in regard to the written form of the contract, the obligation of the surety to bring in the debtor when a suit is brought against him, the allowable objections of the debtor and surety, the obligation of the debtor to inform the surety when he performs an obligation, as well as the periods when a creditor may realize his rights in relation to the surety).

In addition to this, guarantee possesses a number of features which distinguish it from surety. Thus, a guarantee is not a joint and several debtor. A bank is initially obliged to write off the money due to it from the account of an economic organization, and a guarantor must pay only the amount lacking from the total. A guarantor does not have a right to bring a suit in subrogation against a debtor. This may be explained by the fact that a guarantor is always a superior organ which is obliged to apportion circulating funds to the subsidiary organ and to look after its state of affairs. Thus, if a bank mistakenly demands a sum higher than the debt from the guarantor, the excess goes into the account of the debtor, not the guarantor. This shows the basic purpose of a guarantee: to help a debtor to improve his financial condition.

6. *Earnest.* Earnest is a *sum of money given by one of the contracting parties to the other party on account of a payment which it is due to pay by contract, as proof of the concluding of the contract and security for its performance* (Art. 209).

From this, it is apparent that earnest fulfills the following basic functions: first, it is a means of payment (if the buyer of a picture worth 2000 rubles gives the seller an earnest of 500 rubles, he must later pay the additional 1,500 rubles for the picture); secondly, an agreement in regard to earnest, regardless of the amount, must be effected in writing, and if a dispute arises later as to whether or not the basic contract was concluded by the parties, the agreement on earnest affirms the presence of such a contract; thirdly, earnest is also simultaneously a means of securing the performance of an obligation. It is given as an additional means to link the parties. If the party which gave the earnest refuses to perform the obligation, it loses the earnest. In the example given above, if the buyer refuses to go through with the purchase of the picture, he may not demand the 500 rubles paid as earnest from the seller. But if the party which received the earnest is guilty of failure to perform an obligation (the seller sold the picture to someone else and delivered it immediately), he must pay back double the amount (the seller of the picture must pay the buyer 1000 rubles).

As a general rule, an agreement for earnest does not prevent a party suffering losses, which are greater than the amount of the earnest, from having a right to demand compensation for the losses from the party responsible for breach of the contract. However, the parties may provide for something else in the agreement on earnest: to prohibit the claiming of losses no matter what their extent may be; this is known as earnest which turns into "smart money".

Earnest must be distinguished from an *advance*. Like earnest, an advance is given beforehand with the intention of future payment. It may also serve as proof of the presence of a basic contract. However, in contrast to earnest, an advance does not fulfill a securing function. In similar situations, therefore, no matter which party is responsible for the failure to perform an obligation, the party which received the advance must return it to the other party (in the example given above, if the contract of the purchase and sale in regard to the picture is not performed, the seller is obliged to return the advance even if the buyer is responsible, as well as when he himself is the responsible party).

Chapter 20. Liability for Breach of Obligations

1. *Concept and meaning. The measures of property sanctions which are established by law against a debtor who breaches an obligation are called liability for the breach of obligations.*

There are two forms of liability for the breach of obligations, otherwise known as property sanctions: first, *compensation for losses caused*, and secondly, *payment of liquidated damages*. The first form is the basic one; it may always be used – when an action provided in an obligation is not performed at all or when it is performed improperly, i.e. when a debtor carries out the necessary activity but does so in relation to an improper party, at the improper time, by an improper procedure and manner, etc. If a design institute violates a contract with a client by not performing its obligation of producing a design, this is called failure to perform an obligation; if the design is completed, but after the fixed time or with defects, it is an improper performance.

Liability in civil legal relationships fulfills two basic functions: it is used to stimulate a debtor before a breach of an obligation occurs (if you perform, you do not have to pay), and if an obligation is breached, the amount received as compensation for losses or as liquidated damages may be used by the creditor to cover the losses which the breach of the obligation has caused him. In relationships between socialist organizations, liability fulfills still another function – it serves as a signal of certain inadequacies in the performance of the enterprise (organization). The amount of fines paid (penalties) is revealed in the annual reports which an enterprise sends to its superior organ and is considered in the evaluation of its activity: if many sanctions are paid, it signifies that the work is poor.

Because of this, existing legislation provides that in relationships between socialist organizations it is not only a right of the injured party (creditor) to demand property sanctions, but also its obligation to the state. It is not only the party, which does not perform or performs contractual obligations improperly which is in violation of state discipline, but also the organization which does not bring an unpunished debtor to face liability. Failure to perform the obligation of claiming sanctions may result in various unfavorable consequences for the creditor. For example: if a buyer does not present a demand to a supplier to pay liquidated damages for a late delivery of products or non-delivery of goods, the amount of liquidated damages due him goes into the buget of the union republic. In addition to this, a buyer who does not claim liquidated damages without grounds is fined 2% of the cost of the non-delivered goods, which also goes into the union republic's budget.

The Party and the government have repeatedly emphasized the necessity of increasing the liability for the violation of obligations, especially in the mutual relationships between socialist organizations. This requirement is written into the decisions of the XXVI Congress of the CPSU, which seeks: "To improve the forms of relationships based on economic accountability and the mutual economic interest and responsibility for the performance of plan tasks and contractual obligations between suppliers and users, as well as between clients and contractors."[1]

2. *Conditions of liability*. Civil liability for the violation of obligations is expressed in the form of a general rule if the following conditions are present: first, the creditor suffers *losses*; second, *unlawfulness of the actions of the debtor*; third, *a causal relationship* between the unlawful actions and the losses; fourth, *fault of the debtor*. If the liability takes the form of payment of liquidated damages rather than of compensation for losses, only two conditions are sufficient: the unlawful actions and fault of the debtor, insofar as such an obligation is not connected with the existence or absence of losses for the creditor.

A. *Losses*. The concept of losses is explained in Article 219 of the Civil Code. Losses include expenses incurred by the creditor and loss or damage to his property, as well as pro-

1. *Materials of the XXVI Congress of the CPSU*, Moscow 1981, p.199.

fits the creditor did not receive but could have made had the debtor carried out his obligation.

Two of these three types of losses – the expenses incurred by the creditor as well as the loss or damage to his property – constitute *material damage* to the creditor: that which he possesses, and, as a consequence of the violation of an obligation, that which he lost. The third type of losses (*the profits not made by the creditor*) constitutes that which he might have, but as a result of the breach of the obligation, did not receive.

A citizen leases a violin and its case. Because of carelessness, the violin is damaged and the case disappears. The expenses of the creditor include, first of all, the money spent to repair the violin and also the cost of the case. In addition to this, it happens that quite a number of people would have liked to lease the violin. Therefore, had the violin been returned on time in good condition and in its complete state, the leasing center could have leased it to someone else and received the corresponding payment. This is the reason why the citizen is also liable for compensation for the value of the lease of the violin for the entire period that it was under repair and for the time it took to buy a case as compensation for income not earned.

The three types of losses listed above must be reimbursed not only in obligations involving citizens but also those involving socialist organizations. In the latter instance, income not received is expressed in the form of profits not received. Thus, if a supplier does not ship raw materials on time, the buyer has a right to demand from him as compensation not only the entire amount he paid to the workers as wages for the enforced idle time, but also the entire amount of profits the factory would have made had the products been made to the extent provided in its plan.

Article 220 of the Civil Code provides for the possibility that legislation may establish limited liability for failure to perform or for the improper performance of obligations. Limited liability as expressed in law takes various forms:

first, the parties may be deprived of the right to demand

compensation for losses. This occurs when exclusionary liquidated damages are established for the violation of obligations (see Chapter 19);

secondly, the parties may claim compensation for losses only in the form of material damages. Thus, according to Article 371 of the Civil Code, a contractor and a client may not claim income not received (profits not received) from each other when a work contract for capital construction is violated;

thirdly, the parties have a right to demand only material damage, and a strictly specified part of that. Thus, a transport organization must compensate for losses only to the extent of the value of a lost or uncomplete set of goods or baggage, and when the goods or baggage are damaged, to the extent that their worth was reduced (Art. 383 Civil Code and also analogous norms of transport charters and codes). As a result, all the other losses of the recipient may remain uncompensated, including those which constitute material damage (for example, when the cost of the repair of the object damaged during the transport is higher than the amount received from the transport organization).

Article 220 of the Civil Code and the legal acts supplementing it have the significance of imperative norms for the relationships between socialist organizations: socialist organizations are prohibited from concluding agreements directed toward limiting their liability if the extent of this is precisely defined by law for a specific obligation. When the court or *arbitrazh* are informed of such agreements, they declare them invalid and act on the basis of the rules prescribed for the parties.

B. *Unlawfulness of actions.* Unlawful actions are those actions of the parties which violate the requirements of law or other rules to which the parties must adhere. A party acts unlawfully if it performs an act prohibited by law or does not execute an action which it ought to perform by law. Insofar as the law makes it obligatory for a debtor to perform an action in the proper manner, actions which constitute nonperformance or improper performance of an obligation are considered unlawful.

The court (or *arbitrazh*) evaluates the actions of a party in regard as to whether they are unlawful. In so doing it compares that which a party did with the requirements of the law, with other acts which are based on the law, as well as with the obligation itself. In regard to this, the court and *arbitrazh* examine the correspondence between the actions performed by the debtor with the requirements established by the terms of the contract, unilateral transaction, or plan act which gave rise to the obligation.

However, it must be borne in mind that not every violation of an obligation is an unlawful action. Under some circumstances, a party may not only be unable but is obliged not to perform an obligation. Thus, the person in charge of a railroad has a right to temporarily prohibit the transport of goods when natural calamities, accidents, or wrecks would hamper movement. In such situations, the obligation provided in the transport plan may not be carried out by the transporter. Under such conditions, the actions of the transporter who refrained from taking goods for shipment must be considered lawful and no liability must be placed on him.

C. *Causal connection.* Article 219 of the Civil Code places a debtor under the obligation to compensate "for losses caused", e.g. for losses which are in causal relationship with the violation of an obligation. In order for a debtor to incur liability, it is necessary to show that the non-performance or improper performance of a contract are the cause and the creditor's losses the consequences.

Sometimes there is a more complex link between the violation of a contract and a creditor's losses, one which includes two or more intervening links. A factory does not supply a building organization with cement. Since it does not have the cement, the building organization is not able to perform the construction work which is provided in its contract with a client. And because the client's workshop was not put into operation in time, he had to pay a significant amount in the form of a fine for not delivering products to his buyers.

In practice it also often happens that one and the same occurrence is causally linked to various circumstances. In the above example, the specific reason for the interruption of work could have been the absence of necessary construction machinery, or the client's delay in delivering the design, in selecting a building site, etc. All these factors present the court or *arbitrazh* with a complex problem: to sclect those factors from all of these which form the necessary causal relationship with the creditor's losses. The key factor in its decision is the Marxist-Leninist position on the objective nature of causality, the interdependence and conditionality of events in nature and society, of necessity and chance, the role of usage in the conception of events and their relationships, etc.

D. *Fault*. Fault is the particular psychological relationship of a party to his unlawful actions and their unlawful consequences. The fault of the debtor is manifest in his relationship to the very fact of the non-performance or improper performance and to those losses which this causes for the creditor.

In civil, as in criminal, law fault manifests itself in the form of *intention* or *negligence*. Someone acts intentionally if he realizes the unlawful consequence, and desires this or at least conscienciously allows it to happen. In order for an act to be considered negligent the following are necessary: a party could have foreseen it, but thoughtlessly hoped to prevent it; or did not foresee the possibility of such consequences although he ought to and could have foreseen them.

In civil law, not only the form but also the degree of fault is taken into consideration. The degree of fault is expressed in the division of negligence into *gross* and *simple*. Gross negligence occurs when the breach of an obligation may be foreseen and prevented by anyone. Someone acts with gross negligence, for example, in packing a glass dish into a container without taking any measures to carefully wrap it in paper. In simple terms, in contrast to gross negligence a debtor may foresee and prevent a breach of an obligation given his experience, knowledge, and professional preparation. Therefore, stricter standards are applied in defining cases of ordinary than of gross negligence.

It must be noted that in civil law, in contrast to criminal law, the form and degree of fault usually do not influence the degree of liability: regardless of whether a debtor acted intentionally or negligently, he must still compensate for losses in their entirety; as a rule, the form and degree of fault also do not influence the decision as to whether or not to impose liability. In exceptional cases, however, the law requires a specific form or degree of fault to impose liability for the breach of some obligations. Thus, if a party which transferred property for safe-keeping does not receive it back after a fixed period, the custodian is liable for the loss, shortage, or damage to the property only if his actions were intentional or grossly negligent (Art. 427 Civil Code). Intention or gross negligence are also necessary in order for a party to be held liable for delivering property for gratuitous use and failing to mention anything about defects in the property (Art. 345 Civil Code). Sometimes it is also necessary to consider the form and degree of fault in evaluating the behavior of the injured party – the creditor (see in greater detail below).

There are exceptions to the general rule that a debtor must be at fault in order for him to be liable. First of all, one may note the so-called liability of the debtor for the acts of a third party. As noted earlier, under certain circumstances a debtor has a right to transfer the performance of his obligation in whole or in part to a third party. This legal construct is widely used in the supply of goods. For instance, a trade warehouse concludes a contract with an industrial trading unit for the supply of cloth, which it receives from its own supplier – a textile factory. In order that it not lose money and time in loading and unloading the cloth, the warehouse instructs the factory to ship the goods in transit directly to the industrial trading unit. In this instance, the factory as the third party fulfills the obligation of the warehouse to the industrial trading unit.

When he transfers an obligation to a third party, a debtor is liable for the actions of the third party (Art. 223 Civil Code). Therefore, in the example given above, the warehouse must pay a fine to the industrial trading unit if the cloth is not supplied either through its own fault (for instance, the warehouse gives the factory the wrong address for shipment) as well as when the factory itself is at fault. As a general rule, a creditor may demand payment of fines, penalties, or compensation for losses only from the debtor. But as an exception, the law sometimes grants him the opportunity of presenting a claim directly to the third party.

The supply of goods of defective quality may serve as an example. If the breach of contract consists in the fact that the industrial trading unit received cloth of poor quality, it has a right to present claims for the payment of sanctions both to the warehouse and to the factory. Depending on which one of them is found guilty of breach of contract, liability will be placed either on the debtor (warehouse) or directly onto the third party (the factory). In exactly the same fashion, it is the manufacturer who is liable to a buyer of an automobile if component parts, the transmission, or drive train break down during the period of the warranty, even though a dealer was party to the contract of purchase and sale.

In regard to certain types of obligations, the law provides that the absence of fault is not sufficient cause to release a debtor from liability for the non-performance or improper performance of obligations. The number of such exceptions is not great. They consist mainly of obligations arising from the plan of transport. A carrier who does not present the number of transportation means which are established by the plan to the shipper, or a shipper who does not use the means of transport allotted to him are liable to each other for their actions regardless of their liability. This type of increased liability is also borne for the loss, shortage in, or damage to property left for safe-keeping with a socialist organization whose functions include safe-keeping; for example, a seaport is responsible for the articles of passengers left in the custody of a check-room (Art. 427 Civil Code).

3. *Circumstances excluding liability of the debtor.* In all instances where liability is based on the grounds of fault, a debtor is released therefrom if it can be shown that the obligation was breached by chance: the result of an event for which he was not responsible. In regard to this, civil law contains the following presumption (supposition): a debtor who breaches an obligation does so through his own fault. Such a supposition enhances the position of a creditor in a dispute, inasmuch as the one who breaches the obligation – the debtor – has the burden of proving lack of fault (Art. 222 Civil Code).

In situations for which increased liability (irrespective of fault) is established, the law usually provides that a debtor may be released from such liability in cases of *force majeure. Force majeure* is an extraordinary and unavoidable event under given conditions. *Force majeure* is usually of elemental nature (a lightning bolt, flood, hurricane, etc.). War and its consequences are also examples of *force majeure.*

The law allows for a debtor's full or partial release from liability if the *creditor* is at fault for the breach of an obligation. Of course, a debtor must be fully exempt from liability if the breach of an obligation occurred solely through the fault of the creditor (a buyer does not deliver the metal which a factory-supplier needs to make forgings). However, often both the creditor and the debtor are at fault for the non-performance or improper performance of an obligation. In this situation the court (or *arbitrazh*) does not release the debtor from liability but only reduces the extent thereof (Art. 224 Civil Code). Depending on the significance which the intentional or negligent acts of the creditor had on the breach of an obligation, the court (or *arbitrazh*) may reduce the extent of the debtor's liability by one-half, one-third, etc. One of the conditions which leads to the reduction of a debtor's liability is specifically set forth in Article 224 of the Civil Code: when a creditor intentionally or negligently acts to increase the extent of the losses caused by the non-performance or improper performance of an obligation, or does not take measures to reduce them.

Sometimes the law contains a special enumeration of conditions whose existence releases a debtor from liability for an admitted violation of obligation. The most extensive listing is provided in transportation charters and codes. For instance, the Charter for USSR Railroads (Arts. 148 and 149) lists nine conditions under which a transport organization is released from liability for the loss, shortage, spoiling, or damage to goods.

4. *Delay of performance.* Both a debtor and a creditor may be in the position of an overdue party. A debtor admits to delayed performance if he does not complete the action provided in an obligation within the fixed period. Delay by a creditor occurs when he refuses to accept the debtor's proper performance or does not perform those acts which must be carried out in order for the debtor to perform his obligation. Some examples of delay by a creditor are: a buyer to whom a supplier ships goods provided as by contract refuses to accept them, citing a lack of free storage areas; a client does not deliver the equipment which a contractor must assemble according to the contract.

Delay by a debtor results in a number of unfavorable consequences for him (Art. 225 Civil Code):

first, he is liable to the creditor for compensation of losses caused by the delay;

secondly, his liability for the breach of the obligation increases above the ordinary: he is liable not only for any failure of performance which was his fault, but also for making impossible any performance which by chance could have occurred during the time of the delay;

thirdly, if performance rendered after the time-period is no longer of interest to the creditor (the artists violate the contract concluded with the Philharmonic Society by appearing as guest performers not on 4 November, but on another day, after the November holidays), the creditor has a right to refuse to accept performance and to demand compensation for losses.

In relationships between socialist organizations, refusal to accept a contract is allowed only when this is provided by law or by contract. The possibility or impossibility of refusal is usually determined by the specific features of the obligation. Thus, after informing the supplier, a buyer has a right to refuse a quantity of goods presented after the time-period, and a client in a work contract for capital construction is obliged to accept a constructed object regardless of the duration of the delay. Likewise, a consignee must accept the goods sent to his address from a railroad at any time, regardless of the duration of a delay or whether he still needs the goods.

Delay in the performance of a monetary obligation leads to special consequences. A debtor must pay a specific sum per annum for the entire period of the delay on the sum overdue. A debtor must pay three percent interest per annum unless otherwise provided by law or by contract (Art. 226 Civil Code). Furthermore, in relationships between socialist organizations, a faulty debtor must pay a penalty of .04% of the value of the unpaid goods and services per day to the creditor.

When the creditor delays performance, a debtor is granted a right to demand compensation for losses caused. Of course, the general principle concerning personal fault and the fault of third parties applies here as well. Article 227 of the Civil Code provides for the necessity of releasing a creditor from such liability if he succeeds in proving that neither he himself nor any third parties who are obliged to accept performance by virtue of law or of his agency are at fault. A special consequence was established for such occurrences in regard to monetary obligations: the debtor is released from paying the three percent annual interest for the entire period of delay by the creditor which was referred to above.

Chapter 21. Discharge of Obligations

1. *Concept of discharge of obligations and its grounds. By discharge of obligations is meant the termination of those rights and obligations which link a debtor with a creditor.* This occurs as a result of conditions set forth by law under the general heading "*grounds for the discharge of obligations.*" The various types of grounds for the discharge of obligations are enumerated in Chapter 20 of the Civil Code. They are: performance, setoff, coincidence of the persons of the debtor and creditor, mutual agreement, alteration of the plan, and impossibility of performance, as well as the death of a citizen or liquidation of a juridical person.

2. *Performance of obligation* (Art. 228 Civil Code). The normal method of discharging an obligation is by performance. Performance must be real and proper. The Civil Code provides a specific procedure for establishing the fact of whether a debtor has really performed an obligation and whether a creditor has accepted performance. Upon the request of the debtor, a creditor must give a receipt which affirms that the obligation was performed (money paid for an article or work, an order fulfilled and delivered, etc.). The debtor's right to demand such a document from a creditor sometimes becomes his duty. A socialist organization which pays for goods or services in performing an obligation based on an oral agreement must demand that the creditor (another socialist organization or a citizen) give it the proper document. This must affirm not only the fact of payment, but also the basis for payment (how much was paid and for what).

Sometimes a debtor gives a document of debt to the creditor certifying that a specific obligation exists between the parties ("I received 100 rubles as a loan from citizen Ivanov"). When the obligation is performed, the creditor must simply return this document to the debtor. The fact that the debtor possesses the document of debt makes it possible to draw the obvious conclusion that the obligation was discharged. A creditor who is not satisfied with this must prove that the return of the debt did not actually take place.

The violation of the obligation to give a receipt or to return a document of debt by the creditor leads to unfavorable consequences for him: the debtor has a right to withhold performance. Naturally, in these circumstances he is not held responsible for the allowable delay of performance. Furthermore, the debtor is granted an opportunity to demand compensation for losses caused by the withholding of the performance from the creditor.

3. *Setoff* (Articles 229-231 Civil Code). Let us posit the following situation: A borrowed 150 rubles from B. On his part, A performed work for B for which he should have been paid 250 rubles. Instead of two payments taking place, another method is possible: A does not have to repay the loan to B, and B pays him only 100, rather than 250 rubles for the work performed (250 rubles minus 150 rubles). In this and similar situations, *setoff* takes place: *the mutual liquidation of counterclaims.*

A *setoff* is a unilateral transaction. Therefore, the will of only one party, either of the creditor or the debtor, is sufficient to conclude a setoff. A special requirement exists for setoff between socialist organizations: it must be carried out through the bank which holds the assets of these organizations. Setoff is subject to only those requirements which are distinguished by three features established in Article 229 of the Civil Code:

1) they must possess the character of *counterclaims* (a creditor in the first obligation is at the same time a debtor in the second obligation, and a creditor in the second obligation is a debtor in the first obligation);

2) they must be *similar.* Only those obligations which have subjects with similar characteristics may be subject to setoff. Usually these are monetary obligations;

3) if a claim contains a time limit for performance, the time limit *must occur at the moment of setoff.* It is not possible to carry out the setoff of a demand on an obligation in January which must be performed only in May of that year. However,

there are no obstacles for a setoff of demands if at least one of these does not have an established time limit or a defined moment of demand.

The Civil Code indicates those claims which may never be subject to setoff (Art. 230 Civil Code). These are claims which are linked with the personality of a creditor: compensation for harm resulting in injury to health or causing death; lifelong maintenance (when a contract of purchase and sale of a dwelling with a condition of lifelong maintenance of the seller is concluded). The setoff of claims for which the period of limitation of action has expired is also not allowed.

As was noted in Chapter 20, the payment of property sanctions signals deficiencies in the activity of an enterprise. Therefore, it is forbidden to carry out the setoff of sanctions for the violations of obligations provided for by law or contract. If a supplier is obliged to pay a fine of 10,000 rubles for the delivery of incomplete sets of goods to a buyer, and a buyer must pay a fine of 5000 rubles to a supplier for delay in accepting goods, each side must settle accounts with the other to the full extent.

4. *Coincidence of the persons of the debtor and the creditor* (Art. 232 Civil Code). An example of this is a situation when a debtor-citizen becomes the heir of the creditor or a debtor-organization joins or merges with the creditor-organization as the result of reorganization. Inasmuch as in these situations the citizen and organization are "in debt to themselves" as a result of this coincidence, the obligation is automatically terminated.

5. *Mutual agreement of the parties* (Art. 232 Civil Code). This most frequently involves the substitution of one obligation with another by the parties in their relationships between themselves. For example, when a buyer still owes part of the money for a purchased article to a seller, the parties may change the obligation arising from a contract of purchase and sale into a contract of loan. To do this the buyer gives a receipt to the seller in which he indicates that he has re-

ceived the given amount as a loan from the seller.

Obligations which may not be discharged by setoff may not be discharged by mutual agreement of the parties (this concerns, in particular, obligations linked with the personality of the creditor).

Obligations between socialist organizations may, in principle, be discharged by mutual agreement of the parties. It is necessary only that such an agreement does not contradict the acts of planning of the national economy. For example, an agreement between a railroad and a shipper to terminate an obligation arising from the plan to transport and, at the same time, to accept and loan rolling stock is not possible.

6. *Alteration of the plan* (Art. 234 Civil Code). This means of discharging obligations is used only when the debtor and creditor are socialist organizations and the relationship itself is based on an act of planning which is binding on the parties. If such an act of planning is altered by an order which is binding on both parties (for example, an order on the supply of goods issued by the USSR State Committee for the Supply of Technical Materials is annulled), the obligation is terminated or altered correspondingly. The procedure for the alteration and termination of obligations in such situations is usuall regulated by normative acts which deal with the appropriate contract (for example, para. 31 of the Statute on the Supply of Products for Production-Technical Purposes, para. 26 of the Statute on the Supply of Consumer Goods, and para. 47 of the Rule on the Work Contract for Capital Construction, etc.).

7. *Impossibility of performance* (Art. 235 Civil Code). This involves a situation when a circumstance arising after the origin of an obligation makes its performance impossible. This happens, for instance, when an individually-defined article which is the subject of an obligation is destroyed (an antique vase is broken while in transit). The discharging of an obligation does not, however, result in making the entire performance impossible, but only that part of it for which a

debtor may not be held responsible. Therefore, if it is found that the vase broke because the seller failed to take precautionary measures, the obligations remains in effect. It only alters its contents; instead of transferring the article to the obligation of the seller, the buyer is now granted compensation for losses caused by the non-performance.

8. *Death of a citizen or liquidation of a juridical person* (Art. 236 Civil Code). The death of a citizen leads to the discharging of an obligation under strictly defined conditions. The death of a debtor discharges an obligation which provides that it must be performed by the debtor personally, and the death of a creditor discharges an obligation which must be performed for the creditor personally. Thus, the death of an author (debtor) discharges his obligation to write a book for a publisher, and the obligation of a factory to pay a certain sum to a citizen for trauma caused is discharged with his death.

If an obligation is not linked with the personality of the creditor or debtor, the obligation remains in existence after the death of one of them. In this situation, the rights and obligations of the parties are accordingly transferred to the heirs of the creditor or the debtor (a son who receives an inheritance after the death of his father has a right to receive everything which was due to his father, and at the same time, he is liable to pay the debts of his father).

When a juridical person is liquidated, an obligation in which it was involved is, as a rule, discharged irrespective or whether it was the debtor or the creditor. However, as an exception, the legislation of the Soviet Union or the union republics may provide that an obligation of a liquidated enterprise must be performed by another juridical person. For example, the Statute on the Socialist State Production Enterprise (para. 111) places an organ which is superior to the liquidated enterprise under the obligation of satisfying the claims of citizens which are linked with injury to health or cause of death, as well as all other claims of citizens which were not presented before the liquidation of the enterprise on valid grounds. Article 470 of the Civil Code establishes that all the payments of a liquidated enterprise for obligations connected with the reduction of a person's capacity to work or with his death must be carried out by the State Insurance Fund. The liquidation commission allocates a given amount to it for this purpose.

Chapter 22. Contract of Purchase and Sale. Contract of Barter. Contract of Gift

I. Contract of Purchase and Sale

1. *Concept and significance of the contract of purchase and sale.* The goods produced by society and the earned wages of citizens are realized through the use of commodity-monetary forms. The basic legal form for the reciprocal realization of goods is the contract of purchase and sale. The contract of purchase and sale is widely used in relationships involving citizens. Soviet citizens daily acquire various commodities in stores and in doing so they conclude numerous contracts of purchase and sale.

In relationships with citizens, socialist organizations – retail enterprises of state and cooperative products – usually act in the capacity of seller. However, socialist organizations may also act in the capacity of buyers in concluding contracts of purchase and sale with citizens. Thus, the procurement organizations of the Central Union of Consumer Societies purchase agricultural products from the personal garden plots of collective-farm workers, other workers and employees, as well as fruit, etc. on the basis of contracts of purchase and sale. In some circumstances, the contract of purchase and sale is also used in relationships between socialist organizations.

A contract of purchase and sale is an agreement under which the seller is obliged to transfer property to the ownership (operative management) of the buyer, and the buyer is obliged to accept the property and to pay for it a defined sum of money (Art. 237 Civil Code). On the basis of the contract, the property is given over to the ownership of another person, and in mutual relationships involving state socialist organizations the owner does not change, but the right of operative management is transferred.

The contract of purchase and sale is *reciprocal* since the owner receives counter satisfaction – the purchase price. The reciprocity of the contract of purchase and sale distinguishes it from the contract of gift, under which property is transferred to the ownership of another person gratuitously. A contract of purchase and sale is considered concluded at the moment that its participants reach agreement on the transfer of an article and the payment of money. It belongs in the category of mutual contracts since each of the parties has both rights and obligations.

Articles 237-254 of the Civil Code are devoted to the contract of purchase and sale. Several norms of the general part of the Civil Code are also applicable to sale relationships – as regards the form of transactions (Arts. 42-47), the moment of the transfer of the right of ownership (Art. 135), the risk of the accidental ruin of an article (Art. 138), etc. The Civil Code contains the general norms concerning the contract of purchase and sale, as well as the norms concerned with its permutations: the contracts of retail sale, credit sale, on the sale of a dwelling.

The particular features of the sale of goods on credit are defined by a decree of the USSR Council of Ministers and by instructions of the Councils of Ministers of the various republics.[1] Of great significance in regulating contracts of retail trade are the normative acts of the USSR Ministry of Trade, in part the Model Rule on the Barter of Goods purchased in the Retail Network, and the rules based on the latter and confirmed by the Ministries of Trade of the union republics, i.e. the Rules for the Sale of Automobiles, as well as the Rules on the Sale of Furniture and Other Goods.[2]

2. *Form and subject of the contract. Rights and obligations of the parties.* The general rules concerning the form of contracts apply to the contract of purchase and sale. Contracts involving socialist organizations irrespective of the amount, and contracts between citizens involving a sale price of more than 100 rubles must be concluded in *written form*. A contract of purchase and sale which is performed as it is

1. *Civil Legislation. Collection of Normative Acts*, Moscow 1974, p.375.
2. *Bulletin of Normative Acts of the Ministries and Agencies of the USSR*, 1973, Nos. 3 and 7; 1974, No. 9; 1978, Nos. 3 and 9.

concluded may be made *orally*. As a rule, a trade organization and a buyer fulfill the obligations to which they bind themselves at the same time as they perform the contract, and a contract of a retail sale is usually concluded orally. Notarial authentication is required only for the contract of purchase and sale of a dwelling (part of a dwelling or vacation house) which is located in a city or a settlement of an urban type (Art. 239 Civil Code).

The *subject of a contract of sale* may be any property, with the exception of the land, its mineral wealth, waters and forests which are the exclusive property of the state and may be given over only for use. In order to acquire certain types of goods, a buyer must observe special rules. A citizen may purchase weapons, strong poisons, and explosives only with special permission (Art. 137 Civil Code).

Under a contract of purchase and sale, a seller must transfer property to the ownership (operative management) of the buyer. Unless otherwise provided by contract or special rules, the transfer of articles by the seller takes place simultaneously with the payment of money by the buyer. The moment of the transfer of the right of ownership to the buyer is defined by the general rules on the transfer of ownership (Art. 135 Civil Code). A buyer-citizen becomes the owner from the moment of the transfer of the object. If a buyer leaves an individually-defined article (automobile, refrigerator, piano, etc.) in a store after the purchase with the agreement of the store, he is the owner of this article, and it is the obligation of the seller to secure the safeguarding of the article.

At the same time that the buyer acquires the right of ownership he assumes the *risk of the accidental ruin* or accidental damage to the purchased article. This means that if the article is destroyed or damaged under circumstances for which neither the seller, the carrier, nor any other person is responsible, the buyer must suffer all the losses (Art. 138 Civil Code). For instance, if an article which is left in a store for safe-keeping burns during fire caused by a bolt of lightning, the buyer has no right to demand compensation of its value from the store.

An important obligation of the seller is to transfer an article of *proper quality*. If the quality is not stipulated in the contract, then its quality must meet the usual requirements. Such an article must conform fully to standards and technical conditions, and in its artistic-esthetic attributes – to the sample, if a sample was made for the article.

The seller is obliged to inform the buyer of the rights of third parties to the sold article. For instance, in the sale of a dwelling, the seller is obliged to inform the buyer of renters living in the house.

Trading organizations are obliged to sell articles at *prices which are established through state procedures*. Exceptions are allowed for the commission trade in which it is possible to sell articles at prices which are lower or higher than the established retail prices. In a contract of purchase and sale between citizens, the price is determined by *the agreement of the parties*. The sale of surplus agricultural products by collective-farm workers is also carried out at a price determined by the agreement of the parties.

The obligations of the seller correspond to the *rights of the buyer*: the right to demand the transfer of the article, to secure the safekeeping of the sold articles, and to its proper quality. The buyer has corresponding obligations, the most important of these being the obligation to accept the article in accordance with the conditions of the contract and to pay the seller the purchase price.

3. *The obligation of the parties for breach of contract*. The refusal of the seller to transfer the article gives the buyer the right to demand either the enforced transfer of the article or rescission of the contract. In choosing one of these legal remedies, the buyer may at the same time demand compensation for losses caused as the result of the withholding of the transfer of the article or the rescission of the contract.

Analogous consequences occur when the buyer refuses to accept the purchased article or to pay its price. In this instance, the seller may demand that the buyer accept the article and pay the price or that the contract be rescinded, as well as demand compensation for losses if these were caused by the improper performance of the contract.

The Civil Code regulates in detail the *responsibility of the seller* for the defective quality of an article. If a buyer is sold a defective article, he may choose to file any of the following claims against the seller, all of which are provided for in Article 246 of the Civil Code. He has the right to demand: a) replacement of a defective object with specific generic traits with the same type of object of proper quality; b) reduction of the price if the article is usable; c) removal of the defects by the seller at his own expense or reimbursement to the buyer for the expense of their removal (if he removes them himself); d) rescission of the contract, i.e. the return of the article and reimbursement of its cost, as well as compensation for losses.

If the buyer discovers defects in an article which the seller had not informed him of, he has the right to bring a claim against the seller (Art. 247 Civil Code). A claim based on defects in an article may be presented as soon as they are discovered, but not later than six months from the day of the transfer of the article. A buyer who discovers defects in a building (dwelling, vacation house) may file a claim against the seller not later than one year from the day of the transfer of the structure. If the seller does not satisfy the claim or refuses to do so, the buyer has the right to bring suit in court. For a suit brought under these circumstances, a six-month period of limitation of action is established, starting from the day of the filing of the suit or, if no claim was filed, from the day that the time period for its filing expired (Art. 249 Civil Code). There are different time periods for the filing of claims and suits which are based on articles sold through retail organizations which have guarantee periods.

Insofar as the right to ownership of the purchased article is transferred to the buyer in a contract of purchase and sale, the seller may be either the owner or the party acting on authority of the owner. For instance, it is posssible that a seller has sold an article which did not belong to him and for the owner of the article to bring suit against the buyer to effect its recovery. The buyer is obliged to inform the seller of such a suit, and the latter must take measures to protect the interests of the buyer. If it is decided that the article must be transferred to the party bringing the suit, even if the seller participated in the court proceedings, the buyer has the right to demand compensation for losses from the seller. If the buyer did not inform the seller, the latter is freed from having to compensate for losses when the buyer brings a suit against him as a result of the disposition of the article, if he can prove that he could have prevented the alienation of the article.

4. *Special features of a contract of retail sale.* The most frequently concluded contracts of purchase and sale are those between a state or cooperative trading organizations on the one side, and a citizen on the other. Such contracts have the purpose of satisfying the constantly increasing needs of citizens and have several features.

A contract of retail purchase and sale is usually concluded orally. The *written form* is used only for contracts which are not performed as they are being concluded. The written form is used to concluded contracts of retail purchase and sale for acquiring goods on credit, the sale of automobiles, and the sale of goods for the mail-order trade. The proof that an oral contract has been concluded consists of a cashier's or commodity receipt, as well as a guaranteeing certificate and the factory tags with the store's notation of the sale date.

Another feature of the contract under examination lies in its conditions that the seller must perform supplementary services for the buyer: for instance, the seller undertakes to deliver the goods to one's home, to notify the buyer of the arrival of the article ordered, to deliver a purchased gift to the person for whom it is intended, etc.

In a contract of retail purchase and sale, the buyer has the right of unilateral withdrawal from the contract before the receipt of the purchased article. In addition to this, the buyer has the right to exchange goods in good condition for analogous ones within a period of 14 days excluding the day or purchase. The article may be exchanged if the factory tag, commodity or cashier's receipt given out by the store at the same time as the article is present.

The sale of industrial goods on the basis of advance orders and samples has some special features. The retail sale of automobiles and motorcycles with side-cars to citizens is carried out through advance orders.[3] An advance order is accepted by a store upon the appearance of the future buyer in the store, and is registered in a record-book by the store. The time period for fulfilling such an order must not exceed one year from the day of its conclusion and 60 days from the day of payment of the price of an automobile (motorcycle). The moment of the conclusion of a contract of purchase and sale for an automobile or motorcycle with a side-car on advance order is the acceptance of the obligation by the store in the established form. If the buyer does not appear within 30 days after being contacted by the store to pay the purchase price, his order is cancelled and, consequently, the contract is considered to have been rescinded. The buyer has the right to refuse to accept the automobile at any moment. Analogous rules are provided for the conclusion of an advance order for furniture.

5. *The responsibility of a retail organization for defects of a sold article.* The subject of a contract of retail sale may be goods which do not have guarantee periods, as well as goods for which guarantee periods are established. *Guarantee periods* are established for goods which are designed for long-term use or duration (televisions, radio receivers, watches, etc.) with the aim of increasing the responsibility of their makers and to safeguard the rights of the consumers.

Guarantee periods are provided for by standards or technical conditions and go into effect from the day of the retail sale of the product. For those goods for which guarantee

3. *Bulletin of Normative Acts of the Ministries and Agencies of the USSR* 1978, No. 9.

periods are established, the seller – the retail trading organization – guarantees and secures the fulfillment of the requirements in regard to the quality of the products provided in the standards and technical conditions and has a higher degree of liability against the buyer for hidden defects in the merchandise.

Under the rules for the barter of industrial goods bought within the retail trading network of state trade and the network of the RSFSR Union of Consumer Societies, which were confirmed by the RSFSR Ministry of Trade on 28 March 1975, different procedures and time periods have been established for the filing of claims regarding the quality of goods for which guarantee periods are established and for goods without guarantee periods.

The buyer of goods for which time periods are established who discovers hidden defects before the guarantee period elapses, i.e. defects which are not apparent in the normal checking of quality, has the right to file a claim against the seller within the limits of the quarantee period. In this instance, the seller is obliged either to remove the discovered defects without charge or to exchange the defective object for one of proper quality, or to take back the object and return the money paid for it to the buyer.

The work of removing the defects in the article is entrusted to a work-shop for guarantee repairs or to the enterprise-manufacturer. After discovering hidden defects within the time period in a television, radio receiver or other radio equipment, bicycle or other merchandise intended for long-term use, the buyer may turn to a work-shop for guarantee repairs or to the enterprise-manufacturer. If the defects are not removed within ten days, as well as in instances when repairs are still needed after the removal of the defects, the buyer has the right to demand the return of the money paid for the product or exchange of the defective product for a new one in the store in which it was purchased or in another store of state cooperative trade in his place of residence.

Another procedure for the filing of claims as regards defective goods is established for products which do not have a guarantee period. The buyer who notices defects in such products (clothes, fabrics, furs, rugs, etc.) has the choice of exchanging them at the place of purchase or returning the goods and asking for the refund of the purchase price. The exchange of defective goods takes place by presenting a cashier's or merchandise receipt of the store together with the object itself; the factory tag must also be shown. Such a claim may be made within 14 days from the moment of purchase.

A somewhat different method and time periods for the presentation of claims are established for the discovery of hidden defects in goods which have no guarantee period whose presence is confirmed by laboratory analyses or by act of a bureau of product expertise. In this instance, the buyer also has the right either to exchange the article for a new one or to demand the refund of the purchase price from the store. But longer time periods are established for filing a claim with the store. A claim may be presented within a period of six months from the day of purchase of the article within the retail trading network. For seasonal goods, the time period for the filing of claims goes into effect from the moment of the beginning of the season involved. For instance, a claim based on hidden defects in women's fur boots bought in the summer may be filed within a period of six months of the arrival of the winter season.

If a store refuses to satisfy claims based on the quality of sold merchandise, the buyer has the right to bring suit in court not later than six months from the day when the claim was filed or, if a claim has not been filed or if the time of its filing is impossible to ascertain, from the day of the expiration of the period established for the filing of claims (Art. 249 Civil Code).

6. *The contract for the sale of goods on credit*. The contract for the sale of goods on credit (payment by installments) is a variant of the contract of retail sale and makes it easier for citizens to acquire expensive items. Article 252 of the Civil Code deals with this contract. Detailed rules on the contract

of sale on credit are also established by instructions adopted by the Council of Ministers of the union republics. A unique feature of this contract is that the buyer pays for the items by regular payments over a relatively long period, although he becomes the owner of the purchased item when the article is transferred to him.

The sale of goods on credit is not possible for all citizens but only for workers and employees who are steadily employed in a town where a store is located; for sergeants who have re-enlisted and officers; and for graduate students studying in that town and pensioners with permanent passports. A list of goods which may be sold on credit is confirmed by the Ministries of Trade of the Union Republics and is supplemented by the provincial executive committees for goods which are available in that specific region in sufficient quantity.

In order to conclude a contract for the sale of goods on credit, the buyer fills out an agreement-obligation in two copies, one of which remains with the store and the other which is sent by it to the place of work of the buyer. On the basis of this document, the organization for which the buyer works withholds the installment payments from his salary and credits them to the store. Non-working pensioners are given the agreement-obligation and must bring the payments themselves. At the time at which the contract is concluded, the buyer pays only a part of the cost of the purchased article, and the remainer is paid by him in installments (from 6 to 24 months depending on the cost of the object).

The sale of goods on credit is carried out at the prices in effect on the day of the sale. Subsequent changes in the price of the goods bought on credit, whether higher or lower, does not result in a new account. In order to be granted credit, the buyer pays a defined interest. If the buyer does not pay the sum owed by him to the store within the established time period, the store may simply demand it through the court. However, the store does not have the right to demand the removal of the object since the buyer acquired the right of ownership to it from the moment the article was transferred to him.

7. *Particularities of the contract of purchase and sale of dwellings.* Special rules are provided for the purchase of dwellings (Arts. 238-239 Civil Code). They are conditioned by the importance of the subject of such a contract. The contract of purchase and sale of a house or part of a house located in a city or a settlement of an urban-like nature must be authorized by a notary if at least one of the parties is a citizen. For a contract of purchase and sale of a dwelling (part of a house, vacation home) located in the countryside, the simple written form is sufficient. In addition to this, the contract of purchase and sale of a house (part of a house, vacation home) must be registered with the executive committee of the district, city or village Soviet of People's Deputies. Violation of the form of this contract renders it invalid. The right of ownership of a house is transferred to the buyer from the moment of the registration of the contract.

II. Contract for the Sale of a House with the Condition That the Seller Is Maintained for Life

1. *Concept of the contract. Parties.* The contract of the sale of a house with the condition of life-long maintenance is concluded in instances when a seller, usually an elderly or ill person, yields the right of ownership of the house belonging to him with the aim of receiving necessary help in being cared for. The rules for such a contract are contained in Article 253-254 of the Civil Code.

According to *the contract for the sale of a house with the condition of support for life, one party (the seller) transfers the dwelling or part thereof to the ownership of the other party (the purchaser) who takes on himself the obligation to support the seller of the house until his death.* Such a contract may be concluded only between citizens. The seller (alienator) of the house is a person who, for reasons of age or health, is not able to work and is in need of material support and care.

This contract is similar to the contract of sale of a dwelling in that it is reciprocal and is the basis for the transfer of the ownership of a house (part of a house). However, the contract of maintenance for life, the basic element is precisely that assistance which the purchaser of the house must render to the seller of the house.

2. *Form of the contract. Rights and obligations of the parties.* The contract of maintenance for life is effected in the same form as the contract for the sale of a house, i.e. it is concluded in simple written form or in the form of notarial authorization with subsequent registration with the district, city or village Soviet of People's Deputies.

The basic *obligation of the purchaser of the house* lies in offering the necessary assistance to the seller of the house until the end of his life. Maintenance for life includes giving material support in the form of living-space, food, and care for the person selling the house, as well as in providing him with any other assistance. The extent of the material maintenance may be defined by the agreement of the parties. The purchaser of the house does not have the right to alienate the house during life-time of the person who sold the house, and must fulfill his obligations even if the house is accidentally destroyed when, for instance, it burns. If the purchaser does not fulfill his obligations, the alienator of the house has the right to rescind the contract. The purchaser may also raise the issue of the termination of the contract if his material situation changes sufficiently to make it impossible to render the maintenance to the alienator which is provided for in the contract. The purchaser may also request that the contract be rescinded if the alienator overcomes his inability to work completely, for instance, if he recovers from an illness. The contract is also terminated in case of the death of the purchaser.

If the contract is rescinded by request of the purchaser of alienator of the house, the house is returned to its former owner-seller (alienator); however, the expenses for his maintenance, which were provided by the purchaser of the house, are not reimbursed. If the termination of the contract is caused by the seller's recovery of his ability to work, the dwelling remains with the purchaser and the seller retains the right to use of the premises granted to him by the contract free of charge for life.

III. Contract of Barter

1. *Concept of the contract.* The contract of barter, like the contract of sale, is used to transfer the right of ownership (operative management) of property. However, it is used less frequently than the contract of purchase and sale, and usually between citizens. Socialist organizations are permitted to use the contract of barter only in specific situations. For instance, a watch repairshop is permitted to barter unrepaired watches for repaired ones for the cost of the repair.

According to Article 255 of the Civil Code, *each of the parties under the contract of barter is obliged to transfer property to the ownership (operative management) of the other party.* Like the contract of purchase and sale, the contract of barter is bilateral and *reciprocal.* The difference between the contract of purchase and sale and the contract of barter is that the counter satisfaction is expressed not in money, but in the transfer of other property.

2. *Rights and obligations of the parties.* In a contract of barter, each of the parties *transfers and receives into ownership property,* and each of the parties is considered to be the seller of the object which he transfers, and the buyer of the article which he receives. Therefore, the rules of the Civil Code in regard to the form of the contract of purchase and sale, the rights and obligations of the seller and buyer, the moment of the transfer of the right to ownership, the risk of accidental ruin, etc., apply to the form of the contract of barter, its content, and the obligation of the parties.

The basic obligation of each party is the transfer of an object into the ownership (operative management) in exchange for the object of the other party. The right of ownership (operative management) arises from the moment of transfer of the object. If no counter satisfaction is given, if a defective object is transferred, or if other violations of their obligations by the parties take place, the same legal consequences result as when the obligations in a contract of purchase and sale are violated (see section I of this Chapter).

IV. Contract of Gift

1. *Concept of the contract. In a contract of gift, one party (the giver) transfers without charge property into the ownership of another party* (Art. 256 Civil Code). A gift is a contract since giving requires not only the desire of the giver, the person giving the property, but also the agreement of the party accepting the object. Like the contract of purchase and sale, a contract of gift is meant to transfer property into ownership. A contract of gift is *unilateral* and does not create obligations for the party to whom the gift is given. It is a true contract and is considered concluded at the moment of the transfer of the object. As a rule, the parties to the contract are citizens, but a socialist organization and the state may also be granted gifts. A contract of gift must be concluded according to the rules provided in Articles 256-257 of the Civil Code.

2. *Form and content of the contract.* The Civil Code establishes special requirements for the form of a contract of gift. A gift of more than 500 rubles must be notarially authenticated. But a contract of gift of property by a citizen to a socialist organization, regardless of the value of the gift, may be concluded in the simple written form. For the gift of a dwelling (part of a house, vacation home) the rules defining the form of the contract for the purchase and sale of a dwelling must be observed.

The party receiving an object as a gift does not bear any legal obligations. A giver does not have the right to direct the recipient of the gift how the property given as a gift should be used. An exception is provided when a citizen gives property to a socialist organization. In such a contract, the giver may give instructions that the property given as a gift must be used for a specified socially useful purpose. For instance, an artist gives a picture to a club and drawings to a museum. The non-performance of the stipulated conditions allows the giver the right to demand the rescission of the contract and the return of the given property.

Chapter 23. Contract of Supply

1. *Concept and significance of the contract of supply.* The contract of supply is widely used in our national economy and is the most prevalent of all economic contracts. It is applied in the sphere of commodity circulation by industrial enterprises and associations in marketing their products, as well as in providing material-technical supply for associations, enterprises, organizations, and institutions in all branches of the national economy. Under the contract of supply, products for industrial-technical use, as well as goods for consumer consumption (from now on the term "production" will be used), are realized.

The contract of supply fulfills important functions in the national economy. With its aid, enterprises and consumers exert their influence in the planning for the production of raw materials, supplies, structures, and consumer goods in the assortment and quantity which will satisfy their requirements.

The decisions of the XXV and XXVI Congress of the CPSU provide for the increased role of consumers in developing the production program of associations and enterprise-producers. The contract of supply allows producers and consumers to detain and make concrete the plan acts on the production and supply of products and goods. This contract is used to gain information on the demand for products and on changes therein, that is, the contract fulfills a role as an instrument for planning industrial production. The contract also functions in fulfilling the plan. It is both the means for organizing economic ties and for coordinating the economic accountability interests of associations and enterprises (suppliers and buyers) and the means for mutually controlling the fulfillment of the plan tasks with the aid of measures of material responsibility for the violation of contractual obligations.

The increase in the economic independence of associations and enterprises, and the increased role and significance of the contract of supply as an instrument of planning and a means of organizing the mutual relations based on economic accountability of socialist organizations causes it to be the basic document in determining the rights and obligations of the parties in the supply of all types of production.

The contract of supply is regulated by general and special acts. The basic norms on the contract of supply are included in the Principles (Arts. 44-50) and the Civil Code (Arts. 258-266). The detailed regulation of relationships based on supply are relegated by the Principles to the Statutes on Supply. The Statutes on the Supply of Production and Technical Goods and of Consumer Goods, which were confirmed by a decree of the USSR Council of Ministers of 10 February 1981, define the bases, procedure, and time periods for the conclusion, change, and rescission of the contract, its essential conditions, the procedure and time periods for its performance, and liability for violation of contractual obligations.

The distinctions in the application of the contract of supply are expressed in a number of other acts which were confirmed by decree of the USSR Council of Ministers. For instance, the supply to military organizations of special-purpose goods is carried out in accordance with the Basic Conditions of Supply of Production for Military Organizations. Relationships based on the supply of products and goods for export and for the state reserve are regulated by special normative acts.

The Statutes on Supply contain the general norms on the contract of supply. The special features on the supply of various types of products or the supply of goods for specific categories of buyers are provided in the Special Conditions of Supply, which were confirmed by the USSR State Committee on Supply and the USSR State *Arbitrazh*. Relationships based on supply are also regulated by other acts of the USSR State *Arbitrazh*, the USSR State Committee on the Supply of Technical Materials, the USSR Ministry of Trade, and other ministries, state committees, and agencies. For instance, the procedure and time periods for accepting products and goods of specific quality and quantity are established by instructions confirmed by the USSR State *Arbitrazh*.

The Statute on Supply, the Special Conditions on Supply, and the norms on the contract of supply contain the provisions regulating relationships between the organs of state management for the various branches of the national economy which arise in the planning of production and the allocation of products and goods, the relationships of associations and enterprises with the organs of management which plan supply, and liability for the violation of state plan discipline.

On the basis of a contract of supply, one organization, called the supplier, binds itself to transfer into ownership (operative management) of another organization, called the buyer, specific products in accordance with the plan act of distribution. The buyer binds himself to accept these products and to pay a set price for them. A contract which is concluded at their discretion between economic organizations by which the supplier binds himself to transfer products which are not distributed according to plan procedure, over a time period which does not coincide with the moment of conclusion of the contract, is also a contract of supply (Art. 258 Civil Code).

From this definition it is clear that the legislation distinguishes between two variants of the contract of supply: a contract which is based on a plan act on the distribution of production (plan act of supply), and a contract concluded at the discretion of the parties. Their legal distinction lies in the fact that in the former a plan act gives rise to the civil law obligation of the supplier to conclude a contract and to a greater or lesser extent it determines its subject. A contract which is concluded at the discretion of the parties does not depend on a plan act and may not be considered as a plan act, although it does play a role in the planning of the production necessary to satisfy society's needs. The plan contracts of supply predominate in our national economy.

The contract of supply differs in a number of ways from the contract of purchase and sale, the procurement contract, and other contracts involving the reciprocal alienation of an article. Only socialist organizations participate in a contract of supply; it is of a plan character; it is characteristic for it that the moment of the performance of the obligation arising from the contract does not coincide with the time period for concluding the contract.

The first characteristic distinguishes the contract of supply from the contract of purchase and sale, in which citizens may participate. The non-coincidence of the time of the conclusion of the contract with that of its performance helps to distinguish the non-plan contract of supply from the contract of purchase and sale. The demarcation of contracts of supply from procurement contracts is made possible by the subject of the contract, the participants to the contract, and the nature of the plan act which is the basis of the contract. In contrast to the contract of supply, one of the parties to a procurement contract is always a collective farm, state farm, or other agricultural organization which sells its agricultural products on the basis of a special plan for the state procurement of agricultural products.

2. *The plan basis of the contract of supply*. The plan acts on supply form the basis for the conclusion of the contract. They are issued in accordance with the state plans for economic and social development by special organizations of the system of the USSR State Committee on the Supply of Technical Materials, the USSR Ministry of Trade, and by other ministries, state committees, and agencies (union supply and procurement organizations, state committees on supply of the union republics, main territorial administrations of the system of the USSR State Committee on Supply, offices of wholesale trade of the union republics, etc.). Plan acts on supply take various forms. There are plans which bind the buyer to the supplier with direct long-term and long-term economic ties; plans for the supply of individual types of products, which include plans of cooperative supply; warrants [*nariady*] and counter-warrants [*raznariadki*]. These plan acts differ from each other in procedure and designated time periods, content, time periods for performance, and in those to whom it is addressed. A plan to bind for supply by direct long-term or long-term economic ties established the general range of supply of products by group types (assortment) for the entire period of the tie (as a rule, for the period of the

duration of the five-year plans) with distribution on a yearly basis. If necessary, such a plan may be refined every year in accordance with the approved yearly plans for the production and distribution of products.

A warrant is a detailed plan task and, as a rule, it is issued to be in effect for a year. A warrant is sent to a supplier and a specific buyer, or a group warrant may be issued which indicates not the buyer, but the fundholder [*fondoderzhatel'*], the State Committee on Supply of a union republic, or the main administration of the USSR State Committee on Supply, which in turn send the warrants (counter-warrants) with instructions to specific buyers.

In all instances, a plan act for supply is an administrative act which gives rise to an obligation of a supplier to conclude a contract. The obligation for providing supply arises from the contract which is concluded on the basis of the plan act for supply.

The Statutes on Supply establish a number of requirements which must be followed in issuing plan acts. In the first place, such acts must be in accordance with the approved production plan of the enterprise-supplier. *Secondly*, the issuing of a plan act for the supply of production which is produced in excess of the plan may be made only with the consent of the supplier. *Thirdly*, the plans of binding for direct long-term and long-term ties must be issued not later than four months, and warrants not later than 45 days before the beginning of the plan year. The Statutes on Supply also provide the bases, procedure, and terms of changing plan acts on supply.

If the procedure and terms regarding the issuing and distribution of a plan act are violated, the supplier has the right to dispute the same before the organ which issued the plan act and to request that this organ change or repeal the act. During that time that the plan act of supply is not repealed or changed according to established procedure, the supplier is obliged to conclude the contract for supply. The refusal or refraining by the supplier to conclude a contract gives the buyer the right to force him to conclude a contract by turning to *arbitrazh* and demanding a penalty for the refusal or refraining from concluding the contract.

The rights of a buyer which are based on a plan act are defined differently. The buyer has the right to refuse to conclude a contract of supply and to accept the production allocated by the plan. Within ten days after the receipt of the plan act, the buyer has to notify the organ issuing the plan and the fundholder, i.e. the organ to which the products are allocated for further distribution to users, which notified the buyer of the products allocated to it, of its refusal.

3. *Parties to a contract of supply. Structure of contractual ties.* The parties to a contract of supply – suppliers and buyers – may be associations and enterprises – manufacturers and users, as well as supply and procurement, government purveying and wholesale trading organizations (general and specialized administrations of the system of the USSR State Committee on Supply, supplying organizations, regional wholesale trading offices, etc.). Contractual ties may be of simple or complex structure, depending on the participants to the contract. *The simple structure* is used to conclude contracts between associations and enterprises – producers and users (and to provide goods – retail trading organizations); when *the complex structure* is used, one of the parties is a supply and procurement, wholesale trading, or other organization which participates in bringing products to users.

The decisions of the Party and government provide for the systematic development of the economic expediency of direct long-term economic ties between associations, enterprise-producers, and users. In addition to this, contractual relationships, including long-term ones with specialized organizations which offer supply and procurement services (storage and sorting of products and goods, preparing material for production use, etc.) as well as selling supplies in small quantities, are of great significance. Therefore, in addition to direct contractual ties, the contract of associations and enterprise-producers with supply and procurement or regional wholesale trading using material resources and consumer goods, are widely used. In turn, such organizations conclude contracts with the recipients of products – with associations and enterprise-users and with retail trading organizations (stores, markets).

In the decree of the Central Committee of the CPSU and the USSR Council of Ministers of 12 July 1979 on the improvement of the economic mechanism, one of the most important forms for securing material resources for users was recognized to be guaranteed combined supply. Under this, the territorial organs of the USSR State Committee on Supply conclude long-term contracts with associations and enterprises – primary users – to organize the material-technical supply, which includes the obligation to provide supply.

4. *Procedure for concluding and changing the contract of supply*. The Statutes on Supply provide several methods for concluding contracts of supply. A contract may be concluded by: 1) drawing up one document which is signed by the parties; 2) written confirmation by the supplier to accept the order of the buyer; 3) exchange of letters and telegrams; 4) acceptance by the parties to carry out the warrant. As a rule, the choice of methods for concluding a contract is left to the discretion of the parties and depends on the content of the plan act of supply, as well as on the nature of the mutual relations.

As a rule, in concluding a contract which consists of one document, the initiative for sending a draft of the contract lies with the supplier, who is obliged to send the draft of the contract within a twenty-day time period from the moment of receipt of the plan act. When necessary (for instance, when the warrant specifies spare parts for a specified sum), the sending of the draft contract by the supplier precedes the presentation by the buyer of data indicating the grouped assortment or technical characteristics of the products within the limits of the quantity and group assortment provided for in the plan acts. In this instance, the supplier sends the draft contract within ten days after receipt of the data on the grouped assortment or technical characteristics.

The buyer is granted a ten-day period (twenty-day period when a long-term contract is concluded) to sign the draft contract. If the buyer raises objections to the conditions of the contract, it compiles up a list of disagreements within the same time period.

When a contract is concluded by the method of accepting the order of the buyer is a form established by the USSR State Committee of Supply or the USSR Ministry of Trade, the supplier returns the signed order within the established period. Objections to individual conditions of the order are given as reasons for objections in the signed order.

The party which receives the contract with the list of disagreements or the accepted order with objections, examines the disagreements and objections and, if necessary, takes measures to resolve its disagreements with the othe party. The remaining unresolved disagreements must be submitted for decision by *arbitrazh* not later than twenty days after the receipt of the draft contract with the list of disagreements or receipt of the order which was accepted by the supplier with objections. If this is not done, the proposals of the party on the conditions of the contract or order are considered accepted.

Tacit acceptance of the plan act is the characteristic method to establish an obligation of supply by acceptance of the warrant for fulfillment, that is, the warrant is considered accepted for fulfillment and acquires the force of a contract where, within a ten-day period after its receipt, neither of the parties expresses its disagreement with the warrant, nor requires agreement on supplementary conditions of supply, or sends a draft contract. This method of concluding a contract is used at the discretion of the parties in instances when the warrant which is given to the supplier and buyer contains all the necessary data for carrying out the supply (quantity, types or assortment, quality of goods, time periods of supply).

As a rule, contracts for supply of consumer goods are concluded at markets during the wholesale sale of the goods. The procedure for concluding contracts at markets is defined by the Statute on the Conclusion of Contracts at Markets, which was confirmed in 1982 by the USSR Ministry of Trade and the Central Union of Consumers' Societies.

The contract of supply is concluded for five years, for one year, or for some other period taking into account the period for the production and supply of products. The decree of the Central Committee of the CPSU and the USSR Council of Ministers of 12 July 1979, No. 695, "On the Improvement of the Planning and Strengthening of the Role of the Economic Mechanism in Raising the Effectiveness of Production and the Quality of Work," provides for the increased role of long-term (five-year) contracts. Long-term contracts are used to carry out state plans for economic and social development and aid in strengthening contractual discipline.

The unilateral refusal to perform a contract of supply or the unilateral alteration of its conditions are not allowed except in instances provided for by normative acts. Such a refusal is possible, for instance, for the supply of individual types of products which were established by the USSR Committee on Supply. This rule secures the stability of the contract. However, flexibility of the contract is also needed in instances when the contract must be changed or rescinded by agreement of the parties or, if such an agreement is not reached, by decision of *arbitrazh*.

Grounds for the change or rescission of a contract may be: the withdrawal or change of the plan act on supply in accordance with which the contract was concluded; the arising of an impossibility and inexpediency of its performance after the contract was concluded due to changes in production requirements, demands of the population, lack of raw resources, and supplies which are necessary to produce goods in the specified assortment, etc. Moreover, the change and rescission of a plan contract of supply does not have to depend on the withdrawal or change of the plan act on supply. A contract may be changed by agreement of the parties or decision of *arbitrazh* even if the plan act on supply is not changed.

The change or rescission of a contract is formalized by the parties by exchanging letters or telegrams or by concluding a supplementary agreement according to a procedure and within a time period which is established by normative acts on supply or by contract. The Statutes on Supply provide one additional method for changing (rescinding) contracts of supply of products and goods allocated by plan procedure in the instance of the withdrawal or change of the plan act. If in the course of ten days after receiving notification of the change or withdrawal of the plan act on supply neither of the parties expresses disagreement with the change or withdrawal of the plan act, the contract is considered changed or rescinded.

5. *Content of a contract of supply. Rights and obligations of the parties.* The Statutes on Supply provide a model list of those conditions of a contract which in their entirety constitute its content. Among the conditions of a contract of supply, the Statute distinguish the two essential ones without which the contract itself could not exist: *the subject of the supply and the price.* In the Statutes, by the subject of a contract is meant the name, quantity, breakdown of types, assortment, and quality. The parties agree on these conditions irrespective of the method by which the contract was concluded. If even one of these conditions is missing when the contract is concluded, the contract is not considered to be concluded. The absence of other conditions, which define more fully the obligations of the parties, results in other legal consequences and does not lead to the contract being considered not concluded. Special rules are established for long-term contracts which are concluded by direct long-term or long-term economic ties. Such a contract is considered not concluded if it lacks the condition regarding the range of supply over a five-year period and on the breakdown of types (assortment) for five years.

The conditions of the contract of supply which constitute its content are set forth either in the text of the contract or in the specifications. *Specifications* consist of a document attached to the text of the contract which contains the data characterizing the subject of supply: quantity, assortment, quality. The specifications are an integral part of the contract.

The quantity of goods of supply which are distributed by the plan procedure is determined by the parties in the form of goods in kind or in money within the limits of the quantity which is given in the plan act on supply. The quantity of products in a non-plan contract of supply is determined by agreement of the parties.

The *subject* of supply is made concrete by the instructions regarding the assortment. The breakdown of types of the products and goods is determined in the contract on the basis of the order (specifications) of the buyer within the limits of the range of supply and grouped assortment established in the plan act, although for some products it is determined by a listing of types approved by the ministry of the supplier. By grouped assortment is meant an aggregate of products and goods of one type which are differentiated by individual features. For instance, a contract provides for the supply of boots not only of a certain style, but also stipulates the color and size of the boots. If a dispute arises as to the agreement in regard to the grouped assortment, preference is given to the demands of the buyer.

The conditions regarding quality are determined by indicating the number and index of the standard, or the technical conditions, as well as the model (standard) which must match the quality of the supplied products. By agreement of the parties, a supply of products of higher quality than that indicated in the standards or technical conditions and models may be supplied.

The time periods for supply constitute an important condition of the contract of supply. Besides the time periods for supply which are specified in the plan act, the parties may provide in the contract for periodic supply – terms (quarterly, semiannual, ten-day periods) during the course of which the entire amount of products provided in the contract is supplied in parts. If no time periods are provided in the contract, the supply of products takes place every month in equal amounts.

In defining the conditions on price in the contract, the parties are obliged to adhere to the price approved through established procedure by the USSR State Committee on Price or by ministries and departments. For some type of products, if the determination of the price is left to the competence of the production association or enterprise, the parties may agree on the price during the process of concluding the contract (contractual price). Beside this, the parties may through their agreement determine additions to or reductions in the centrally established prices. In the contract, the parties also define the procedure and form of the accounts for the dispatched products (for accounts, see Chapter 32).

The basic obligation of the supplier is to transfer the products to the buyer within the determined period in the quantity, assortment, and proper quality agreed upon with the buyer. The buyer is obliged to accept these products and to pay by the method provided for in the contract. In case of delayed delivery of supply, the supplier is obliged to supply the unsupplied quantity of goods if the buyer does not refuse to accept delayed supply of products.

The buyer (recipient) accepts the received products by quantity and quality. In acceptance on the basis of quality, the buyer (recipient) verifies whether or not the received products conform to the requirements of the standards, technical conditions, or approved models. The acceptance of the products by quantity is carried out by the buyer by verifying whether or not the products actually received conform to the quantity which the supplier has indicated in the shipping invoice or other accompanying documents. The procedure and terms for accepting products and goods by quantity and quality are defined by instructions approved by the USSR State *Arbitrazh*.

6. *Responsibility of the parties for non-performance or improper performance of the contract of supply*. For a violation of the obligations in a contract of supply the parties must pay liquidated damages (fine, penalty) established by the Statutes on Supply, the Special Conditions on Supply, other

normative acts, or by the contract, and is also liable for the *losses* caused by the violation of the contract.

The Statutes on Supply are based on the principle of full compensation for losses caused by a violation of a contract of supply. The party incurring the losses has the right to demand from the party which perpetrated the violation any expenses incurred as a result of the non-performance or improper performance of the obligation of supply, as well as the profits which it did not make. The liquidated damages established for the violation of the contract of supply are of a calculated nature, i.e. losses in excess of the liquidated damages are recoverable. Only liquidated damages for the supply of products of improper quality or in incomplete sets are of a fine character, and they are paid by the supplier (producer) in excess of the compensation for losses which were caused by such a supply. The compensation of losses and payment of liquidated damages does not free the party which violated the contract from actual performance of the obligation.

The decisions of the Party and government have repeatedly emphasized increased responsibility for violating contractual discipline. The other party has an obligation to file a claim placing responsibility on the party violating the contract. The Statutes on Supply establish that associations and enterprises must apply the sanctions provided in normative acts on supply or in contracts for the violation of contractual discipline in the proper manner, without mutual concessions. The consequences of the violation of this obligation are also established. If a buyer does not file a claim against the supplier for payment of liquidated damages for failure to supply or for delayed supply, the amount of liquidated damages is recovered and placed into the union budget. Moreover, if by *arbitrazh* it is established that the refusal by the buyer to recover liquidated damages is without grounds, it recovers a fine of 2% of the value of the undersupplied products from it (the buyer) to be placed into the union budget.

One of the most significant violations of contractual discipline is failure to supply or delayed supply of products. In case of failure to supply or delayed supply, the supplier must pay liquidated damages in the amount of 8% of the value of the non-supplied *products*, and 5% of the value of non-supplied *goods*. Liquidated damages for non-supply or delayed supply of products are calculated on the basis of the individual designations of the assortment, and not on the basis of the overall value of the non-supplied products. Taking into account the importance of the prompt supply of products, the Statutes on Supply provide for an increased rate of liquidated damages for non-supply for a specified category of buyers. For example, liquidated damages of one-half above the rate (12% and 7.5%) are recoverable for the non-supply or delayed supply of products in provinces of the Far East, and for early supply on other provinces.

The Statutes on Supply make distinctions in regard to responsibility for supply of products of improper quality. The extent of the fine to be paid depends on the nature of the violation. Different consequences are provided for supply of defective goods, products with defects which may be removed without returning them to the producer, etc.

If the goods supplied do not correspond to the standards, technical conditions, or models (i.e. substandard supply), the producer (supplier) must return to the buyer the money paid (for the products) and pay a fine of 20% of the value of the defective products. If the defective products had been awarded the "Symbol of Quality", the producer (supplier) must pay a fine of 30% of the value. The fine for substandard supply is recoverable from the account of the producer automatically (without his consent).

If the buyer notices defects in the products which may be eliminated on the spot, the buyer has the right either to remove the defects on his own, but at the producer's expense, or to demand that the producer (supplier) remove the defects. In the supply of consumer goods, the buyer also has the right to refuse to accept goods from which defects were removed. A fine is recovered from the supplier whether he removes the defects himself or whether they are removed by the buyer.

The amount of fines in cases where it is possible to remove the defects without returning the products to the supplier is significantly lower than the fines which are paid by the producer for supplying defective products. In this instance, the producer pays a fine in the amount of 5% of the value of the *products* and 2% of the value of the *goods* from which defects were removed.

Strict liability is also provided for the violation of payment discipline (for the groundless refusal of the acceptance of or delay in payment, etc.).

The extent of liquidated damages (fine, penalty) is established by the Statute on Supply and the Special Conditions on Supply and may not be changed by agreement of the parties. Another rule is in effect in cases of increased responsibility. At the conclusion of the contract, the extent of the liquidated damages may be increased by the agreement of the parties. The parties also have the right to provide for liquidated damages in a contract of supply in case of the non-performance or improper performance of obligations for whose violation no liquidated damages are provided in legislation.

Chapter 24. The Procurement Contract for Agricultural Products

1. *Concept and significance of the procurement contract.* The Procurement contract is used for the state procurement of agricultural products, and it serves as a means of organizing the economic ties of agricultural enterprises (collective farms, state farms, and their associations) with processing organizations and industrial and trading enterprises. The procurement contract aids in the sale of agricultural products, and agricultural production concentrated under state authority for subsequent plan distribution. The role of the procurement contract is constantly being increased in our national economy because this contract will aid in fulfilling the Food Program, which was adopted at the May (1982) Plenary Session of the Central Committee of the CPSU.

The procurement contract is regulated by general and special normative acts. Articles 267 and 268 of the Civil Code are devoted to it. Instructions on the procedure and time periods for concluding a procurement contract, the conditions for payment, and delivery terms for agricultural products are contained in the decree of the USSR Council of Ministers of 23 April 1970 on the organization of state purchases of agricultural production.[1] In developing this decree, the USSR Ministry for State Procurement confirmed the Statute on the Procedure for the Conclusion and Performance of Procurement Contracts for Agricultural Products.[2] In addition, the USSR Ministry for State Procurement and the Ministries for State Procurement of the union republics have confirmed model procurement contracts for individual types of agricultural products, as well as instructions on the method of purchase of individual types of products, which define the conditions for accepting and storing of processed products and the payments for them.

Under a procurement contract, a collective farm, state

1. *Collection of Decrees of the USSR* 1970 No.8 item 63.
2. *Collection of Normative Acts on Economic Legislation*, Moscow 1979, p. 431

farm, or other agricultural enterprise (association) binds itself on the basis of the plans of state procurement to transfer agricultural products in a specified quantity, assortment, and in proper condition to the ownership (operative management) of a processing organization (enterprise), and the processing organization (enterprise) binds itself to accept the products and pay a set price, as well as to render assistance to the collective farms, state farms, or other agricultural enterprises (associations) in organizing production and in selling the agricultural products.

The parties to this contract are *socialist organizations*. One of the parties, called "the economic unit" in a number of normative acts, is always a state-farm organization. The other party to the contract is called the procurer or processor. The procurer may be a specialized processing organization, trading enterprise and organization, or an industrial enterprise.

The procurement contract is a plan contract, which is concluded in accordance with the State Plan for the Procurement of Agricultural Products. Moreover, relationships which are outside the scope of sale based on the plan are also formalized by a procurement contract. The *subject* of the contract consists only of products resulting from agricultural production.

2. *The plan basis of the procurement contract.* The procurement contract is concluded on the basis of plans of sale to the state of agricultural products which were approved by collective farms, state farms, and other agricultural enterprises and associations. The general plans for the sale of products in their natural state for the entire period of the five-year plan is established for agricultural enterprises (associations).[3] The plans for purchase are established for the processing organizations. The organs which plan state purchases also determine the extent of purchases in excess of

3. The procedure for planning the production and processing of agricultural productions was established by the Decree of the Central Committee of the CPSU and the USSR Council of Ministers of 14 November 1980 (*Collection of Decrees of the USSR* 1981 No.1 item 1).

the plan. However, the plan for purchases in excess of the plan is not presented to agricultural organizations but only to processing (organizations).

Besides the plan target for state purchases of agricultural products, another plan basis for the contract is an act issued by a district administration for agriculture which binds an economic unit to a processing organization. In instances when the plan of sale is established for the production association in agriculture, the production association ensures the transmitting of the plan targets to the collective farms and state farms which constitute the association. In this way, the basis of the procurement contract consists of the plan target for state procurement and for the sale of agricultural products, and the act binding the procurer to the economic units which defines the concrete parties to the contract.

3. *Procedure and time-periods for concluding and changing the contract.* The Statute on the Procedure for Concluding and Performing Procurement Contracts for Agricultural Products provides for the conclusion of a procurement contract by agricultural enterprises either with a specializing processing organization which realizes the supply of agricultural products to industrial and trading organizations (for instance, with a grain delivery center or an organization of the Central Union of Consumers' Societies), or directly with industrial and trading enterprises. In the first instance, when the agricultural products are dispatched as transit by industrial and trading enterprises, a complex structure of contractual ties is involved, while in the second instance, direct contractual relationships are involved. Regardless of which organization is the processor, a procurement contract is concluded directly in that economic unit to which the processor (procurer) sends its representative. The contracts must be concluded separately with each procurer not later than 1 January of the relevant year. After it has been concluded, a procurement contract must be registered with the District State Inspector for the Procurement and Quality of Agricultural Products. The contract is transferred to the administration of the agricul-

tural industry of the district executive committee, as well as to the State Inspector for Procurement. These organs exercise supervision over the conclusion and performance of procurement contracts.

If a dispute arises concerning the conditions of the contract, and one of the parties is a collective farm, the contract is given over to decision by the court. In other instances, pre-contractual disputes are decided by the proper *arbitrazh.* A procurement contract is concluded for one year or for several years. A long-term contract is refined every year by the parties by taking into account the procurement plans for each year.

In the process of performing the contract, the parties have a right to change the assortment of the procured products by mutual consent. The parties must inform the District State Inspector for Procurement and Quality of Agricultural Products of the changes which they have made on the contract. Other changes may be made in the contract by agreement of the parties. Changing the conditions of a contract is allowed only in instances when this does not contradict the established extent of the procurement of agricultural products.

4. *Content of the procurement contract. Rights and obligations of the parties. The contents of the contract* consists of its conditions in regard to the quantity, assortment, and quality of the procured products, time periods of sale, price, and other conditions included in the contract in accordance with the Model Procurement Contract for the corresponding product.

The quantity, assortment, and time periods for sale in the contract are established on the basis of the State Plan for Procurement of Agricultural Products. In addition, the quantity of the procured products is listed in the contract by taking into account the products provided in the plan, as well as those procured in excess of the plan. The quantity and assortment of the products in excess of the plan are determined by agreement of the parties. By agreement of the parties, the contract may define the procedure for delivery (distribution) of the products, schedule for delivery (distribution), the procedure and form of payment, etc. Beside specifying the price for the procured products, the parties also determine the additional payment for products sold in excess of the plan.

The parties to a contract *must* observe all the conditions of the Model Contract. However, they do have the right to include supplementary conditions in the contract based on local particularities (for instance, conditions on packing or on the schedule for delivery). Any terms of the contract which contradict the Model Contract are considered invalid.

A contract which has been concluded defines the *obligations of the economic unity and the processor*. The economic unit is obliged to ensure that the quality of the procured goods corresponds to the standards or technical conditions, and to transfer (deliver) the agricultural products to the processor wihtin the time period established in the contract. In this connection, it is recognized as expedient to change over to the acceptance of agricultural products directly in the economic unit (most importantly vegetables, fruits, berries, potatoes, milk, etc.) (para. 11, Decree of the Central Committee of the CPSU and the USSR Council of Ministers of 14 November 1980).

The processor is obliged to accept the products provided for in the contract, including those in excess of the contract, on time, and to pay the established price for them. The processor pays for the products sold by the collective farms by giving an advance of 30% or, if the collective farm is economically weak, 40% of the value of the products. These advances are taken into account when the processor pays for the products.

The processor is under the obligation to provide the economic unit with standards, technical conditions, wrapping and packing material. At the request of the economic unit, it may be provided in the contract that the processor give instructions and render advice to the workers of the economic unit on questions concerning the receiving of the products, evaluation of their quality, and the payment for them.

5. *Responsibility of the parties according to the procurement contract.* If they breach an procurement contract, the parties must bear *responsibility*. The economic unit pays a penalty or liquidated damages to the processor for the delay or non-supply of products within the determined time periods. The penalty is calculated at the amount of 0.1% of the value of the non-supplied products for each day of delay. If the delay is more than ten days, the calculation of the penalty is terminated and the economic unit pays liquidated damages consisting of 2% of the value of the non-supplied products in addition to the penalty. The penalty and liquidated damages are recovered on the basis of the individual designations of the assortment provided in the contract. The economic unit is absolved from responsibility for a breach of the contract if the obligation to sell is not performed as a result of natural calamity or through the fault of the processor. For example, the economic unit does not pay liquidated damages for failure to provide the products if the harvest was destroyed (ruined) by hail.

The responsibility of the processor arises from the refusal to accept the products provided for in the contract or for delay in accepting them, as well as for late payment for the accepted products. For the refusal to accept or for delayed acceptance of the products, the processor must pay liquidated damages to the economic unit in the amount of 3%of the value of the products which were not accepted or accepted only after a delay. Late payment for goods causes the processor to pay a penalty of 0.1% of the value of the accepted products for each day of the delay. If the delay is longer than ten days, liquidated damages of 2% of the late payment are recovered in addition to the penalty.

Whether or not liquidated damages (fine, penalty) are paid, the guilty party must reimburse the other party for any losses caused as the result of breach of the contract. However, the payment of liquidated damages does not free the party from the performance of its obligations.

Chapter 25. Contract of Property Hire. Contract of the Gratuitous Use of Property

1. *Concept and significance of the contract of property hire.* The contract of property hire (sometimes called contract of property lease of rent) is used to assign property for temporary use and is used in the mutual relationships of citizens as well as of organizations. Many citizens hire a vacation home for the summer and a garage for the winter. There are organizations which specialize in hiring out property needed by other organizations (tools, machinery) and to citizens – articles of domestic and everyday use (refrigerators, baby carriages, musical instruments, movie cameras, tents, etc.). Socialist organizations lease buildings (accommodations) and other property.

Under a contract of property hire, *one party – the owner or possessor of the operative management (the lessor) – transfers the property for temporary possession and use in return for payment to another party (the lessee) who must return it when the contract is terminated* (Articles 275, 291 Civil Code).

The lessee uses the property of the lessor in return for payment. The contract of property hire is reciprocal; that is its essential condition. If the use of the property is not paid for, the contract takes on a different character: use of the property becomes gratuitous (Section 5).

Under a contract of hire, organizations and citizens receive property for use without spending any means to acquire it for ownership (operative management), and if structures and buildings necessary to the organizations are involved they do not have to turn to superior organs for a decision as to whether they will be taken into the account of the other organization.

This contract is also beneficial to the other party. It gives the party which does not need the property for a certain period the opportunity to receive payment for hire (lease) which enables him to recover the expenses for capital repair on the property and also frees him for that period from a number of operating expenses for the upkeep of the property being used.

The contract of property hire is regulated by Article 275-294 of the Civil Code. Individual norms concerning this contract are included in the Statutes on the Enterprise (para. 16), Combine (para. 44), and Scientific-Production Association (para. 44). The rates of payment for the leasing (hiring) of non-residential accommodations (buildings),[1] as well as the Model Contracts on Leasing Domestic Articles[2] were confirmed by the governments of the union republics.

There are departmental normative acts on the contract of property hire. They include the Model Contract for the Lease of Non-residential Accommodations (Buildings), which was confirmed by the RSFSR Ministry for Communal Economy, the Rules for Assigning Articles for Household Use, Musical Instruments, Sport and Tourist Inventory, and Equipment and other Property to Citizens for Temporary Use, which was confirmed by the RSFSR Ministry of Everyday Services to the Population.

2. *Subject of the contract of property hire. Parties.* The *subject* which is provided in the contract always consists of *individually-defined articles.* The character of the property transferred for hire and its purpose, influence the *price of the contract* – the extent of the hiring (leasing) price. Use of the hired property must not lead to the complete exhaustion of the property; at the end of the time period of the contract, it must be returned to the lessor. Property which has generic features such as fuel, grain, or potatoes may not be the subject of a contract of hire. If the object to be returned is not the same but of the same type as the transferred object, the parties conclude a contract of loan (see Chapter 32).

The lessor must be the owner of the property or possess the right to its operative management. A change of owners

1. *Collection of Decrees of the RSFSR* 1965 No.17 item 105.
2. *Collection of Decrees of the RSFSR* 1965 No.1 item 2, 1982 No.2 item 4; 1969 No.15 item 76.

(possessors of the right of operative management) does not affect the validity of the concluded contract of hire and it retains the obligations for the new owner (possessor of the right of operative management).

Such natural objects as land, minerals, waters, and forests, which are the exclusive property of the Soviet state, may not be subjects of the contract of hire. They are assigned for use to organizations and citizens in accordance with the rules of land, water, and forest legislation, and the legislation on minerals.

The parties to a contract of property hire may be enterprises, institutions, and organizations, as well as citizens. Thus, on the basis of this contract, leasing centers of the system of the USSR State Committee on Supply transfer tools, machinery, and other technical means to associations, enterprises, and educational institutions for temporary use in return for payment. Socialist organizations often lease means of transportation, equipment, and machines.

In the mutual relationships between socialist organizations, the contract of property hire may be of plan character. The conclusion of a contract for the lease of non-residential premises (buildings) precedes the plan distribution of the premises (buildings) by the executive committee of the local Soviet of People's Deputies between future users which was agreed earlier with the possessors of these premises (structures) and formulated by an order (decision) of the executive committee. In this instance, the lessor does not have the right to violate the decisions of the executive committee of the local Soviet and is obliged to conclude a contract with the juridical person which is indicated in the order.

Socialist organizations conclude contracts of property hire in the *written form*. Between citizens, this contract must be concluded *in writing* if it is for a period of more than one year (Article 176 Civil Code).

A socialist organization, which has properly performed the obligations imposed on it by the contract, has a prior right to renew the contract as against other parties when the validity of the contract expires.

The general period for a contract of property hire must not exceed ten years, for the hire of buildings (non-residential accommodations) it must not be over five years, and the period for the hire of equipment or other property must not exceed one year. If the parties do not indicate the period in the contract, it is considered to be concluded for an indefinite period, but no longer than the periods specified above (Article 277 Civil Code).

3. *Rights and obligations of the parties. Their responsibility*
In a contract of property hire, the lessor is obliged: a) to transfer into the possession and use of the lessee the property specified in the contract in the condition conforming to the terms of the contract and the purpose to which the property is to be put (Article 281 Civil Code). The condition of the transferred property is usually stipulated in the contract, which also note the defects in the property; b) to carry out capital repairs on the property transferred for hire. Socialist organizations use the money received for hiring out (leasing) property for this purpose. However, it may be established by legislation or the agreement of the parties that capital repairs are to be carried out by the lessee (Article 284 Civil Code); c) to render assistance to the lessee in using technically complex property, for instance, an automobile. The procedure for such assistance is usually defined in the appropriate rule and may be refined in the concluded contract.

The *obligations of the lessee* in a contract of property hire are more numerous and consist of: a) using the hired property in accordance with the contract and the purpose of this property (Article 283 Civil Code); b) prompt payment of the agreed price for the hired (leased) property (Article 286 Civil Code). Usually such payments take place periodically. For instance, in the contract for the lease of non-residential accommodations (buildings) from the state or communal housing stock, payment for the lease must be quarterly, prior to the commencement of the next quarter; c) maintaining the hired property in good condition and carrying out current repairs at his own expense unless otherwise provided by legislation or the contract, and paying the expenses for its maintenance (Article 285 Civil Code). For example, organizations which occupy non-residential accommodations in residential houses, in addition to paying for the lease and central heating and communal services, must share with all the other possessors of the house the expenses from the residential fund, proportionate to the space of the accommodations which it occupies; d) returning the property in the same condition in which it was received, taking account of normal wear and tear, or in such condition as may be provided in the contract (Article 291 Civil Code).

Upon agreement of the lessor, the lessee has the right to transfer the hired property for temporary use to another party (sublessee) for payment (Article 287 Civil Code), while remaining responsible for the use of the property by the sublessee according to the terms of the first contract.

The lessee has the right to carry out capital improvements of the hired property, which is the obligation of the lessor, if such repair is necessary, or if the repair was stipulated in the contract and the time period for its performance, as provided in the contract, has expired (Article 284 Civil Code). In this instance, the expense of the capital improvements which is the obligation of the lessor but which were carried out by the lessee, may be recovered from the lessor or deducted from the hiring (leasing) fee.

If the lessor fails to perform his obligations of giving the property stipulated in the contract to the lessee, the latter has the right to demand this object from the lessor or to renounce the contract (Article 282 Civil Code). If the lessor refrains from transferring the object or from accepting it from the lessee (when the latter returns it to the lessor), the lessee has the right to demand compensation for losses caused.

If the lessee is unable to use the hired object in accordance with its purpose, he has the right to demand a corresponding reduction in the hiring price, taking into account the actual use of the object, or to rescind the contract prematurely (Article 290 Civil Code).

If, in exceptional circumstances, the lessee has to carry out capital repairs of the property according to law or the contract and fails to perform them, the lessor has the right to prematurely terminate the concluded contract (Article 289 Civil Code).

With the permission of the lessor, the lessee may improve the conditions of the hired property, adapting it to his needs, and has the right to demand compensation for the necessary expenses incurred for this purpose, except in some circumstances. The lessee has the right to remove any improvements which are carried out without the permission of the lessor if they can be removed without causing harm to the property and if the lessor does not agree to pay for their worth (Article 293 Civil Code).

If the lessee intentionally or negligently damages the condition of the property, this serves as grounds for the premature termination of the contract by the lessor (Article 289 Civil Code). Non-performance of such an important condition of the contract as payment of the hiring (leasing) fee for a period of three months also gives the lessor the right to demand premature termination of the contract (Article 289 Civil Code).

4. *Contract of domestic hire.* This contract is a variant of the contract of property hire and therefore is subject to the general rules in regard to it. However, because of its function, the contract of domestic hire has some specific features.

The lessor of articles for household use, musical instruments, sports equipment, automobiles, and other articles for personal use is always a specializing socialist organization (leasing work-shop), and not a citizen.

A citizen may lease out an article for domestic use to another citizen if this is not of the nature of the prohibited handicraft and artisan business. However, in this instance the relationships are not those of domestic hire, but of property hire and the rules of that contract are applied.

Domestic hire is formalized by a written contract in accordance with an existing model form or an issuing receipt. In order to conclude a contract for hiring means of transportation, it is necessary to show a driver's license. Beach equipment is issued after the deposit of a monetary pledge, the extent of which must not exceed the worth of the hired property. The price of a contract of domestic hire is determined by a confirmed price list.

The duration of the hiring period must not exceed the periods given in the rules for hire. Objects of significant material value are hired for a period not exceeding one year. The contract of domestic hire which is concluded without a time period is considered to terminate within the period established by the proper model contract or the rules on leasing.

The lessor is obliged to verify the good condition of the hired object in the presence of the lessee (Article 281 Civil Code), and then to maintain the property in proper technical condition so that it may be used. The lessee has the right to demand the removal of defects in the property or its exchange. An organization is also obliged to instruct the lessee on the rules concerning the use of technically complex objects or give instructions on the procedure for use.

A citizen concluding a contract of domestic hire may renounce the contract at any time (Article 290 Civil Code). He does not have the right to sublet the hired property (Article 287 Civil Code), since this contradicts the purpose of the contract.

If the property is damaged, the price of the repair as fixed by the price list is paid to the lessor. If it is impossible to repair the article, an article of the same type must be returned or its value must be paid. If the lessee does not have the object of hire he is obliged to pay for the losses caused. In addition to this, a fine is recovered from the lessee in proportion to the worth of the lost article of hire unless he can show that he is not guilty of the loss of the property.

5. *Contract of gratuitous use of property. Under a contract of gratuitous use of property, one party binds itself to give a specific property to another party without charge for temporary use, and the latter binds itself to return the given*

property. The contract is used in relationships between citizens as well as in the use by them of library material.

Sometimes an organization concludes such a contract. The manager of an organization has the right to give buildings (accommodations) for gratuitous use to departmental medical and children's institutions, to young workers' schools, vocational and technical schools, courses for raising qualifications, schools for advanced technical training, dining halls, and trade union committees. Models of machines, machine tools, instruments, and other equipment, samples which were dismantled from exhibitions may be offered without charge to higher educational institutions for instructional purposes, and objects of material worth – to schools, vocational and training schools and children's non-school institutions for the purpose of training and to equip work-shops and academic offices. If he receives permission from a ministry or department, the manager of an organization may also give equipment and means of transport to other state organizations for use free of charge, and equipment and instruments for scientific-technical work to scientific institutions and higher educational institutions.

The contract of gratuitous use is similar to the contract of hire in its content, and therefore many of the rules on property hire apply to it (Article 342 Civil Code).

In particular, such a contract must be made in writing between citizens if its term of validity is longer than one year. The property granted by the contract and its use must agree with the conditions of the contract. The party which receives the property must carry out its current repairs and is responsible for the expenses for the maintenance of the property. The owner of the property (possessor of the operative management) is responsible for capital repairs.

In addition to this, the contract of gratuitous use has a number of special features, inasmuch as the relationships of the parties in these circumstances are of *gratuitous character*.

As a general rule, the time period of such a contract between socialist organizations must not exceed one year, irrespective of the subject of the contract, even if the contract is concluded without any indication as to the time period. After a period of one year, the contract is considered to be terminated.

A party giving property for gratuitous use is liable only for those defects in the property which it intentionally or through gross negligence fails to mention when it transfers the property (Article 345 Civil Code). There is no liability if the defects were not mentioned as a result of simple negligence.

The party giving the property for gratuitous use has the right to demand the premature rescission of the contract if the user uses the property so as to contradict the contract or the purpose of the property, or if he intentionally or through negligence worsens its conditions, or gives the property for use to a third party without consent.

Chapter 26. Contract for Lease of Living Accommodations

1. *Concept and significance of the contract for the lease of living accommodation. The housing fund of the USSR.* Among the basic socio-economic rights of Soviet citizens which are listed in the USSR Constitution is the right to housing (Article 44). This means that citizens of the USSR are guaranteed permanent and constant use of housing, that they are not under the threat of having to live in the streets, and that those in need of housing are guaranteed an improvement in housing conditions.

The USSR Constitution provides the fundamental material and legal guarantees for realizing the right to housing. These include: the development and protection of the existing state and public housing fund, state assistance for cooperative and individual housing construction, public supervision of the allocation of living space, and provision for those who have a need for improved housing by carrying out the program for the construction of well-equiped housing, through low rent and by payment of their communal services.

The basic housing fund of the USSR belongs to the Soviet state (Article 44 USSR Constitution). The housing fund of the country, which consists of the totality of all dwellings, as well as living accommodations in non-residential buildings, is divided into four parts:

a) *the state housing fund* which includes the dwelling and living accommodations which belong to the Soviet state. This, in turn, includes the fund under the management of the local Soviet of People's Deputies (the housing fund of the local Soviets), and the fund under the management of the ministries, state committees, and departments (departmental housing fund);

b) *the social housing fund* which includes dwellings and other living accommodations belonging to collective farms, cooperative organizations, their associations, trade unions, and other social organizations;

c) *the housing fund or housing construction cooperatives* which includes houses belonging to individual citizens by right of personal ownership, and apartments within the houses of individual builders in the housing construction cooperatives.

Permanent use by citizens of living accommodations in the houses of the state and social housing fund is realized by payment according to a contract of lease with a housing-management organization. The contract for the lease of living accommodations is the basic legal form for the use of the state and social housing fund.

The *parties* to such a contract are: the family which received an order for residence in the apartment, and the housing-management organization. They must agree on the conditions and on the observance of the sanitary and technical requirements supplied for the use of the living accommodations and its equipment (water-pipes, sewer system, central heating, gas stove, etc.). At the conclusion of the contract, the citizens and the housing-management organization, as legally equal participants to the contract, make concrete those rights and obligations provided in law which are connected with the condition and safe-keeping of the living accommodations provided for use. In a contract for the lease of living accommodations there must also be agreement on the conditions on which, by request of one of the parties, an agreement must be reached. For instance, at the request of a party settling into a communal apartment which is used in common by several families, the management organization must define the rights and obligations of the new resident in regard to the use of the auxiliary areas of the apartment (kitchen, bathroom, etc.).

In houses which belong to housing construction cooperatives or to citizens on the basis of the right of personal ownership, the contractual form of the lease of living accommodations is utilized by those citizens who lease housing for payment from members of a cooperative or the owner of a dwelling (part of a house). In this instance it is necessary to attain agreement on a wider range of concerns than in a state or social housing fund (in particular, agreement must be reached on the rent for the house, the period of its use, etc.), since the legislation does not define all the conditions of the lease of housing in the fund of housing construction cooperatives and individual housing funds.

2. *Legislation on the contract for the lease of living accommodations.* The conditions of the contract for the lease of living accommodations are defined by the legislation of the Soviet Union and the union republics. The most important acts are the Principles of Housing Legislation of the USSR and the Union Republics (from now on these will be referred to as Principles of Housing Legislation), adopted by the USSR Supreme Soviet in 1981. All other acts of housing legislation of the USSR and the union republics must agree with these. The basic republic acts in this area are the housing codes. The Housing Codes of the Azerbaidzhan SSR, Lithuanian, and Armenian SSRs went into effect in 1983.

The amount of rent for houses of the state and social fund is established by decree of the USSR Council of Ministers. The Councils of Ministers of the union republics confirm model codes of the lease of living accommodations, the rules for the use of living accommodations, and the maintenance of dwellings and their surrounding grounds.

Housing legislation also includes normative acts which are issued by the Ministries of the Housing-Communal Economy of the republics.

In the absence of necessary instructions in the Principles of Housing Legislation, the housing codes, or in other acts on the relationships resulting from the contract for the lease of living accommodations, the relevant rules of civil legislation are utilized. For example, to recognize as invalid through court proceedings a contract for the lease of living accommodations as a result of the annulment by the court of the order for living accommodations, the three-year period of limitations of actions, provided in Article 16 of the Principles of Civil Legislation, is applied.

3. *Contract for the lease of living accommodations. Under a contract for the lease of living accommodations, one party*

(the lessor) is obliged to give to the other party (the lessee and members of his family) living accommodations (living space) in return for payment for use, to maintain the dwelling properly, to carry out capital repairs on the living accommodations (living space) in a timely fashion, and to secure the uninterrupted functioning of apartment equipment, and the lessee and members of his family are obliged to treat with care the leased accommodations and its equipment, a dwelling and the grounds appurtenant thereto, to pay the rent on time and in the instance that the accommodations are vacated, and to return them to the lessor.

A contract of lease which is concluded by a citizen with a member of a housing construction cooperative, who is the basic user of the living accommodations in a cooperative house, does not contain a number of the most important enumerated rights and obligations of a hirer of living accommodations in a state or social housing fund. The rights and obligations of the hirer in this instance equal the rights and obligations of the sublessee (see Section 6).

The contract of lease allocates to the lessee and his family rights and obligations for the use, possession and, to a specified extent, the disposition of the living accommodations which differ significantly in their content from the rights and obligations of the lessee in a contract of property hire which was examined above (Chapter 25).

The essential condition of the contract is its *subject* – the concrete, defined living accommodations (living space), its sanitary and technical conditions, which secure its suitability for permanent residence of the family involved. Repeated settlement in barracks and cellars which are not suitable for residence is not allowed by the contract for the lease of living accommodations.

Living accommodations which are supplied to citizens in houses from the state and social housing fund must have public services and amenities constructed according to the conditions of that specific population area. The contract of lease is concluded specifically for living accommodations which consist of one or several rooms. The separateness of the accommodations is the decisive feature of the residences in these funds. A part of a room, a through-passage adjoining room, and particularly auxiliary rooms (storage room, corridors, kitchens) may not be independence subjects of a contract.

The requirements in regard to the subject of a contract for the lease of living accommodations in houses belonging to housing construction cooperatives and individual housing funds are different. The separateness of the accommodation is not required in these houses, and the subject of the contract may be a part of a living accommodation – floor space or even auxiliary premises. In these circumstances the presence or absence of the defined minimal public service amenities in the living accommodations (plumbing, sewage system, central heating, gas, etc.) or the auxiliary areas have no legal significance for concluding or renouncing a contract.

Apartment rent in the USSR is as low as possible. Payment for communal services (water, gas, electricity, thermal energy) is in addition to the rent and is also low.

The amount of the rent for living accommodations in a house of a housing construction cooperative which is paid by the lessee to a member of the cooperative according to an agreement between them may not exceed the operating expenses collected from the member of the cooperative which are spent on these premises. The payment based on contract for use of living space in houses belonging to citizens by right of personal ownership is also determined by an agreement of the parties, but may not exceed the defined rates established for such houses by acts of the government of the union republics.

The basis of the housing legislation is established by the *permanence* of the contract of lease for houses in the state or social housing funds. This rule creates stability for the permanent use of leased living accommodations.

In the individual housing fund, the period of a contract acquires particular legal significance since the preferential right of the lessee to continue using the space is linked with it. The lessee who concludes a contract with the owner of a dwelling for one year with an obligation to vacate the premises at the expiration of this period is not given a preferential right to renew the contract of a new period. The lessee who concludes a contract for the lease of living accommodations for a period longer than one year with the owner of a residential house acquires such a preferential right, which is defended in court if the other party violates it. His right to use of the residence is more secure.

In houses belonging to a housing construction cooperative, a period which is stipulated in the contract for the lease of living accommodations also gives the lessee a definite advantage: he may not be evicted by a member of the housing construction cooperative before the period expires. The lessee who does not have a contractual agreement in regard to the period of residence in a cooperative apartment may be evicted within three months after being given notice.

4. *Procedure for allocating living accommodations. The order.* In the houses of the state and social housing funds, the conclusion of a contract for the lease of living accommodations is preceded by the acceptance by an organ of management, with the participation of a social organ, of a decision on allocating living accommodations to a citizen who is registered for the improvement of the living conditions at his place of residence or place of work with observance of priority on a waiting list. In order to be registered, a citizen – who is in need of improvement in his living conditions – files a claim for living space with the executive committee of a local Soviet or the administration or trade union committee of the enterprise, institution, or organization at his place of work, which will make a decision orally and collegially. Such a decision is not necessary to conclude a contract of lease in other housing funds.

If the living space was constructed from central state capital investments for housing construction, the joint decision of the administration and the trade union committee of the enterprise, institution, or organization is submitted for confirmation by the executive committee of a local Soviet with the participation of a social agency on housing problems. If the residence is built from funds belonging to an enterprise or organization, the joint decree of the administration and trade-union committee of the enterprise, institution, or organization is only reported to the executive committee of the local Soviet for receipt of an order.

In a social housing fund, living accommodations are granted on the basis of a joint decision of the organ of the relevant organization (management of a collective farm, union of consumer cooperatives) and the trade-union committtee, with a subsequent report to the executive committee of a local Soviet.

On the basis of the decision taken according to the procedure described above, the executive committee grants to a citizen a single-form *order*. The order is issued with the agreement of all the members of the family on receiving the living accommodations. An order is the instruction of the organ of management which provides for the obligation of the lessor to conclude a contract for the lease of living accommodations with the holder of the order. The order is not a substitute for the contract and is not a document allowing use of the living accommodations. It is temporary, not a permanent, act in force. An order is an administrative act for the distribution of vacant apartments in state and social housing funds which has the significance of a juridical fact for the creation of civil legal relationships for the use of living accommodations.

Use of and residence in living accommodations is not possible without concluding a contract of lease. The order gives the holder of the order the right to use of the residence on the basis of a contract. It grants to a citizen the right to enter into contractual relations with a housing-management organization. The organization does not have a right to conclude a contract with a citizen who does not have an order for a given living accommodation. The holder of an order has the right to refuse to conclude a contract without having to give a reason for the refusal. For instance, one of the members of the family refuses because the apartment is above a store. By giving the order to a housing-management organization, he indicates his agreement to settle into the quarters which were provided. A housing-management organization does not have a right to refuse to conclude the contract and supply him and his family with the living accommodations.

After the expiration of the short period of its validity, the order loses force and may be extended only by the executive committee which issued it. Without an extension of the term of the validity of the order, a contract of lease may not be concluded.

The executive committee which issued the order may not unilaterally declare it invalid. The order is declared invalid by court only in the following instances: 1) violation of the procedure established by law for providing living accommodations; 2) violation of the civil rights of other citizens or organizations to the living accommodations indicated in the order; 3) providing citizens with incorrect information as to the necessary improvements in the living conditions; 4) illegal actions performed by officials in making a decision on providing living accommodations; 5) other instances of the violation of the procedure and conditions for providing living accommodations. The violation of the procedure in providing living accommodations may be, in particular, a unilateral decision of officials, a decision by an organ of management without involving society in the decision of questions on the personal provision of space, or one made exclusively at the decision of the trade-union committee but without the participation (or with the objection) of the organ of management. If a contract for the lease of living accommodations is concluded on the basis of an order which is declared invalid, the court also declares the contract invalid. The parties occupying living accommodations or the owner are subject to sanctions by the procurator according to administrative procedure, without being provided with other living accommodations.

5. *Parties to a contract for the lease of living accommodation. Conclusion and form of the contract.* As a rule, the organ of management which provided the living accommodation does not coincide with the organ which has the right to conclude a contract for the lease of the living accommodation. Only a juridical person which has the residential houses on its accounts

has the right to conclude a contract in regard to houses in the state and public housing fund. Therefore, the *lessor* in a contract is either a specializing housing-management organization (housing-management office, a production housing-repair association, a production housing-repair trust), or an enterprise, institution, or organization which has the dwellings on its account.

A contract for the lease of living accommodation is concluded with a *holder of an order* and the members of his family who are indicated on the order. As a result of the concluded contract, the holder of the order becomes a *lessee*, a party to a contract. With the acceptance of the order by the housing-management organization, an agreement is reached with the holder of the order to provide him with the living accommodation indicated in the order for residential use.

In regard to houses of a housing construction cooperative, the lessor is a member of the housing construction cooperative who is granted the right to lease an occupied cooperative living accommodation (see Chapter 27); in an individual housing fund, the lessor is a citizen who is the owner of the dwelling (or a part of it, or of an apartment in a house of a collective of individual builders).

The lessee is empowered by law to represent the members of his family and persons, who cease to be members of his family but who reside in the leased quarters, to carry out legal acts in their name in relations with the lessor (for instance, to reserve the accommodation). In doing so, the lessee does not have any advantage over the members of his family. Other members of the family may be substituted for the lessee if certain established procedures are observed.

In most instances, certain legal acts must be carried out to change or terminate the contract personally by everyone who lives jointly with the lessee in the residence (the agreement of everyone to re-plan the living accommodations, exchange it, etc.).

Two categories of family members and relatives who live with him jointly constitute the *members of the lessee's family*. Spouses, children of the lessee and his spouse, and parents (adoptive and adopted) are always considered members of the family. In order that other persons be recognized as members of the family, it is necessary that they maintain a common household with the lessee. The law does not establish a minimal period of common residence of these parties with the lessee, nor the conditions for maintaining a common household. These other persons may include either relatives or dependants who are not capable of working (in the majority of the union republics).

In houses of the state and public social fund, a contract of the lease of living accommodation is concluded in the *simple written form.*

In houses which belong to housing construction cooperatives or to citizens by right of personal ownership, the law does not provide that the contract must be in the written form and this question is decided at the discretion of the parties.

6. *Rights and obligations of the parties.* The lessor is *obliged* to transfer to the lessee the accommodation in a condition which conforms to the established sanitary and technical requirements, to carry out the necessary capital repairs provided for in the rules for the use of living accommodation, to secure the uninterrupted functioning of the equipment of the house (accommodation), and to remove without delay any disrepair or damage in the apartment or house, as well as to ensure the proper security and technical maintenance of the entrance house and adjoining grounds. The lessee and members of his family, regardless of age, as well as persons who have ceased to be members of his family but who are continuing to reside in the leased accommodation, have equal rights and obligations. Each one of them in a house in a state or social fund has *the right*: to use the living and auxiliary accommodation (corridor, kitchen, bathroom, etc.) indefinitely, to bring in other persons according to the proper procedure, to protect his right to retain the living accommodation during a temporary absence, and to request alteration of the contract in circumstances provided by law, as well as to be transferred to another living accommodation of smaller scale in case

there is surplus space in the occupied living accommodation, and to terminate the validity of the contract of lease at any time.

A person who lives jointly and continuously with the lessee has the right to turn to the court in the instance of a dispute between them in regard to the use of the living accommodation. The court, in deciding such disputes, considers the established mutual relationships of the parties and other pertinent reasons and interests of the parties residing in the living accommodation. For example, if no agreement was reached among the members of the family in regard to exchanging the living accommodation, then any of them has the right to request the enforced exchange through judicial proceedings of the living accommodation for residence in various houses (apartments).

None of the residents may be evicted from the leased living accommodation or limited in his right to use the living accommodation except on grounds and by procedure provided by law.

The lessee of the living accommodation is *obliged* to treat with care the accommodation transferred for his use and its equipment, as well as the entrance area, residential house, and the objects constituting the public services and amenities of the house; to observe the rules for the use of the living accommodation and the maintenance of the house and the surrounding area; to carry out current repairs in the apartment systematically in accordance with the list of repair work indicated in the rules for use of living accommodation; to render payment on time for the use of the living accommodation and communal services; and when vacating the premises, to yield them in proper condition. A living accommodation may not be used by citizens for personal gain, to obtain income not based on wages, or for other mercenary aims, as well as to the detriment of the interests of the state and society.

The legislation of the union republics establishes the norms securing the living space needed for the permanent residence of one person in state and public housing funds, which are based on socio-hygienic calculations for satisfying the living space requirements of everyone. In most union republics the norm for the living space of one person is established at nine square meters; in the Georgian, Azerbaidzhan, Lithuanian, and Armenian SSRs, it is twelve square meters. In a number of republics when the nine-meter norm for living space is in effect, specific cities may have a norm of greater than twelve square meters.

The norm for living space is applied in calculating the surplus area in a leased living accommodation, the rent, the settlement of other persons, sub-leasing space, and in other instances established in the legislation. The norm for living space may not coincide with the amount of the actual living area provided for citizens.

Supplementary living space is an additional norm of living area which is necessary for certain categories of citizens because of their professional work or as a result of severe forms of certain chronic illnesses, as well as that given to some persons because of specific services to the state and society. These persons include: Hero of the Soviet Union, Hero of Socialist Labor, personal pensioners, those having the rank of a colonel or higher, scholars, members of various creative unions, those suffering from certain severe, chronic illnesses, executives, etc. As a general rule, those who have a right to supplementary space in excess of the norm avail themselves of it in the form of a room or a living accommodation of ten square meters. Those who are ill and those who need additional space because of the conditions and character of their work may occupy supplementary living accommodations of larger size.

A surplus of living accommodation above the norm, even if it consists of one separate room, may not be withdrawn. The lessee who has surplus living accommodation has a right, with the consent of the family, taking into account to whom the dwelling belongs, to request that the executive committee or the enterprise, institution, or organization supply him with smaller accommodations instead. In most of the union republics, the lessee and members of his family pay for surplus living accommodation at three times the standard rate.

In the instance of the temporary absence of the lessee, a member of his family, or all of these persons, the right to use the living accommodation is, as a rule, reserved for those who are absent for six months, and in some circumstances (if called into military service, treatment at health care facilities, etc.) – for the entire period of absence. If one of the persons is absent for more than six months on valid grounds, the lessor may prolong the period, and if there is a dispute in regard to the right of the absent party to occupy the leased living accommodation – the court may prolong the period.

In the instance of an official trip abroad, a business trip to the regions of the Far East or to areas comparable to the Far East, and in other instances (to enterprises which are particularly important for the national economy, to construction jobs) which are provided in legislation, the living accommodation is reserved (retained) for the lessee or, as the case may be, for members of his family for the time necessary for the job. The housing organs of the local Soviets issue a certificate (securing warrant) to those who are leaving to secure their right to the use of their living accommodation. A lessee and members of his family who are temporarily absent retain the rights and obligations of the contract, and the living accommodation which is reserved during the temporary absence of citizens is not considered to be surplus. Only a court may declare that a person has lost the right to use his living accommodation by remaining away longer than the time period established by law.

The conditions of a contract may be changed only with the mutual agreement of the parties (lessor, lessee, the members of his family and the persons who have ceased to be members of his family but are continuing to reside in the lessee's dwelling). However, the law provides for instances when the desire of one party, or even one member of a family, to change the contract is obligatory for all. For instance, a member of a family is entitled, with the consent of all the other members of the family. to request that a contract be concluded with him separately for his share of living space if this area constitutes a separate living accommodation. In other words, a person is entitled to request to be recognized as an independent lessee (for instance, in case of the death of the primary lessee or at the same time as another lessee).

Citizens living in one common apartment who use the living accommodation therein on the basis of separate contracts have the right to demand that one contract be concluded with them if they are united into one family using all their living accommodations. The refusal of the lessor to realize the rights of a member of one or more families in these instances may be contested by them in court.

A lessee has the right to exchange the living accommodation he occupies for premises occupied by another lessee with the mutual transfer of the rights and obligations under contracts of lease or to exchange with a member of a housing construction cooperative on condition that he becomes a member of a cooperative.

A written agreement of the adult members of the family living with the lessee, including those temporarily absent for whom the right to the living accommodation is being reserved, is required for exchange.

The lessor gives permission for the exchange of living accommodation in houses in institutional or social housing funds, as well as for houses which belong to citizens by right of personal ownership. Permission for the exchange of houses belonging to housing construction cooperatives is given by the executive committee of the local Soviet. A lessee may contest a refusal to exchange by judicial proceedings. It is not possible to contest in court a refusal to exchange by a citizen-owner of a dwelling or by a collective farm.

The law establishes conditions under which the exchange of living accommodations is not permitted. Specifically, exchange is not allowed if the house is under threat of collapse or is subject to demolition, if the exchange is of speculative (sham) nature, or if the lessee is subject to eviction on grounds provided by law and has filed a suit in court relating thereto, as well as where the living accommodations are work areas or are located in a hostel. An exchange is formalized by exchanging the orders which are issued by the executive committee. A refusal by the local Soviet to issue an order may be contested in judicial proceedings during a period of six months. When the parties receive exchange orders, it means that the contract is in effect in a judicial sense.

By observing the rules of registration a lessee has the right to house his spouse, children, parents, and other members of his family in the leased accommodation. The housing of minor children with their parents does not require the consent of the members of the family. Newly housed members of the family acquire the same right to the use of the living accommodations as the lessee and other members of the family. However, an agreement may be made between earlier residents and newly settled members of a family on the use of the leased living accommodation according to various proportions. For example, a new resident may be allocated a room which is larger or smaller in area than that of the other members of the family.

A lessee and the members of his family living with him may, by mutual agreement, permit other citizens who are not members of the family to live termporarily in the living accommodation under their occupancy without accepting any payment for the use of the accommodation. If the temporary residents remain in residence for longer than one and a half months, the norm of living space established in the republic must be observed for each resident.

If the temporary residence by persons who are not members of the family is paid for, it is a residence under a sublease. A lessee has the right, with the agreement of other members of his family as well as that of the lessor, to sublet part of the accommodation leased by him or, if he is temporarily absent, his entire accommodation, while remaining responsible to the lessor for the terms of the contract. It is necessary to observe the norms for living accommodation for each resident. Temporary residents and sublessees are obliged to vacate the accommodation at the request of the lessee, and in case of a dispute – upon the decision of the court.

In houses of house-building cooperatives, the rights and

obligations of the lessee are equal to the rights and obligations of sublessees in the state and social housing fund. For example, a lessee in a housing construction cooperation fund has a right to residence, but not to permanent use, of the cooperative apartment.

In an individual housing fund, the lessee has the rights and obligations of a lessee in a state or social housing fund with several particularities and exceptions: use of the accommodation by the lessee does not have to conform to the norms; other members of the family may be housed in the occupied space with the consent of the lessor except in instances of the lessee's housing his minor children, and if he occupies a separate living accommodation – his spouse and parents who are unable to work; the lessee is deprived of the right to request that another living accommodation be given to him while capital repairs are carried out (see also section 7).

7. *Responsibility of the parties in a contract for the lease of living accommodation. Termination of the contract.* A lessee and adult members of his family and person who have ceased to be members of the family have joint and several property responsibility for obligations arising from a contract for the lease of living accommodation. This means that the lessor has a right to request payment for compensation of losses, as well as liquidated damages for late payment of rent and communal services from all the members of the family jointly or from any one member of the family in particular (Article 181).

The lessee of a living accommodation has the right to terminate the contract at any time. If the lessee and the members of his family move to another place for permanent residence, the contract is considered annulled from the day of their leaving. A contract of lease for houses in house-building cooperatives and in the individual housing fund is terminated at the expiration of the period of the validity of the contract.

A lessee in the individual housing fund is granted a preferential right to renew a contract the term of which has expired, except in the instance where it was concluded for one year. The owner of the house has the right to refuse to renew the contract if the accommodation is needed to house him and his family, as well as if the lessee has systematically failed to perform his contractual obligations.

The constitutional right to housing allows the unilateral annulment of the contract at the initiative of the lessee and consequent re-settling of the lessee and those living jointly with him. One group of grounds for termination of a contract is linked with the fault of the users in not observing its conditions. In this instance, they re-settle without being provided with another living accommodation. This measure fulfills the role of sanctions against those users who do not observe the constitutional obligation to treat housing with care (Article 44), who do not act properly as recipients and users of a living accommodation, *i.e.* who violate the requirements of housing legislation. A citizen has a clear right to the use of housing if he adheres strictly to all the obligations prescribed by housing law. And, conversely, he may be deprived of the right to the use of living accommodation through gross violation of his obligation as a user.

A lessor has the right to cancel a contract and cause the residents to move only on grounds established by law and, as a rule, by judicial procedure. An exception to judicial procedure exists when a lessee moves according to administrative procedure with the sanction of the procurator when his house is in danger of collapse (Article 36 Principles of Housing Legislation).

In regard to houses belonging to citizens by right of personal ownership, a lessor has the right to request premature annulment of the contract and the re-settling of the lessee and his family without providing them with another living accommodation in cases of systematic non-payment of rent by the lessee.

Lessees and other users of living accommodations in houses in the state, social, and individual housing funds who systematically destroy or damage them, use them counter to their purpose, or violate the rules of socialist communal life making it impossible for others to live with them in one apartment or house, and where measures of warning and social action prove to be ineffective, may be evicted without being supplied with another living accommodation for being unable to live communally, or may be obliged to exchange their accommodation for another accommodation suggested by a party interested in making the exchange.

A specific example of the prohibition of joint residence under law (Article 38 Principles of Housing Legislation) is when the court forbids persons who have been deprived of parental rights to live jointly with their children.

A contract is abrogated as invalid and the users evicted without being supplied with another living accommodation if the court annuls the order on grounds based on the fault of the persons who received it (see section 4). However, if the space which they had occupied earlier in houses within the state and social housing fund are still vacant, they may move into that space. Citizens who receive an invalid order through no fault of their own are housed according to judicial proceedings by being provided with another living accommodation.

Another set of grounds for the unilateral demand to cancel a contract at the initiative of the lessor stems from the state of social needs or the housing needs of the citizen-owner of the house. These grounds, which are listed by law, are not connected with the behavior of the lessee or those living jointly with him. Those who are evicted because of state or social requirements are provided with another well-built accommodation in houses within the state and social housing fund; if the eviction is for the purpose of satisfying the housing needs of the owner of the house, no other living accommodation is provided.

Grounds for cancelling a contract in houses of the state and social housing fund for state or social requirements are: demolition of the house, danger of the collapse of the house (living accommodation), or restructuring a residential house (living accommodation) into a non-residential one. A reason for the cancellation of a contract in houses within the individual housing fund because of state or social needs is the decision by the executive committee to demolish a residential house which is under personal ownership in order to requisition of the plot of land. These grounds for cancelling a contract are used sparingly, since the basic task is to safeguard and increase the housing fund as a whole.

The accommodation which is provided for the evicted party must be concretely named in the court decision on the cancellation of a contract and the eviction of the lessee or in the relevant decree of the procurator.

Besides these general grounds for the cancellation of a contract, special instances are established for the institutional fund in regard to eviction from the houses belonging to the enterprises and institutions of the most important branches of the national economy, a listing of which is established by the Council of Ministers of the USSR and the Councils of Ministers of the union republics. Workers and employees may be evicted from houses of these enterprises and institutions by being provided with another living accommodation, which conforms with sanitary and technical requirements, when they terminate work relationships at their own desire without valid reasons of when they violate labor discipline or commit crimes.

Officers, ensigns and warrant officers, servicemen on reenlistment in the Armed Forces of the USSR, and persons of equal rank who were discharged from actual military service to retire or to go into the reserves, as well as persons living with them, may be evicted from the living accommodations occupied by them to military settlements with the provision of another well-built living accommodation.

Collective-farm workers who receive living accommodation in houses belonging to the collective farm can be evicted upon being provided another living accommodation which meets sanitary and technical requirements if they are excluded from membership in the collective farm or leave the collective farm at their own volition (Article 37 Principles of Housing Legislation).

In the individual housing fund, the contract of lease may be prematurely cancelled with the lessee and he and those living with him may be evicted without being provided with another living accommodation if the court establishes that the accommodation is needed for the personal use of the owner of the house and the persons living jointly with him. Such an eviction is also permitted by law when the dwelling occupied by the lessee and his family is destroyed as a result of the re-furnishing or re-building of the house by decision of the executive committee of the local Soviet, and accommodations of equal value are not available for him.

8. *The contract of lease of employment living accommodation. Use of accommodations in hostels.* The housing rights of citizens who use employment living accommodations and space in hostels have important legal particularities. Employment living accommodations and hostels are accommodations and buildings for a special purpose, The users of these accommodations do not have the guarantee of permanent use and legal authority which is given for permanent use of living accommodation.

Citizens who because of the nature of their labor relationships must live at the location of their job or close to it (a list of such categories of workers is established by the legislation of the USSR and union republics) are provided with employment living accommodations on the industrial territory or close to it by decision of the administration of the enterprise, organization, or institution where they work, and are transferred on the basis of special orders issued by the executive committee of the local Soviet. Such space is given to state-farm and collective-farm workers, janitors, mailmen, etc. As a rule, such an accommodation is divided into separate apartments. The decision on the apportionment of the accommodations among the number of workers is made by the executive committee.

In houses belonging to collective farms, the ratio of living accommodations to number of employees and the establishment of the list of workers for whom such apartments are provided is carried out by decision of the general assembly of the members of the collective farm (assembly of representatives) and confirmed by the executive committee. Service apartments are also provided for certain categories of military personnel (ensigns, warrant officers, for the first five years of service).

A housing-management organization concludes a contract of lease of living accommodation for a period covering the entire work time of the lessee. The right to use the employment accommodation terminates at the same time as the employment contract, and when he terminates the relationships, a worker (member of a collective farm, soldier) together with all the persons living with him, is subject to eviction without being provided with another living accommodation after being notified by the lessor. The eviction is carried out by judicial proceedings. However, a large group of persons who receive service accommodations are granted the right by law to be provided with another living accommodation by the lessor (Article 40 Principles of Housing Legislation) which meet sanitary and technical requirements).

The range of powers of a lessee of an employment living accommodation is significantly narrower than that of a lessee in houses within the state and social housing fund. It is impossible to reserve or sublease employment living accommodations, to exchange them for other living accommodations, etc.

A single or married worker may be provided with space in a hostel for a period of work or study by decision of the administration, trade union committee, or committee of the Young Communist Organization. Only specially constructed or re-equipped dwelling houses are provided for the purpose of hostels. Separate apartments in residential houses may not be allocated to be used as hostels. Ministries and institutions have confirmed the Statutes on Hostels which establish the procedure and condition of the use of living accommodations in them.

Seasonal and temporary workers and persons working on the basis of an employment contract for a fixed period (for instance, those working in regions of the Far East) who terminate their jobs, as well as persons studying at educational institutions who leave, are subject to eviction according to judicial procedure without being provided with another living accommodation within the hostel, since they were provided with one (only) in connection with their work or study. Workers who are released at their own desire without having valid reasons, or for violating labor discipline or committing crimes are subject to eviction without being provided living accommodation. However, persons who are released for other reasons (as well as a wide range of persons who have worked in an organization for not less than ten years; single persons who live jointly with their minor children, etc.) may be evicted only where they are provided with another living accommodation which meets sanitary and technical requirements (Article 41 Principles of Housing Legislation).

Chapter 27. Use of Accommodations of a Construction Cooperative

1. *Housing-construction, dacha construction, and garage construction cooperatives.* There are many housing-construction cooperatives (*ZhSK*) in the USSR. The satisfaction of the housing requirements of citizens through their participation in a housing-construction cooperative is one of the most important legal forms for the realization of the constitutional right of Soviet citizens to housing. Cooperative dwellings constitute an independent housing fund in the USSR (see Chapter 26).

There are also cooperatives to meet the housing and professional requirements of citizens, which are linked with the construction and utilization of non-residential structures and buildings (for instance, a cooperative for the construction of work-shops for artists).[1] For summer vacations, dacha cooperatives (*DSK*) and cooperatives for the construction and management of multi-story pensions for country retreats have been formed.

The satisfaction of the needs of individual owners of means of transportation is met by cooperatives for the construction and operation of collective garages (*GSK*) for cars and motorcycles; individual owners of boats (motorboats) form cooperatives for the construction and operation of mooring-areas for boats and motorboats, individual owners of dwellings form cooperatives to bring natural gas into their houses.

Any type of cooperative which is established with the permission of a district (city) executive committee consisting of the requisite minimum number of citizens is a socialist organization which secures the realization of a defined material interest uniting a group of persons: housing, vacation, parking and servicing of individual means of transportation and (other things), attaining a material result in economic activity.

Cooperatives are voluntary associations of citizens in urban or rural population centers which are formed on the basis of

1. *Collection of Decrees of the RSFSR* 1980 No. 3 item 19; 1981 No. 19 item 128.

place of residence under housing-communal organs of the executive committee of People's Deputies (*ZhSK, DSK, GSK*) or on the basis of place of employment under associations, enterprises, and organizations (*ZhSK*), which have a permanent judicial connection with the cooperatives. The number of members in a cooperative may not exceed the number of cooperative accommodations. The procedure for the registration of citizens who desire to join a housing construction cooperative, as well as the conditions for accepting citizens into a (*ZhSK*) are established by the Principles of Housing Legislation adopted by the Council of Ministers of the USSR on 19 August 1982, "On the House-Construction Cooperative", as well as by republic housing codes and other legislation of the union republics. The acceptance of citizens who desire to join a *ZhSK* for registration is carried out by decision of the executive committee at the place of residence of these citizens or if at the place of employment – by the joint decision of the administration and trade union committee.

The conditions for joining other types of cooperatives are regulated by republican legislation in some union republics, and by decision of the local Soviets of People's Deputies in others.

Citizens accepted into cooperatives must be eighteen years of age, in need of the relevant living-condition requirement, and meet the established criterion of need. For example, a person who possesses a house (part of a house) by right of personal ownership which is suitable for permanent residence and which the owner may use to fill his needs does not have the right to become a member of a *ZhSK* or a *DSK*. Associations, enterprises, and organizations are allowed to use the assets of the incentive funds for gratuitous material assistance and to partially liquidate the bank credit to build cooperative apartments for workers with a specific length of service at the particular enterprise, association, or organization. For example, for newly-married couples, two years of service is sufficient to make them eligible for this privilege.

Housing-construction cooperatives are granted long-term credit (for 25 years) for the construction of residential houses in the range of 70% (80%) of the construction costs of the house. *Dacha* and garage construction cooperatives do not receive credit to carry out construction.

A cooperative has the rights of a juridical person from the time of the registration of its charter, which is adopted by the general assembly of the founders, with the executive committee of the local Soviet of People's Deputies. The buildings and structures built by the cooperative belong to it by right of cooperative ownership and are used by it on the basis of self-repayment. The structures of a cooperative may not be removed, sold or given away either in their entirety or in part (as accommodations) by juridical persons or citizens, with the exception of the transfer which occurs as the result of its liquidation.

The cooperative, through the general assembly of members of the cooperative as the highest organ of cooperative management, chooses the administration for carrying out current matters, carries out the admission of members into the cooperative and expulsion from membership of the cooperative, distributes accommodations among the members of the cooperative and also allows the transfer of the share of a member of the cooperative to other members of his family, establishes the amount of the entrance, share, and other dues of the members of the cooperative which constitute its assets, and has the right to claim prompt payment of the dues as well as adherence to other obligations provided in the charter.

The executive committee of the local Soviet of People's Deputies supervises the activity of the cooperative, the state of its operation, and the carrying out of repairs on the buildings belonging to it. The district (city) executive committee has the right to revoke a decision of the general assembly of the members or the administration of the cooperative if it contradicts existing legislation.

The securing of living accommodations in the houses of a *ZhSK*, the basic rules of the organization, and the activity of a *ZhSK* are regulated by the Principles of Housing Legislation and the decree of the Council of Ministers of the USSR of 19 August 1982, "On the Housing Construction Cooperative".

The legislation of the union republics establishes the procedure for the organization and activity of cooperatives and the rights and obligations of its members. These relationships are regulated in detail by the housing codes of the republics and the model charters for house, *dacha*, and garage construction cooperative. The Model Charter for the *ZhSK* was confirmed by decree of the Council of Ministers of the RSFSR on 24 September 1958; the Model Charter for the construction and operation of collective garage-parking for automobiles of individual owners – by decree of the Council of Ministers of the RSFSR on 24 September 1960. This act was made applicable to cooperatives for the construction and operation of collective garage-parking for motorcycles and motorscooters by the decree of the Council of Ministers of the RSFSR on 12 July 1971.[2]

2. *Obligations arising from membership in a construction cooperative*. In a cooperative, the obligations to provide a citizen the use of an accommodation arises from his memership in the cooperative and the payment by him of share fees, and in this they are different from the obligations arising from the contract of the lease of living accommodations or of property, which are agreements between a lessor and lessee.

A share is a monetary sum which must be paid by a member of the cooperative to carry out cooperative construction and is equal to the construction cost of the accommodations which is provided for use. A member of a housing-construction cooperative has to pay a share of between 30% and 20% of the cost of the apartment before the beginning of construction, depending on the region and category of citizen; a member of a *dacha* or garage construction cooperative has to pay the full amount.

A member of the cooperative who is accepted into the general assembly *of shareholders of the cooperatives* and who pays his share has the right to demand that the cooperative

2. *Collection of Decrees of the RSFSR* 1965 No.23 item 144; 1958 No.13 item 154; 1960 No. 33 item 160; 1971 No. 13 item 100.

provide for his use accommodations determined by the general assembly which are under the ownership of the cooperative.

Acceptance into the membership of a cooperative and the distribution of accommodations among the members of a cooperative is under the competence of the general assembly as the highest administrative organ of the cooperative. The decision of these questions is under the supervision of the local Soviet. For example, in accordance with the Principles of Housing Legislation the decision of the general members of the *ZhSK* on accepting a citizen into the membership of the *ZhSK* is subject to confirmation by the district (city) executive committee, and the settling of apartments in a house of the *ZhSK* is carried out on the basis of orders of the district (city) executive committee (Article 43). Citizens who are included on the list confirmed by the executive committee of the local Soviet as entering into an organized *ZhSK* are considered to be members from the time of the registration of the charter of the *ZhSK*. Citizens who are accepted into an existing *ZhSK* are considered to be members from the moment of the confirmation of the decision of the general assembly by the executive committee. The revoking of the decision of the general assembly of members of the *ZhSK* by the executive committee of a local Soviet in the process of supervision violates or calls into question the membership of the person into a cooperative, and the refusal to be given an order to use the cooperative accommodations violates or calls into question his right to use the accommodation which arose on the basis of his membership in the cooperative and the payment of share fees.

Therefore, in such instances a citizen has the right to turn to the courts for protection of his violated or disputed right – to remain a member of a cooperative or to secure his right to use the cooperative accommodation. The executive committee of a local Soviet acts as a third party in a judicial investigation of civil law disputes between a cooperative and a shareholder. In deciding such disputes, the judges consider the reasons for the decision taken by the executive committee of the local Soviet. For example, a court takes into consideration the right of the executive commitee of the local Soviet to revoke the decision of the general assembly of shareholders into whose membership the citizen was accepted if he already has a *dacha* in his personal possession.

3. *Rights and obligations of a member of a cooperative. Rights of members of the shareholder's family.* A member of a cooperative has the *right*:

a) to receive a separate cooperative accommodation for an indefinite period in accordance with the amount of his share;

b) to use jointly with his family the accommodation provided for the entire period of the existence of the cooperative and to retain the right to it regardless of the reasons for and length of absence from the place where the building was located;

c) to leave the cooperative voluntarily of his own volition at any time before the beginning and during the process of the construction and operation of the cooperative structure;

d) to transfer his share to an adult member of his family with the agreement of the general assembly of the membership of the cooperative by adhering to specific conditions (for example, in a garage construction cooperative, the member of the family has to have his own registered automobile);

e) to sublease a part or the entire accommodation in his use for payment with the agreement of the administration of the cooperative (this right of a member of a *GSK* is not provided in the Model Code of the *GSK* of most of the union republics);

f) to receive his share accumulation upon leaving the cooperative in the established amount and procedure.

The range of rights of members of the *ZhSK, DSK*, and *GSK* have some differences because of the different purposes of the accommodation. For example, in most of the union republics only a member of a *ZhSK* has the right, with the permission of the executive committee of a local Soviet, to exchange the cooperative accommodation which he uses for a living accommodation in the use of another shareholder or apartment renter upon condition that a person who exchanges his living areas is accepted into the membership of the cooperative in accordance with the rules established by the Model Charter of the *ZhSK*.

A member of a cooperative has a so-called perferential right as against a new member of the cooperative to receive an available apartment in the cooperative building through the procedure for an improvement in living conditions or to be given for use an available room in the same apartment in which he is living.

In this instance, a member of a cooperative has the right to bring a suit in court to protect his preferential right as against a new member of the cooperative.

Enterprises, institutions, and organizations under whom *ZhSK*'s are organized may recommend workers who have a need for improved living conditions for membership in a cooperative and available living accommodations only if no members of a *ZhSK* are in need of improved living conditions.

The members of the family of a shareholder and persons who have lost family ties do not have an equal range of rights as the members of the cooperative. However, the termination of family relationships with a member of a cooperative does not give rise to the right to evict the persons who are living with him. The members of the shareholder's family who are living with him have the right to use an accommodation in a cooperative building during the time that they retain their membership in the cooperative. The spouse of a shareholder may have the right to a share of the share accumulation if this property was acquired by the spouses during the marriage, but this in and of itself does not grant to the spouse of the shareholder membership in the cooperative. A family is represented in a cooperative by one member of the cooperative. Since membership in a cooperative is connected with a specific person, the rights of a member of a cooperative may not be transferred by him to another member of the family unless the shareholder leaves the cooperative.

Besides rights, the members of a cooperative also have *obligations*: to pay share fees on time, to bear a portion of the expenses of the management and operation of the structure as determined by the general assembly of shareholders, to partake in other payments constituting the assets of the cooperative, to carry out decisions adopted by the general assembly, to treat the used accommodation and cooperative structure with care, and to perform any other obligations provided by the charter.

A member of a cooperative may not simultaneously be a member of another cooperative of the same type except in exceptional instances which are provided for a member of a *ZhSK* for a determined time period.

A share paid from means which are the joint property of spouses, may, as the means of the cooperative, be apportioned upon their divorce only in the *ZhSK* and *DSK*, if each one of the spouses is allocated a separate housing (*dacha*) accommodation. The spouse who is entitled to a part of the share has the right to use the accommodation allocated to him in the instance that he becomes a member of the cooperative, since he has right of preference.

4. *Responsibility of the parties for obligations arising from membership in a construction cooperative.* Members of a housing construction cooperative which receives credits from the state for the construction of dwellings have an obligation to the cooperative to pay off the loan promptly. Payments are made from the current account of the cooperative, which applies the entrance fees of the members of a *ZhSK* toward the liquidation of the regular payments on the debt. The members of a cooperative pay 0.5% annual interest for the use of the credit; the bank adds 3% interest annually for late payment of debt.

A cooperative has the right to recover debts from its members through nondisputable proceedings on the basis of an executive warrant by a state notarial office upon showing the actual obligation of the debtor and documentary proof of late payment of his debt.

Such a procedure is also used to recover the debt of members of a cooperative who are more than two months behind in their payments in order to cover the expenses for the maintenance and operation of the house (*dacha*).[3]

A member of a cooperative (*ZhSK, DSK*) may be expelled from a cooperative and at the same time be evicted by judicial proceedings together with all the persons living with him without being provided with another living space (vacation accommodation) on grounds provided by the Charter: in particular, such grounds consist of the following: non-adherence to the Charter, non-performance of the obligations established by decision of the cooperative, systematic violation and destruction of the accommodation, and in a *ZhSK* and *DSK* also the violation of the rules of socialist community life, making it impossible for others to live jointly with him in one apartment (house, *dacha*), after measures of warning and social influence prove ineffective. In the latter case, the living (*dacha*) accommodation remains in the use of the members of the family if one of them becomes a member of the cooperative. A member of a *GSK* who is expelled from the cooperative on grounds listed in the Charter is deprived of the right to use the garage (parking space).

It is possible to evict members of the family of a shareholder in a cooperative accommodation in instances provided for in the Model Charters of the *ZhSK* and *DSK*, as well as on the grounds of the systematic destruction and damaging of the accommodation or the violation of the rules of socialist community life. Minor descendants of a deceased member of a cooperative who had lived jointly with him in a cooperative living accommodation or *dacha* may not be evicted from the occupied accommodation for reasons of the absence of the deceased from membership in the cooperative by cause of death.

The termination of obligations arising from membership in cooperative living accommodation also occurs when the activity of the cooperative terminates. When a building belonging to a housing construction is demolished as a result of the requisition of a plot of land for state and social purposes, the cooperative is given a building equal in value to the demolished

3. *Collection of Decrees of the RSFSR* 1976 No.7 item 56.

building, and the obligations based on membership are therefore not terminated.

5. *Obligations arising from membership in a garage construction cooperative*. The difference in the obligation of members of a *GSK* from the obligations arising from membership in a housing/*dacha* construction cooperative are based on the purpose of the garage construction cooperative.

Citizens living in a given area who are owners of individual automobiles, motorcycles, and motor scooters which are registered according to established procedure may be members of a garage construction cooperative. Like members of other cooperatives, a members of a *GSK* is obliged to contribute monetary assets in the amount of a full share before the start of construction.

The particular features of the legal regime of a *GSK* include: the right of a shareholder to transfer his share by established procedure only to that member of the family who has his own car (motorcycle, motor scooter); the preferential right of heirs who are members of the family of a deceased member to continued use of the garage facility if one of the members of the family who is the owner of an automobile (motorcycle, motor scooter) becomes a member of the cooperative. Division of a share in a *GSK* is not possible. A member of this type of cooperative does not have a preferential right to better parking conditions for his means of transportation when another garage accommodation or parking space in the cooperative falls vacant.

It is possible to expell from membership in a *GSK* persons who have sold or given away their automobiles if they use the accommodations (parking spaces) which have been provided in order to gain earnings not based on work, or if they move to another permanent place of residence at another population point.

Chapter 28. Work Contract

1. *Concept and significance of the work contract.* This is a contract for the performance of various types of work. It is widely used in relationships involving the participation of citizens, above all in the area of the everyday needs and services of the population (sewing of a single item or repair of sewn articles, footwear, the building of customized furniture, making jewelry, repair and construction of dwellings on the order of citizens, etc.). The work contract also is frequently used in the national economy (repair of buildings, structures, equipment, means of transportation; creating customized products, repair jobs which are not connected with capital construction). There are organizations for which the obligations of the contractor are the basic type of economic activity (prerelease testing trusts, repair-construction organizations, etc.).

The conditions of the work contract are defined in Articles 350-357 of the Civil Code. The rules regarding specific types of work contracts between organizations, as well as the rules concerning work contracts for the securing of the everyday needs of citizens (household orders), are established by the legislation of the USSR and of the union republics.

Thus, the Councils of Ministers of the union republics confirm Model Contracts for Household Orders (for example, model work contracts for the repair or construction of a dwelling);[1] the republic Ministries for Services of Everyday Needs devise and confirm the conditions for the acceptance and delivery of citizens' orders (for example, the Uniform Rules for Services for Everyday Needs were confirmed in the RSFSR), and the Ministries for Housing and the Communal Economy have confirmed a model (sample) contract for the provision of technical services by state housing operation organizations for cooperative dwellings.

In a work contract, one party (the contractor), binds himself to perform at his own risk defined work on the instruc-

1. *Collection of Decrees of the RSFSR* 1965 No. 1 item 2; 1982 No. 2 item 4; No. 11 item 71.

tions of the other party (the customer), with either the customer's or his own material, and the customer binds himself to accept and pay for the finished work (Art. 350 Civil Code). In this manner, the work contract is a contract for the performance of work, the result of which is a defined material thing, during the process of which the material of one of the parties is used.

The work contract differs from the labor contract, which it resembles. In a work contract, the subject is the performance of defined work and the material result of this activity; in a labor contract, it is the performance of the work itself, the process of the work activity (Art. 15 Labor Code). A contractor performs work independently. A worker in a labor contract is included in the collective of an organization and is obliged to observe the labor routine established for it.

As was already mentioned, a contractor performs jobs at his own risk. The *risk* of a contractor lies in the unfavorable consequences for him, as provided by law (he does not receive payment for his work) because of possible accidental circumstances which may occur during the performance of the work which are not the fault of the parties.

A contractor incurs the consequences of the accidental destruction of the subject of a contract before it is delivered to the customer or the accidental impossibility of the performance of the obligation (Art. 363 Civil Code). If the contractor supplies his own materials, he bears the risk of their accidental loss or damage (Art. 357 Civil Code). Finally, a contractor must also take upon himself the risk of the possible increase of his expenses above the agreed firm price or estimate (Art. 352 Civil Code).

When the thing which is produced from the materials of the contractor is transferred to the ownership (operative management) of the customer, the work contract is similar to the contract of purchase and sale or the contract of supply. The work contract differs from the contract of purchase and sale in that its purpose is to produce a thing and therefore the moment of the conclusion of the contract does not coincide with its performance; also, the subject of the work contract has individually-defined features. In addition to this, the work contract differs from the contract of supply in its participants: in a work contract, the parties may be either citizens or organizations, while only organizations may be parties to a contract of supply.

2. *Rights and obligations of the parties. Their responsibility.* A contractor usually performs the work stipulated by the contract with his own materials and resources. However, there are exceptions to this general rule.

Socialist organizations have the right to transfer their materials and equipment to other organizations for the manufacture of products. In this case, the norms for the use of the materials, dates for the return of the remnants and substantial scraps, as well as the liability of the contractor for non-performance or improper performance of his obligations must be provided for in the contract (Art. 355 Civil Code).

In performing a work contract, socialist organizations may charge other organizations ñ subcontractors ñ with the performance of specialized jobs, while remaining liable for the performance of the contract as a whole. Citizens are not allowed to engage sub-contractors and must perform contracted jobs through their own labor and the labor of members of the family if those jobs are not connected with prohibited trades.

The period of the contract (including intervening periods) is determined by agreement of the parties, and if a plan task or established norms for the completion of work form the basis of the contract, the period is in accordance with the periods which are indicated by the documents mentioned above.

The work performed by a contractor is paid for at established rates; in the mutual relationships between citizens, the payment is based on agreement of the parties. A firm or approximate estimate may be established for the performance of a particularly difficult job if it is not possible to determine all future expenses beforehand. Payment for the work of a contractor must be made when the work is finished, but in contracts in which citizens participate, payment may be made beforehand, and an advance may also be given.

A contractor has the following *obligations*:

a) to perform the work in strict conformity with the order of the customer and within the established period;

b) to safeguard the property entrusted to him by the customer: materials, remnants, etc. (Art. 356 Civil Code);

c) to notify the customer promptly of the unsuitability or defective quality of the customer's materials, of the fact that following the customer's instructions may lead to the complete unfitness of the thing or to its being defective, or of the presence of other circumstances, having nothing to do with the contractor, which threaten the fitness or strength of the work performed (Art. 358 Civil Code),

d) to give the finished work to the customer and inform him of the use of the customer's materials and to return the remnants to him (Art. 354 Civil Code).

The customer has the right to supervise the performance of the work during the time it is being performed by the contractor (Art. 360 Civil Code).

The customer is *obliged* to accept a performed work of suitable quality and to pay for it. If deviations from the terms of the contract which lessen the quality of the work, or other defects become apparent during the time of acceptance, the customer is entitled to claim their gratuitous removal or a corresponding decrease in the amount paid for the work.

The defects in a performed work may be obvious or hidden. *Obvious* defects may be discovered in the usual manner of acceptance. *Hidden* defects are impossible to detect by examining the performed work at the time of acceptance. For example, a workshop sewed winter boots in the summer the defects of which were revealed only when they were worn in winter. The customer must inform the contractor of obvious defects immediately, and of hidden ones as soon as they are discovered (Art. 361 Civil Code).

Reduced periods of the limitation of action for a suit on the grounds of a contractor's liability has been established at six months, and for hidden defects – within one year of the discovery after the acceptance of the work (for buildings and structures – within three years if one of the parties is a citizen). If notification regarding defects is made within a guarantee period which may be provided for in the work contract, the limitation period begins on the day of the notification, and in relationships between socialist organizations – from the day that the defects in the work are discovered (Art. 365 Civil Code).

A work contract may be terminated unilaterally by the notification of one of the parties, if such possibility is provided for by law (Arts. 359, 360 Civil Code).

3. *Contract of household orders.* A work contract for the services of everyday needs of citizens (the contract of household orders) has several special features. In these cases, the customer is always a citizen, and the contractor is always an enterprise (studio, workshop, etc.) for everyday needs. As a rule, performance of the contract must be formalized by a document of established form.

Enterprises for everday needs perform orders in accordance with existing standards and technical conditions for everyday services and within the established period, taking into account the desires of the customer. The performance of the order may have to be urgent, in which case supplementary payment is requested. The value of the materials given to the contractor by the customer is determined by agreement of the parties, based on their value but not exceeding the retail price.

The material may also be supplied by the contractor and paid for by the customer at the retail price. Material provided according to a list of goods may be supplied to a customer who permanently resides in a specified population center on credit, e.g. by installment payments (see Chapter 22). The performed work, depending on the subject of the order, is paid for by the customer at the confirmed price either after the performance of the order or when the contract is concluded, in the form of an advance or in full.

Guarantee periods for the quality of the performed work are determined by the standards and technical condition. For seasonal items (summer, winter things), the guarantee periods are calculated from the beginning of the relevant season whose time frame is established according to local conditions. If major deviations from the contract are discovered in the item which was produced from the material of the customer, the latter may, at his discretion, request either that another item be produced from the same type of material of the same quality, or the rescission of the contract and compensation for losses suffered.

Chapter 29. Work Contract for Capital Construction

1. *Meaning and significance of the contract.* Capital construction is one of the most important branches of material production. More than 100 billion rubles are expended annually on capital construction in the USSR. This makes it possible to start the construction of hundreds of new enterprises, thousands of various types of buildings and structures, and to build more than two million new apartments every year.

One of the two methods is used to organize capital construction – the economic or the work contract method. In the first case, a production enterprise or association carries out construction-installation and other related jobs through its own resources. In the second case, construction is carried out through the labor and resources of specialized contracting organizations, and the results of their work are accepted and paid for by a customer. As a rule, construction is carried out more quickly and is cheaper under the contractual method. Therefore, it has long been the favored method in our country. At the present time, 92% of the general output of construction-installation work is carried out under this method.

Construction-installation organizations perform essential jobs based on the work contract for capital construction which they conclude with customers. This contract is of the same type as the usual work contract, but has a number of important distinctive features. Because of this, the work contract for capital construction has for many years been regulated as a special type of civil law contract by legislation.

The Civil Code devotes a separate chapter (Arts. 368-372 Civil Code) to the work contract for capital construction. The USSR Government issued two important acts in regard to the development of these norms: Rules on the Work Contract for Capital Construction, and the Rules on Financing Construction. The first define the procedure for concluding and performing work contracts, and the second defines the mutual relationships of the parties in a work contract with the bank which finances the construction, as well as the procedure of payment for the performed work. In addition to this, the Statute on the Mutual Relationships of Organizations, i.e. between general contractors and subcontractors, was confirmed on 31 July 1970 by an instruction of the administration of the USSR *Gosstroi* jointly with the USSR *Gosplan*.

In some of the union republics, Special Rules on Work Contracts for Capital Construction on Collective Farms have been issued. Such Special Rules were specifically confirmed by the RSFSR *Gosstroi* on 29 January 1972.

Under the work contract for capital construction, the contractor organization binds itself to construct with its own labor and resources and to turn over to the customer organization an object provided for in the plan, in accordance with the approved design and estimate documents, and within a fixed period of time, and the customer is bound to supply the contractor with a building site, to deliver to him the confirmed design and estimate documents, to ensure timely financing of the construction, to accept the completed structures, and to pay for them (Art. 368 Civil Code).

This contract is *bilateral, compensatory*, and *consensual*. Only state, cooperative, and other social organizations may be parties to it. The work contract for capital construction is based on plan acts which are binding for both parties. In contrast to the contract of supply, this contract is always a plan contract. Adherence to this requirement is supervised by a bank in making payments for the work performed.

2. *Plan prerequisites for the contract.* In order that a work contract for capital construction may be concluded, not only a single – as is the case in the contract of supply – but several plan acts are necessary. The structure which is the reason for the conclusion of the contract must be included beforehand in a confirmed *letter of allocation* [*titul'nyi spisok*] as well as in the *work plan* of the contracting organization, and must have the necessary *design and estimate* documentation.

The basic form of state planning consists of stable five-year plans for capital construction. These plans (for five-year periods with tasks allocated by year) are confirmed by the Council of Ministers of the USSR for each ministry and department of the USSR and for each union republic. They define the tasks in exploiting production capacities and basic funds, the amount of capital investment and the construction-installation work for each structure.

The Council of Ministers of the USSR also confirms the five-year plans (with buildings allocated by year) of contracting work. They provide for the amount of work to be performed for each ministry and department acting as contractors, and for the union republics, ministries, and departments acting as customers, as well as define the district where the construction is concentrated. After this, these plans are brought to the actual performers – contracting organizations and customers of organizations.

A *letter of allocation* defines the place, cost, and duration of the construction, as well as other data for each construction. A letter of allocation is an unalterable plan act. It has this generally obligatory significance: it controls not only the parties to the work contract, but also the plan, financial, banking, and supply organizations. The inclusion of structures in a letter of allocation signifies that the necessary juridical basis for the construction has been obtained. Letters of allocation are compiled on the basis of another plan act – an itemization of the construction projects. Depending on the importance and the estimated cost of the structure, the itemization of the construction projects is confirmed either by agreement of the Council of Ministers of the USSR (the most important structures) or the ministries and departments of the USSR and the Councils of Ministers of the union republics with the agreement of the USSR *Gosplan* (if the estimate is three million rubles or more), or directly by the ministries and departments of the USSR and the Council of Ministers of the USSR (if the estimated cost is less). The parties to the work contract must agree on a letter of allocation for interim work. This defines the amount and the type of the construction-installation work which must be performed during a specific year.

The design and estimate documents include the designs, work documentation, and estimate calculations. If the construction is carried out on the basis of a model or recurring design, a single document is worked out – the working design with the total estimate calculations. From the moment that these are confirmed by a ministry (department) of the USSR, the Council of Ministers or a ministry (department) of a union republic, and in the case of the largest and most important enterprises, buildings, and structures – by the Government of the USSR, the designs and estimates acquire the significance of a plan act which is binding on the parties.

3. *Parties and procedures in concluding a contract.* Often several organizations participate in the construction of a thing, each one of which specializes in a distinct aspect of the job (erecting the foundation, installation of electrical or sanitary-technical equipment, etc.). In these cases, the customer concludes a contract with one defined organization only – the general contractor. The general contractor, in turn, enters into contractual relationships in his name with the other specializing organizations – the subcontractors. The general contractor organizes the totality of the work on the thing and is responsible before the customer for the actions of each of the subcontractors. This rule, however, has two exceptions.

First, by joint decision of the ministries (departments), the customer, and the contractor, there may be not one, but two or more general contractors. In this situation, the customer must conclude a work contract with each of the general contractors for the relevant aspect of the job. The system of dual general contractors is used in the construction of large-scale production, housing and cultural/domestic complexes, roads, main electrical lines, and several other things.

Secondly, it is possible to have situations when the customer concludes a contract for the performance of installation or other specialized work (for example, laying down a railroad track or a road, etc.), with one organization directly and bypasses the general contractor. Such a contract (which is called a direct contract) may be concluded by the parties only with the consent of the general contractor.

If it is a matter of new construction which must be completed after the limits of a calendar year, the parties conclude a *general work contract* for the entire construction; and in addition to this, they also conclude supplementary agreements for every year of the construction except the first. Therein, the conditions of the general contractor are made concrete, refined, and detailed. When construction does not extend beyond the limits of a calendar year, a contract for one year – rather than a general contract – is concluded. Like all contracts made by such organizations, a work contract for capital construction must be concluded in the *written form*. The USSR *Gosstroi* and the USSR *Stroibank* jointly confirm the model general and yearly contracts, as well as the supplementary agreements. Special conditions for regulating the mutual relationships of the parties in greater detail are usually attached to them.

The contractor must be the initiator in the conclusion of the contract. However, it is necessary that he receives from the customer beforehand (not later than fifteen days from the moment of the confirmation of the state plan for economic and social development), a letter of allocation for the construction, a letter of allocation for the interim construction, and several other documents without which the draft of the contract may not be worked out. In the course of one month from the moment they are received, the contractor is obliged to send the draft contract which he has drawn up to the customer, who has ten days in which to sign and return the contract. If the customer does not agree with some conditions of the contract, he must draw up a protocol of disagreement which he sends to the contractor along with the signed contract. Now the contractor must settle the disagreements between him and the customer within ten days. If he does not succeed in doing this, the customer must institute a pre-contractual dispute. Pre-contractual disputes involving a letter of allocation for internal construction are decided by the ministry (department) of the customer and contractor. All other pre-contractual disputes are decided by the organs of *arbitrazh*.

4. *Rights and obligations of the parties*. Contractors and customers are subject to a wide range of obligations. The *obligations* of the contractor include:

a) to construct the thing provided for in the plan by his own labor and resources in accordance with the design-estimate documentation. It is possible for the contractor to make slight deviations from the design if they do not change the basic outcome and do not affect the strength of the details of the structure. He is obliged to indicate all such deviations in the working drawings;

b) to participate jointly with the customer in securing the material and technical goods for the structure. As a general rule, the contractor is obliged to secure the building materials (cement, bricks, etc.) and the customer, the equipment;

c) to deliver the thing within the established period. Two special commissions – one a working commission, and the other a state commission – are set up to attain this aim. The working commission carries out the acceptance of the thing for operation. The working commission includes: representatives of the general contracting and subcontracting organizations, which participate in the construction and will participate in the operation of the thing. The state acceptance commission for the acceptance for operation of unique and especially important things (for instance, nuclear power stations, etc.) is appointed by the Government of the USSR, and for other things, depending on the estimated cost of the structure – by the ministries and departments of the USSR, the Councils of Ministers of the union republics, etc;

d) to remove promptly any imperfections or defects which are exposed as the result of the acceptance of the work. Moreover, the state commission has the right to accept things meant for production (plants, factories, mines, etc.), as well as dwellings only after the contractor removes the imperfections exposed by the working commission.

In contrast to the contractor, the customer has some obligations even prior to the conclusion of the contract. The customer is *obliged*:

– to deliver the design and estimate documentation. The customer must deliver the approved design to the contractor even before the conclusion of the contract – before 1 July of the year preceding the plan year, as well as the working drawings for the entire thing, or at least for the work to be done during the plan year;

– within a fifteen-day period after the date of the confirmation of the State Plan for Economic and Social Development for the USSR to deliver to the contractor a letter of allocation for the entire construction, as well as for the construction during the first year (the latter is called an interim construction letter of allocation), drawings of the equipment and material, tests as to the strength of the individual parts of the building and structures subject to reconstruction during the first year of construction.

All other obligations of the customer arise only upon the conclusion of the contract. Among a number of contractual obligations of the customer are the following:

a) to deliver to the contractor an available building site. Persons living on the site of the construction in houses subject to demolition must be re-settled in another accommodation prior to a period indicated in the contract;

b) to secure prompt financing of the construction in the full amount. In order to accomplish this, the customer must deliver to the financing bank (this function is carried out by either the *Stroibank* or the *Gosbank*) the necessary plan documents, as well as the contract concluded by the parties. The bank verifies the documentation with care, and if it considers that it meets the established requirements, makes a decision on whether to give financing;

c) to render various services to the contractor in carrying out the work. The contractor has the right to use the erected buildings and structures before they are delivered at completion without payment for his production needs, to accommodate his workers, etc. The customer must grant the contractor the opportunity of using the services of his shops and other factories. If the contractor does not have a housing fund, the customer is obliged to provide temporary living accommodations for his workers in houses and hostels which belong to him. The customer is obliged to treat the workers of the contractor on an equal basis with his own workers and employees and to provide a club, dining room, medical center, pioneer camp, etc. If the customer has his own leased railroad tracks and rolling stock, he must give the contractor the opportunity of using them for payment within the limit of the established estimate price;

d) to exercise supervisory and technical control over the progress of the construction, paying special attention as to whether the work performed conforms with the designs and estimate by such indicators as amount, cost, and quality. In order to do this, he is granted the right to examine the progress and quality of the equipment. However, the customer may not give any instructions to the workers of the contractor or interfere in any way in his operating-management activity;

e) to accept the finished things of the construction and to pay for them. The decree of the Central Committee of the CPSU and the Council of Ministers of the USSR of 12 July 1979, "On the Improvement of the Planning and Strengthening of the Economic Mechanism for the Increased Effectiveness of the Production and Quality of Work", provided that the customer must settle with the contractor in full for the completed construction and the enterprises delivered for operation, the starting complexes, lines, and things prepared for the output of production and the provision of services at the rate of the estimated cost of the commodity production. At the present time, this system of settling accounts, under which the contracting organization receives money from the customer all at once, is ever more widely used. The customer pays after he definitely accepts the enterprise which was constructed and prepared for the output of production.

In a subcontractor's contract, the general contractor has the role of customer, and the subcontractor, that of contractor. Therefore, the general contractor has the obligations of the customer enumerated above in relation to the subcontractor, and the subcontractor has the obligations of a contractor in relation to the general contractor.

5. *Obligations of the parties.* The Rules on Contracts for Capital Construction, as well as the Statute on the Mutual Relationships of Organizations, provide for liquidated damages, varying in amount and type, which are levied when the parties violate pre-contractual and contractual obligations.

A fine of 50 rubles per day of delay is levied against the customer for delayed delivery of the design of the contract and for delay in signing for the design.

The contractor is liable for delayed completion of the construction of the thing – enterprise, its individual lines, release complex, buildings, and structures. In this case, a fine of .05% of the estimated cost of the construction-installation work for each day of delay, but not more than 500 rubles per day, is levied against him.

It is possible to have instances when the customer discovers defects after the acceptance, during the process of using the thing. The rules provide that the contractor is liable for this only under two conditions:

first, if the defects are discovered within the limits established by the Rules on Guarantee Periods (these provide for one year for construction in general, and two years from the day of acceptance of buildings and structures for use in the case of dwellings; for systems of central heating of dwellings, it is one heating season after the delivery for use, etc.); secondly, if the defects occurred because of the contractor's fault.

The discovered defects must be defined in a statement drawn up by both parties (if the contractor does not respond to the request of the customer to draw up a statement on the defects, the customer draws up a statement unilaterally which has the same force as a bilateral one). The statement must include the periods for removing the defects; their discovery also makes it necessary for the contractor to pay liquidated damages in the amount equal to delayed removal of the defects, which are discovered during the acceptance of the things (liquidated damages of 100 rubles per day). After the expiration of the periods established in the statement, the customer may remove the defects by his own means. In this case, the contractor must compensate him for expenses incurred, and in addition, pay him one-half of the sum as a penalty for violating the obligation of removing the defects.

A customer who has not delivered the entire documentation or is late in delivering the documentation necessary for the conclusion of a contract (letters of allocation, etc.) must pay a penalty of 50 rubles for each day of delay, and a fine of 250 rubles per day for delayed delivery of the design-estimate documentation.

The customer is liable for delayed delivery of installation equipment or for its incomplete state. For a delay of up to ten days in the delivery of equipment, a penalty of 3%, and for a longer delay, one of 8% of the value of the delayed equipment, is levied; the penalty for the delivery of incomplete equipment is 20% of its value.

The customer must also pay liquidated damages if he refrains from accepting the things constructed during the construction (an enterprise, its individual lines, buildings, and structures). The liquidated damages are established as .05% of the estimated cost of the construction-installation work for each day of the delay, but not more than 500 rubles per day.

Besides liquidated damages, the aggrieved party has the right to request compensation for losses not covered by the liquidated damages (penalty, fine) from its unpunctual contractor. However, Article 371 of the Civil Code limits the liability of a party in a work contract for capital construction: he must compensate only for those losses which may be expressed in the expenses of the other party, in the loss of or damage to his property. Thus, as an exception to the general rule, the party which violates a work contract for capital construction is freed from the necessity of compensating for losses which are in the form of income not received by the aggrieved party, income which he would have received had the obligation been performed by the debtor (profits not received).

Chapter 30. Contract of Carriage

1. *Transportation legislation of the USSR.* There are five basic types of transportation in the Soviet Union (railways, motor, air, sea, and inland waterways), which in their entirety form one transportation system. It is based on socialist ownership of the means of transportation, and its activity is planned within the limits of state plans for economic and social development.

Transportation enterprises for all types of transportation carry out transport of goods as well as of passengers. The basic rules in regard to the contract of carriage, which apply to all types of transportation, are established in the Principles (Arts. 72-77) and in the Civil Code (Arts. 373-385).

A more detailed procedure for carrying out transport functions, including the performance of contracts of carriage, is regulated by the transportation statutes and codes which are issued for the various types of transportation. At the present time, the Statute on Railroads of the USSR [*Ustav zheleznykh dorog SSSR*], confirmed by decree of the Council of Ministers of the USSR of 6 April 1964 (hereinafter abbreviated as *UZhD SSSR*), the Statute on Inland Water Transport of the USSR [*Ustav vnutrennego vodnogo transporta SSSR*], confirmed by decree of the Council of Ministers of the USSR of 15 October 1955 (hereinafter abbreviated as *UVVT SSSR*), the USSR Merchant Shipping Code [*Kodeks torgovogo moreplavaniia SSSR*], confirmed by edict of the Presidium of the Supreme Soviet of the USSR of 17 September 1968 (hereinafter abbreviated as *KTM*), are in force. Air transportation is carried out in accordance with the USSR Air Code. The regulation of motor transport is under the authority of the union republics. The RSFSR issued a Statute on Motor Transportation, confirmed by decree of the Council of Ministers of the RSFSR of 8 January 1969 [*Ustav avtomobil'nogo transporta RSFSR*], (hereinafter abbreviated as *UAT RSFSR*).

On the basis and development of these statutes and codes, the transport ministries issue rules of transportation for the relevant type of transport which define in greater detail the conditions and particularities of various types of transport (for example, perishable, liquid, or dangerous goods, goods in containers, etc.). The rules of transport, as well as the rates of payment for transport, are published in collections which are called tariff manuals, or simply, tariffs.

Besides this, a number of decrees on transportation issues were adopted by the Council of Ministers of the USSR, and there are also the normative acts of the transportation ministries, which are issued jointly or with the agreement of other ministries and departments.

The norms of legislation on transportation, particularly those on the transport of goods, are for the most part imperative since transportation activity is massive in nature and must be carried out regularly; therefore, deviations favoring owners of goods must not be allowed. The norms of maritime law are exceptions to this, as they include many dispositive norms. This may be explained as a particularity of transport by sea, as well as by the fact that maritime transport often carries goods of foreign organizations and must observe their special requirements.

2. *Planning the carriage of goods*. The plan is the basis for all types of transport activity in the USSR, and, as a general rule, contracts for the transport of goods are concluded on the basis of plan tasks which are obligatory for the parties. The procedure for planning transport is defined in the Basic Statutes on Yearly and Quarterly Planning for the Carriage of Goods, which were confirmed by decree of the Council of Ministers of the USSR of 4 January 1970. Supplementary directives concerning this issue are contained in the transportation statutes and codes, as well as in the rules on the carriage of goods which exist for each type of transport.

The plans for the carriage of goods are a component of the state plans for the economic and social development of the Soviet Union. They are worked out by taking into account the plans for production, capital construction and material-technical supply, and on the basis of applications by organizations which require transport. Yearly and quarterly plans for transport are confirmed, and superior organs give out monthly plan tasks to shippers and carriers which are binding on them.

Shippers and carriers are obliged to agree upon (make concrete) the procedure for carrying out the monthly plan for transport established for them, taking into account the special features of a specific type of transport. With this aim, a shipper submits 10-day applications for carriage to a carrier of railway, internal waterways, or motor transport. In addition to this, special yearly contracts in regard to the organization of transport are concluded for motor and internal waterways transportation: these are called navigation agreements for inner waterways. For maritime transport, monthly plans arranging for ships are established, while for air transport, contracts for the organization of transport are concluded if the shipper sends goods regularly; in other cases, applications are submitted.

After performing these actions, a shipper has the *obligation* to present the goods, and the carrier – *to make available* means of transportation. The performance of these actions serves as the conclusion of the *contract of the carriage of goods*. The contract is formalized by relevant transportation documents, which specify the goods and the place to which they are to be transported.

If the obligations to present the goods and to deliver the means of transport are not fulfilled, the shipper and carrier pay a penalty in an amount defined for railway transport according to the quantity of goods (non-delivered railway cars and containers), and for other types of transportations – in an amount defined on the basis of the payment due for the transportation (from 20% to 100%). Such a penalty has the character of exceptional liquidated damages, and it is not possible to demand an amount in excess of this for losses incurred. Moreover, when it is not his fault a carrier who does not supply means of transport is obliged in addition to provide such means to transport the non-delivered goods during the course of the following month at the request of the shipper. This involves the principle of *real* performance in relation to the obligation of carriage (Chapter 18).

Goods which are not provided for in the monthly plan or which are presented in excess of the plan are transported on the basis of advance applications by the shipper which are accepted by the carrier. For some types of goods (agricultural products, goods for export), a general transport obligation is established to accept them for carriage. If obligations for the transport of goods in excess of the plan or on the basis of the plan which are accepted by a carrier upon the application of the shipper are not performed by the parties, the above-mentioned liability provisions in the form of a penalty are applied.

3. *Contract for the carriage of goods. Transport documentation. Under a contract for the carriage of goods, a transport organization (carrier) binds itself to deliver goods entrusted to it by a shipper to the point of destination and to deliver them to the person authorized to receive the goods (recipient), and the shipper binds himself to pay the established rate for the transport of the goods (Art. 373 Civil Code).*

In this manner, the contract of carriage is related to *bilateral compensatory contracts*. Only a *socialist organization* may be a carrier, but both socialist organizations and citizens may act as *senders* of goods (shippers). The special features of the contract of carriage is that it involves a third party – the recipient of goods (consignee), who does not participate in the conclusion of the contract but is nevertheless granted the right to request that the carrier deliver the goods to him. Therefore, a contract for the carriage of goods belongs to the group of contracts in favor of third parties (on this contract, see Chapter 17).

Depending on the type of transport, contracts of carriage are divided into those for railway, motor, air, sea and inland waterways transport. The transport of goods abroad is called international (for maritime transport – foreign) carriage.[1]

1. International transport has significant special features and is not examined in this text.

Carriage by sea within the country is called cabotage. A distinction is made between *petit* cabotage (transport between ports located in one sea basin), and *grand* cabotage (transport between ports of several sea basins). Within the limits of this general classification of the means of transport, it is necessary to distinguish among the various types of contracts of carriage reflecting the particularities of the individual mode of transport.

When only one transport organization acts as carrier, the contract is called *contract of carriage for local carriage*. When on the side of the carrier there are several carriers (for example, the transport of goods from Moscow to Sverdlovsk is performed through the resources of three railways – Moscow, Gor'kii, and Sverdlovsk), and when one contract of carriage is concluded and one transport document delivered, this is known as *direct* carriage (carriage by direct service). The transport of goods by direct service may also be performed by carriers of various types of transport, in which case it is called *direct mixed service*.

Depending on the quantity of the goods transported by railway, *small* (up to ten tons) and *railway car* shipments are distinguished; bulk goods (coal, oil, lumber, etc.) may be transported by *unit* trains. Small shipments (up to two tons) are also accepted for motor transport. When an entire ship, a part of it, or specified areas are used for maritime transport, a *contract of charter* is concluded, while a *bill of lading* [*konosament*] is the basis for the conclusion of a contract of carriage by sea.

In addition to these, for carriage by sea or inland waterways, where the technical organization of the transport operation has a number of special features, two contracts – which have a special purpose but may also be used for the transport of goods – are also utilized. These are the contracts of *towage* and of *time-charter*.

Under a contract of towage, the owner of one ship binds himself to tow, for compensation, another ship or other type of floating vessel to a specified mooring either within a definite time or to execute maneuvers. The towing of lumber on rafts is subject to the rules on the contract of carriage, and special rules are established for other types of towing since the problems of towing are different and require special regulation.

Under a contract of time-charter, which is used in sea transport and which is also called contract of affreightment for a period of time, the owner of a ship binds himself to present a ship for compensation (leasing payment) to the other party – the charterer (lessor) – for a defined period for the transport of goods, passengers, or for other purposes (Art. 178 *KTM SSSR*). Thus, the contract of time-charter may be used either for the transport of goods or for other purposes, and to be precise, it must be considered a variant of the contract of property hire (lease).

All types of contracts for the carriage of goods belong to the category of transactions for which the written form is prescribed. The *goods order* [*gruzovaia nakladnaia*] is such a form, and for sea transport – the *bill of lading* [*konosament*]. The goods order and the bill of lading are filled out according to the established procedure and contain the basic data in regard to the freight (name, number of pieces, weight), as well as information on the carrier, shipper, and recipient of the goods. To confirm his acceptance of the goods, the carrier gives a goods receipt, or a copy of the bill of lading or goods order, to the shipper.

Transport documents must be formalized attentively and in detail, since the information contained in them defines the rights and obligations of the participants in a contract of carriage. The data contained in a transport document may be disputed if it contains false information; however, in order to do so, it is necessary, as a rule, to present convincing proof in writing.

4. *Rights and obligations of the participants in a contract for the carriage of goods*. The parties in carriage – the carrier, the shipper and the recipient of the goods – are accorded a wide range of mutual rights and obligations, and for individual types of transportation, such rights and obligations have

significant particularities reflecting the nature of the activity of a specific type of transport. A carrier has the following basic *obligations*: to supply for transport technically sound means of transport (railway cars, containers, ships); to secure the safety of the goods while they are in transit; and to deliver the goods at the point of destination within the established periods. A special feature of railway transit is that when the shipper himself loads the goods onto the railroad cars, as well as for transport in containers or cylinders, the shipper makes the decision as to the suitability of the means of transport for the goods (Art. 46 *UZhD SSSR*).

The periods for the delivery of goods are determined by the rules of transport, depending on the distance of transport; the carrier is also under obligation to deliver the goods by the shortest route. In railway transport, depending on the type of goods and the nature of the transport (small, railway car, or *unit*) of the goods, freight or high speed is used. High speed is used for goods which are perishable, for liquids, and for several other goods; the periods of delivery are shorter than for freight speeds.

The basic *obligations* of the shipper are: to present the goods for transport in suitable packing which ensures their complete safety in transit and to pay the transportation fee for the delivery of the goods. Comments regarding unsuitable packing are not allowed in the transport document, and in such a situation the carrier is obliged to refuse to accept the goods for transport. As a general rule, the transportation fees must be paid before the commencement of the transport.

Various procedures were established for the performance of the operation of loading and unloading goods for the different types of transport. For railway transport, such operations are performed at the location of general use of railways (e.g. at railway stations), and in other situations (at the entrance way of the clientele), as well as in the loading and unloading of dangerous, perishable, heavy, extremely large, bulky, or liquid goods – by the shipper and recipient of the goods. In motor transport, as a general rule, the loading and unloading is carried out by the shipper and recipient, and in air transport, by the carrier alone. In general, in maritime transport, such operations are carried out by sea ports.

For all types of transport, defined periods are established for performing loading and unloading operations which are meant to stimulate their fastest performance. A penalty must be paid for the violation of such periods in an amount which depends on the length of the delay and which is calculated on a progressively increasing scale. Such a penalty is levied by non-contentious procedure (see Chapter 32) and in maritime transport is called *demurrage*.

The weight of the transported goods is also determined differently for different types of transport. For railway transport, the rule is that the party shipping the goods determines the weight; for motor transport, the shipper and carrier jointly determine the weight; for air transport, the carrier determines the weight. As a general rule, in sea and inner waterway transport, the weight is determined by the shipper, but specific methods are used to establish it: the weight may be calculated by the draught of the ship.

When the goods are delivered to the point of destination, the carrier is obliged to inform the recipient of the arrival of the goods, and the latter is obliged to accept the goods delivered to his address. It is possible to refuse to accept the delivered goods only when the quality of the goods is so changed as to make impossible their full or partial use.

At the time of the delivery of the goods to the point of destination, the recipient is obliged to verify that their condition accords with the rules in effect for the specific types of transport and, in relevant situations, to draw up a special document establishing the non-preservation of the arrived goods. Such an act is called a *commercial act*, and it is necessary in order to present claims against the carrier by the recipient. For most codes of transport, the non-preservation of the delivered goods may also be certified to by making a notation in the bill of lading.

In maritime transport, if during the period of navigation or mooring of the ship a situation occurs which could be grounds for the presentation of property claims against the carrier, the captain must carry out special actions which have the purpose of securing proof and which are called *sea protest* (Chapter XVIII *KTM SSSR*).

A claim of sea protest is made by the captain in the ports of the USSR to a notary and must contain a description of the circumstances of the event and the measures taken by the captain. On the basis of such a claim, as well as on the examination of the captain, and, if possible, of not less than two witnesses from the commanding crew of the ship, and two witnesses from the ship's company, the notary draws up an act of sea protest. Such an act will be used in the future as proof for the resolution of possible disputes (on the non-preservation of the goods, the collision of two ships, etc.).

5. *Responsibility of the parties in a contract for the carriage of goods*. From the moment that the goods are accepted to the time that they are delivered to the point of destination, the goods are under the jurisdiction of the carrier, and he bears responsibility in cases of the non-preservation of the delivered goods, which may manifest itself in their loss, shortage, or damage. The responsibility of the carrier for all types of transport, in accordance with the general principles of Soviet civil law, arises only if the carrier is at *fault* which, as a general rule, is the assumption (on the concept of fault, see Chapter 20).

In the transport statutes and codes, the fault of the carrier is characterized not in the general form given in Article 222 of the Civil Code, but in the form of a model list of circumstances the presence of which absolves the carrier from responsibility (Arts. 148 and 149 *UZhD SSSR*, Arts. 132 and 133 *UAT RSFSR*, Art. 160 *KTM SSSR*, Arts. 191 and 192 *UVVT SSSR*). The carrier has the right to prove that the non-preservation of the goods occurred as a result of these and analogous circumstances, which he could not prevent or eliminate, and if this can be established, the carrier must be considered not guilty and be freed from responsibility.

Moreover, there are rules in effect for many types of transport under which the party presenting a claim carries the burden of proving the fault of the carrier for the non-preservation of the goods if sp‹ ial circumstances are established which presume the lack of .ault of the carrier. In such circumstances, to state it in a legal fashion, there occurs a shift in the burden of proof. This rule has great practical significance since it requires the presentation of necessary proof not by the carrier, but by the owner of the goods.

To situations of the shifting of the burden of proof of the carrier to the owner of the goods are related the issues of whether the goods are kept in railway cars, motor cars, or containers which are in good repair and have the repair seal of the shipper; the non-preservation of the goods as a result of natural causes due to their being transported in open rolling stock or open deck; the failure to take necessary measures by the person attending to the goods on behalf of the owner of the goods, as well as the short-fall of goods within the limits of the norms of natural loss, established for a specific type of transport.

In case of the non-preservation of the goods, the carrier is liable to the extent of the actual value of the lost or incomplete stock of goods, or for the sum by which the value of the goods is decreased owing to the damage to the goods. Moreover, in the loss or shortage of goods, the transport payment is returned if it is not included in the price of the goods.

The carrier does not have to compensate the owner of the goods for other possible losses in excess of the amounts listed above. However, when the goods are sent off, their value may be declared, and in such instances the carrier pays the owner of the goods compensation within the limits of the declared value. If the carrier uses the goods for his own purposes, he must compensate for the value of the goods at twice the amount (Art. 151 *UZhD SSSR*, Art. 135 *UAT RSFSR*, Art. 195 *UVVT SSSR*).

If the goods are not delivered within the established period, the carrier, as a general rule, must pay a fine the amount of which is calculated as a defined percentage of the sum of the transport fee and which depends on the length of the delay. For instance, railway and motor transport organizations pay a fine of 12% to 75% of the transport fee, and air carriers – 5% for each missed 24-hour period, but not more than 50% of the transport fee. It is not possible to levy losses incurred in addition to this fee. A fine for delay in delivery is levied when the carrier is at fault, and he is entitled to prove his lack of fault for the delay in the delivery of the goods.

If the goods do not arrive at the point of destination for an extended period of time, they are considered lost after the period of time defined in the transport statutes and codes elapses, and the carrier is obliged to pay compensation for the goods. However, if the goods are subsequently delivered, the shipper is obliged to accept the goods and to return the compensation received for them earlier to the carrier.

In maritime transport, when the goods are transported in foreign carriage, special rules regarding the liability of the sea carrier are established which are based on international agreements. First, the carrier is not liable for the non-preservation of the goods if this occurred as a result of the actions or oversight of the captain, other persons of the ship's crew or the pilot in carrying out the navigation or management of the vessel (Art. 161 *KTM SSSR*). This rule is commonly known as release from liability for a navigation error. Secondly, in the transport of goods by bill of lading, if the value of the goods is not declared and included in the bill of lading, compensation for lost or damaged packages or customary freight unit may not exceed 250 rubles (Art. 165 *KTM SSSR*).

The transport statutes and codes also provide for instances of liability of the shipper and recipient to the carrier. A shipper's liability takes the form of a fine for giving incorrect information about the goods in the transport document, and, in addition, he must compensate for losses caused because of this, as well as pay a fine for any damage to the means of transportation during carriage. In case of the tardy removal of the goods from the station (port) of destination, it is possible to demand a higher payment for the safekeeping of the goods from the recipient.

In the course of maritime transport, losses may occur as a result of exceptional expenditures made deliberately and reasonably or as sacrifices for the purpose of saving the ship, freight, and transported goods from common danger. For example, some of the goods or ship's equipment may be thrown overboard to lighten the weight of the ship, or a sinking ship is deliberately brought to rest on a shoal, or measures are taken to prevent a fire from spreading on board, etc.

The reason for such action is that the sacrifice of a part will result in saving most of the rest – the ship itself and most of the goods. The losses resulting therefrom form a special institution of maritime law called *general average* [*obshchaia avariia*], and such losses are apportioned among the ship, freight, and goods in proportion to their value (Art. 232 *KTM SSSR*).

The value of the ship, freight, and goods, the amount of the losses incurred, and their apportionment is established by specialists called *average adjusters*, and the legal document created by them is called an *average statement*. In the USSR, average adjusters are in the service of the USSR Chamber of Commerce and Industry. In accordance with Article 249 of the Merchant Shipping Code of the USSR, interested parties may dispute in a Moscow city court the average statement which was drawn up.

6. *Contract for the carriage of passengers and baggage. Under this contract, the carrier binds itself to transport passengers to their point of destination, and if the passengers register baggage – also to transport baggage to the point of destination and to turn it over to a person authorized to receive the baggage; passengers bind themselves to pay the established fare for carriage, and if they register baggage – also to pay the freight for the carriage of the baggage* (Art. 374 Civil Code).

Although outwardly similar, the contract of carriage for the passengers differs significantly from the contract for the carriage of goods both in its content and in the conditions of the liability of the carrier to the passenger.

The *written form* is established for the contract for the carriage of passengers. The conclusion of the contract and the fact of payment for carriage are confirmed by the passenger's holding a ticket of established form. If baggage is registered, a baggage receipt containing information on the number of pieces and the weight of the baggage is given to the passenger.

The special features of the conclusion of the contract for the carriage of passengers is that, as a general rule, the transport enterprise is obliged to enter into such a contract. The carrier has the right to refuse to sell a ticket to a passenger only if there are no more spaces, or if the passenger seems to be in a state of illness. After concluding the contract, the carrier is obliged to provide the passenger with space in accordance with his ticket, and if this is not possible, to provide him with other space if the passenger consents.

A passenger is granted a number of rights under the contract of carriage which reflect the purpose of the contract and the special circumstances which may arise in regard to particular groups or even individual passengers. A passenger has the right to bring with him one child under the age of five without compensation if it does not take up separate space, to carry hand luggage within the established weight limit, to extend the period of validity of the ticket because of illness, or to return the ticket to the carrier and receive its price less an established deduction. The passenger also has the right to make stops during the trip.

A passenger has the right to register objects and things to be transported as baggage to be conveyed to the place of destination in special baggage accommodations in the custody of the carrier. As a rule, baggage is conveyed by the same means of transportation as is the passenger. Specific types of transport may have maximum weight per piece of baggage. A passenger has the right to declare the value of the registered baggage by paying an established sum.

Articles belonging to a passenger which remain with him during the carriage and are under his observation are called hand luggage. The general weight of such articles for specific types of transport must conform to established limits which are from 50 to 80 kilograms. Some things (flammable, toxic, or those which would soil the rolling stock or clothing) are not allowed as hand luggage.

The carrier is liable for the non-preservation or the shortage of baggage unless it can show that it was not at fault. The extent of the carrier's liability depends on the declared value of the baggage, and if it was not declared, the extent of the compensation is defined by the rules of transport and by tariffs. A fine in the amount of a defined percentage of the transport fee is paid for late delivery of the baggage.

There is a rule in effect that it is the passenger's obligation to look after and safely keep hand luggage, and that he must be responsible for it. However, if the non-preservation of hand luggage is caused through the carrier's fault (for instance, the occurrence of a collision), the carrier is under obligation to compensate for losses caused to the passenger.

According to Article 385 of the Civil Code, the liability of the carrier to the passenger himself is decided differently. If the passenger's death or injury to his health occurs during the course of transport, the liability of the carrier is defined not by the rules of transport legislation but in accordance with the provisions for liability for causing harm (Chapter 40 Civil Code), and the carrier is liable as the source of increased risk (see Chapter 35). This solution is directed towards strengthening the protection of the health of the passengers since the liability for causing harm is stricter. Increased liability to the passenger, which is referred to in Article 385 of the Civil Code, is provided for air transport. Air transport carries liability even if the harm caused to the passenger is by *force majeure*.

7. *Filing of claims and suits in carriage cases.* When claims arise against a carrier, shippers and passengers must file a written claim. As a general rule, failure to file such a claim results in the loss of the right to turn to *arbitrazh* (or a court) with the suit. The establishment of an obligatory claim

procedure in carriage cases may be explained by the large numbers of carriages and the necessity of giving the carrier the opportunity of carrying out relevant investigations in order to explain the reasons for the non-performance of the concluded contract.

Claims must be accompanied by the following documents: goods orders, bills of lading or baggage receipts (in the original), acts on the non-preservation of goods (commercial acts), data on the value of the goods (baggage), and estimates of the amount of the claim. If all the goods are lost, the carrier is subject to an inquiry concerning their disappearance.

As a general rule, both the shipper and the recipient are authorized to bring a claim against the carrier for lost goods, but only the recipient may bring a claim for the shortage of or delay in the delivery of the goods. But the recipient may transfer the right to bring a claim to the shipper, as well as to his superior organs and transport expediting organizations. Claims must be addressed to a sea or river steamship authority which performs the carriage, and in railway, motor, or air transport, they must be addressed to the carrier of the place of designation of the goods. In cases involving the carriage of passengers and luggage, claims may be presented to the carrier at either the place of departure or destination, at the passenger's preference.

Uniform time periods are established for presenting claims and for their examination by the carrier for all types of transportation. Claims must be presented within six months, and claims for the payment of fines and premiums must be presented within 45 days. The carrier must examine the claim and notify the applicant within three months whether he will satisfy or reject it; in the case of claims based on the carriage of direct mixed service, within six months, but claims for the payment of fines, within 45 days (Art. 384, para. 2, Civil Code).

In cases of denial of a claim or of non-receipt of an answer within the period established therefore, a suit may be brought against a carrier within the period of the limitation of actions, which is two months (Art. 384, para. 3, Civil Code). In regard to this, it must be kept in mind that the periods for presenting claims may not be interrupted nor renewed; however, the rules of suspension, interruption, and renewal are applied to the limitation of action periods (Chapter 9).

A six-months period of limitation of actions is established for the presentation of claims arising from carriage by the carrier against the shipper, recipient, or passengers (Art. 384, para. 4, Civil Code). In these cases, the obligatory preliminary declaration of written claims is not necessary. However, if a dispute arises between socialist organizations, a claim must be declared inasmuch as obligatory, pre-*arbitrazh* settlement is prescribed for such disputes.

Chapter 31. State Insurance

1. *The nature of insurance and its organization in the USSR.* State insurance in the USSR has as its aim: to compensate citizens and organizations for losses which are caused by various natural disasters, accidents, and other similar circumstances from an insurance fund which was created through payments by interested persons. There may also occur circumstances in the life of citizens which do not result in property damage, but which nevertheless result in the need for supplementary funds for the insured party or his relatives (death, old age, coming of legal age, marriage, etc.). Insurance may be used in these circumstances.[1]

The insurance fund forming the state insurance of the USSR is created by payments of individual interested organizations and citizens, which are collected by a special state insurance organization – *Gosstrakh* [State Insurance], which manages this fund and carries out the payment therefrom when circumstances such as those listed above, occur.

In compensating for the harm caused to a socialist organization or to a citizen, Soviet state insurance helps in the safeguarding of socialist and personal property, and the restoration and development of productive forces. The assets collected by *Gosstrakh* are also used in the fight against accidents and natural calamities through financial expenditures to prevent their occurrence (on fire prevention measures, against epizootic outbreaks, on veterinary matters, etc.).

In each republic, insurance matters are carried out by a special state organization – The *Main Administration for State Insurance* [*Glavnoe upravlenie gosudarstvennogo strakhovaniia*] of the union republic (*Gosstrakh*) which is under the management of the Ministry of Finance of that republic. The main administrations (managements) of state insurance are organized on the basis of economic accountability, are juridical persons, and act on the basis of the Model Statute on the

1. This chapter does not examine the social insurance of workers, employees, and collective farm workers, which is set forth in texts on labor and collective farm law.

Organs of State Insurance in the union republics, which was confirmed by decree of the Council of Ministers of the USSR.[2] They carry out their activity through provincial (territorial) administration, district inspectorates, and a far-flung network of insurance agents. Some insurance operations are carried out by the all-union organization, the USSR *Gosstrakh* which is under the administration of the Ministry of Finance of the USSR.[3] Insurance operations which are connected with foreign trade and foreign shipping are carried out by a special state organization – *Ingosstrakh* [Foreign State Insurance]. In accordance with Article 3 of the Principles of Civil Legislation, relationships based on state insurance are regulated by all-union legislation.

2. *Types of insurance.* Insurance is divided into two basic types depending on its subject: *property insurance* and *personal insurance.* The compensation received as a result of insurance is always in the form of property; however it is not only property, but the personality itself, its personal, non-property assets – life and health, which may be insured. When a person's health is damaged he may lose his ability to work, and consequently, his wages, based on work, while death deprives dependants of the deceased of the source of their livelihood. However, property loss may not occur if, for instance, the insured party did not work or if the deceased was insured so that the benficiaries are persons who did not receive their support from him while he was alive and had no right to receive it.

Both property and personal insurance, may, in turn, be sub-divided into types depending on the concrete subject of insurance (building insurance, insurance of household property, and agricultural crops, etc.; life insurance and employment insurances, insurance of children, mixed insurance which includes payments of money when a certain age is attained, etc.).

2. *Collection of Decrees of the USSR* 1958 No.17 item 137.
3. *Collection of Decrees of the USSR* 1967 No.22 item 157.

Insurance may also be divided into *voluntary* or *obligatory* depending on the basis of the insurance relationship. Voluntary insurance arises as the result of the conclusion of an insurance contract, while obligatory insurance does not depend on the will of the participants in an insurance legal relationship and arises either as the result of factors provided in law, for instance, as the result of the fact that a piece of property which enters the economy is subject to obligatory insurance, or as the result of the conclusion of another type of contract (the obligatory insurance of passengers in transport).

3. *Concept and general characteristics of insurance legal relationships. Basic insurance concepts (terms).* An insurance legal relationship is one of the types of obligation (see Chapter 16). *Under an insurance legal relationship (obligation), one party (the assured) binds itself to pay a defined sum (or sums) to the other party – the insurance organization (the insurer), while the insurer binds itself, in the case of property insurance, to compensate the assured or a third party for losses suffered when an event which is provided for in the contract or in law occurs within the limits of the amount established by law or contract (insurance sum), or, in the case of personal insurance, to pay the insurance sum.*

Thus, besides the assured and the insurer, who are essential participants (parties) to this obligation, one other party may participate in it – the party who is the beneficiary of the insurance. When the insured event occurs, this party has the right to receive insurance compensation or the insurance sum, depending on whether property or personal insurance is involved; consequently, it is a creditor in relation to the insurance organization. In this instance, the insurance contract – which was concluded by the assured and the insurer – may be considered a particular instance of a contract in favor of a third party (see Chapter 17).

In instances provided in the rules of the Ministry of Finance of the USSR, the life or health of another person, and not of the assured himself, may be insured by personal insurance. In this situation, the insured party as such is not a party to the insurance legal relationship since it has no rights or obligations which are linked to the insurance. But insurance may also be concluded in favor of the insured party, and in this case it simultaneously becomes the beneficiary with all the juridical consequences thereof. This occurs, for instance, when parents insure their children or when an organization insures workers who perform dangerous work (involving fire, etc.). In the latter situation, the insurance is concluded in favor of the insured if he loses the capacity to work, and in favor of his family if he dies.

Let us now examine some basic insurance concepts (terms), which are found in the rules on insurance:

a) *insurance risk* is the danger the consequences of which give rise to insurance (natural or other calamities, fire, death, etc.);

b) *insurance event* is the occurrence of the event which gives rise to the payment of the insurance (death of the insured party, injury to his health as the result of an insurance risk under the relevant legal relationship, loss of damage to property resulting from the same reasons, etc.). An insurance event is the concrete situation wherein the insurance risk appears;

c) *insurance interest* has significance only in property insurance. It is the loss which the insurer may suffer as the result of an insurance event. Property insurance precedes compensation for losses, but insurance in favor of the given party may not take place if he does not have an insurance interest or an interest in excess of the insurance. This rule excludes the possibility of using insurance for the purpose of profits.

In order for an insurance interest to exist, a mere subjective interest in safeguarding a certain property is not sufficient, for instance, a desire to secure the safeguarding of the property of relatives or close ones. It is necessary that the persons in whose favor the insurance is paid has some kind of right or obligation in regard to the insured property. Such a person may be not only the owner of the property but also the renter, holder of a pledge, commission store, carrier, etc., to the extent that they are responsible for the safeguarding of a particular property and, consequently its loss or damage may cause an additional property obligation for them. Several parties may have insurance risk in the same property at the same time, and each one of them may be its insurer and beneficiary within the limits of this interest, e.g. of the losses which it may cause when an insurance event occurs;

d) *insurance evaluation* is the monetary evaluation of the insured property. It reflects the maximum damage which may be the cause of an insurance event. The existing rules allow the insurance of household property without insurance evaluation, that is, in this situation the insurance sum may be established by the insurer at his own discretion without verification;

e) *insurance sum* (insurance cover) is the maximum sum that the insurer binds himself to pay when an insurance event occurs. In property insurance it may be lower than the insurance evaluation or may equal it, but it may not be higher;

f) *insurance payments* (also called insurance fees or insurance premiums) are the payments which the assured makes to the insurer. An insurance contract is usually considered to be concluded from the moment of the first insurance payment;

g) *insurance compensation* is the sum which the insurer for property insurance pays as compensation for the harm caused when an insurance event occurs. The extent of insurance compensation is defined in various ways. If the insured property is completely destroyed, the extent of the insurance compensation is equal to the insurance sum (insurance cover) irrespective of whether the insurance sum equals the insurance evaluation or the actual value of the insured property or whether it is lower than these if the property was not insured for its full value. If the property which was not insured to its full value is partially destroyed or damaged, the extent of the insurance compensation may be defined in two ways: by the *system of proportionate liability* or by the *system of first risk*.

In most types of insurance, in particular in obligatory insurance, the system of proportionate liability is used (Art. 387 Civil Code). Under this system of determining the extent of insurance compensation, the insurance sum decreases proportionately to the saved value of the insured property, i.e. the extent of the insurance sum is related to the extent of the loss caused by the insurance event as the insurance sum is related to the insurance evaluation. For instance, if property worth 5000 rubles is insured for 2500 rubles and one-half is destroyed as a result of an insurance event, the insurance compensation is 1250 rubles, that is, one-half of the insurance sum.

Under the system of first risk, a loss is compensated for in full, but only within the limits of the insurance sum. Thus, if the loss is less than or the same as the insurance sum, it is compensated in full; if it is higher, then the part of the loss which exceeds the insurance sum remains uncompensated. Under this system, in the example given above, the insurance compensation is 2500 rubles, that is, the full value of the destroyed part of the insured property, inasmuch as the damage does not exceed the insurance sum. The system of first risk is used in insurance contracts for household property.

The assurer is *obliged*: to pay the insurance payments to the insurer according to established rates and within established periods; to maintain the insured property in proper condition; to take all possible measures to prevent the destruction of or damage to the insured property; to take measures to save the insured property during and after a natural calamity or accident and to take measures to prevent further damage; to show to the inspector of the insurance organization all the property remaining after the occurrence of the insurance event; to notify in writing the insurance organization of the occurrence of the insurance event within an established period; and in a personal insurance case, to produce all documents necessary to receive the insurance sum.

The insurance organizaton is *obliged*: to commence a determination of the amount of insurance compensation within an established period; to draw up a document concerning the destruction of or damage to the insured property; to pay the insurance compensation (which may not exceed the actual loss and insurance sum) or the insurance sum in cases of personal insurance (in cases of insurance for loss in general of the capacity to work, where only a partial loss occurs, the insurance sum paid is proportional to the loss of capacity to work).

If the insurance event occurred as a result of the intent of the assured or the insured party (in personal insurance cases) or as the result of a crime committed by the latter, the insurance sum (insurance compensation) is not paid.

According to Article 399 of the Civil Code, the right is transferred to the insurer who has paid out insurance compensation, within the limits of the amount paid by him, to demand compensation for harm caused by an insurance event which the assured (or other person who received the insurance compensation) has against the party who is responsible for the harm. This rule is applied only in property insurance. In personal insurance, the party which receives insurance compensation retains for itself all property rights which arise from the occurrence of the insurance event: the right to make a claim against the party responsible for harm (see Chapter 35), the right to receive benefits, pensions, etc.

4. *Obligatory insurance*. Obligatory insurance arises in circumstances which are especially provided for in law. Thus, it is used in relation to specific property (buildings, agricultural crops, domestic livestock) of collective farms in accordance with the Decree of the Presidium of the Supreme Soviet of the USSR of 28 August 1967, "On State Obligatory Insurance for Collective Farm Property".[4] The Decree of the Presidium of the Supreme Soviet of the USSR of 2 October 1981, "On State Obligatory Insurance for Property Belonging to Citizens",[5] provides for obligatory insurance for buildings

4. *Gazette of the Supreme Soviet of the USSR* 1967 No.35 item 481.
5. *Gazette of the Supreme Soviet of the USSR* 1981 No.40 item 1111.

and domestic livestock belonging to citizens.

The legislation contains a list of damages (insurance risks) whose occurrence effectuates insurance. The extent of insurance payments are established by legislation and vary according to type of property, the district in which it is located, and the category of assurers. As a rule, insurance is given for the full value of the property, and if only a part thereof is destroyed, compensation is on a proportionate basis.

All property which is subject to obligatory insurance is considered to be automatically insured from the moment of its entry into the economy. The insurance is in effect during the time that the property is in the economy, irrespective of whether or not insurance payments are made.

The conditions of obligatory state insurance for the property of collective farms are applied to state farms within the system of the Ministry of Agriculture of the USSR.

State structures which are leased out and, under certain circumstances, property belonging to other owners which is given to state organizations, also falls under obligatory insurance. In accordance with Article 198 of the Civil Code, state pawnbrokers must insure the property which they take as pledge at the pledgor's expense.

The insurance of passengers against accidents is an obligatory private insurance which is provided for by Decree of the Presidium of the Supreme Soviet of the USSR of 4 June 1982, "On the obligatory state insurance of passengers on air, railway, sea, inland waterway and motor transport".[6] The insurance is in effect for accidents which occur during transport (flight) or at a station, port, terminal, or wharf. It is not in effect for passengers on all types of transport for international service, local service, maritime and inner waterway transport for casual outings and excursions, inner waterway transport for inner-city service and ferry crossing, motor transport on city routes or inter-city routes within one province, territory, autonomous republic, or union republic

6. *Gazette of the Supreme Soviet of the USSR* 1982 No. 23 item 412.

which is not divided into provinces. In case of injury to health the insurance is paid to the insurance party himself, and in case of death – to his heirs. The insurance sum is 1000 rubles for death or for complete and permanent incapacity to work, and in case of partial loss of capacity to work, the insurance is decreased in proportion to the loss. The insurance fee (amount) is paid at the time of the sale of the travel documents; passengers who have a right to travel free of charge are considered to be insured without payment. All passengers are considered to be insured from the time of declared readiness to board until the time they leave the premises of the terminal, port, or airport of the point of destination indicated on the ticket.

For some categories of workers, obligatory insurance of pensions is established. In this situation, the organization for which the insured work is the assured, and the occurrence of an insurance event gives rise not to a one-time payment, but to the granting of a pension.

5. *Voluntary insurance.* The definition of the insurance contract is given in Article 388 of the Civil Code. Voluntary insurance may be concluded independently of obligatory insurance or in addition to it. However, in regard to property insurance, this is allowed only in instances when the insurance sum for the obligatory insurance is lower than the actual value of the insured property and then only for the amount not covered.

There is a large number of various types of voluntary insurance, both property and personal, each regulated by special rules issued by the Ministry of Finance of the USSR on the basis of the relevant decrees of the Council of Ministers of the USSR. Within the sphere of property insurance are the voluntary insurance of the property of cooperative and social organizations, insurance of structures and livestock of citizens (above the sum of the obligatory insurance for their full value), insurance for household goods, and insurance for the means of transportation belonging to citizens. The sphere of personal insurance includes insurance for death and loss of the capacity to work, individual insurance against accidents, insurance for pensions, insurance for children, insurance of workers at the expense of an enterprise, institution or organization, and mixed life insurance.

There is also the so-called accumulated personal insurance. Its special feature consists of the fact that it is concluded not with the view that an insurance event causing harm will occur but for the accumulation of assets to be available at a defined period or situation. This type of insurance includes insurance for marriage. Elements of this insurance are included in mixed life insurance (the insurance sum is also paid when a certain age is reached) and in insurance for children (the insurance sum is also paid when a child comes of legal age).

An insurance contract is concluded in the *written form*. It is formalized when the insurer gives a special document to the assured which, depending on the type of insurance, has various names (insurance proof, insurance receipt, insurance policy, insurance certificate).

Chapter 32. Contract of Loan. Accounting and Credit Relationship

I. Contract of Loan

1. *Concept and significance of the contract of loan.* The contract of loan is used in relationships between citizens involving the comradely lending of money or things. This contract is the basis for the mutual relations of citizens and state savings banks, other banks, pawnbrokers, and mutual assistance benefit funds. Socialist organizations are allowed to conclude contracts of loan between themselves without the participation of a bank only with regard to things. Thus, a collective farm has the right to give to another one grain or fodder as a loan in a gesture of comradely assistance. The contract of loan is prevalent in the mutual relationships of socialist organizations with banks and in the granting of bank loans and credit.

Articles 269-274 of the Civil Code are devoted to the contract of loan. The loan relationships of citizens with savings banks, pawnbrokers, and mutual benefit funds are regulated by normative acts, in particular the Statute on State Savings Banks of the USSR[1] and Model Statute on the Pawnbroker.[2] Loan relationships with banks are defined by the statutes on Banks of the USSR and their instructions regarding the granting of credit.

Under a contract of loan, one party transfers money or generic goods into the ownership (operative management) of another party which must return the sum of money or an equal quantity of goods of the same kind and quality (Art. 269 Civil Code). The party which transfers the money or things is called the *lender*, and the party which receives them is the *borrower*.

Under a contract of loan, money or things are transferred into the ownership (operative management) of the borrower and the money received is spent by the borrower and the

1. *Collection of Decrees of the USSR* 1977 No. 21 item 131.
2. *Collection of Decrees of the RSFSR* 1968 No. 10 item 54.

things are used. Therefore, the borrower does not return the same money which he borrowed but its equivalent. This is the difference between the contract of loan and the contract of property hire, under which an individually-defined thing is given for use and must be returned.

As a general rule, a contract of loan is *gratuitous*. Citizens do not have the right to receive interest for money which they lend. Under a contract of loan, interest may be added only in circumstances provided in law. Interest is charged in the lending operations of mutual benefit banks and of pawnbrokers. Thus, a pawnbroker has the right to receive payment for the loan given out for the pledge of things. Banks charge interest when giving credit to socialist organizations.

2. *Form and content of the contract of loan*. The form of a contract of loan depends on the amount of money lent or the value of the things given for loan. The written form is prescribed for a contract involving more than 50 rubles. If the parties violate this requirement, then each of the parties in a dispute loses the right to plead transfer (non-transfer) of money or things by witnesses' testimony (Arts. 296, 271 Civil Code). A contract of loan is considered concluded from the moment that money or things are transferred. It is usually formalized by a promissory note which is given by the borrower when he receives the money or things.

The borrower is obliged to return the money or things within the period agreed to in the contract. If a period is not agreed on, the borrower must return the money or things within seven days from the day that the lender asks for their return. The obligation of prompt performance arises when this is given in normative acts or in the content of the contract. For instance, a savings bank is obliged to return a deposit promptly at the request of the deposition. In the instance of a delay in the return of money, the borrower is obliged to pay 3% annually for the period of delay on the amount overdue if another amount of interest is not established by law or contract (Art. 226 Civil Code). This rule is also applied when the loan itself is gratuitous. The party which lends the money or things has a right to claim losses which were caused by the delay in performing the obligation, against the borrower. If the borrower does not perform his obligation of returning the borrowed money, the lender has the right to turn to court or *arbitrazh*.

3. *Special features of contract concluded with pawnbrokers or mutual assistance funds*. In accordance with the Model Statute, pawnbrokers lend money (give loans) to citizens for a period of from one to three months. A loan is given only for the pledge of a thing, and the amount of the loan given by a pawnbroker must not exceed 75% of the amount at which the pledged object is valued. The articles are valued by agreement of the parties. A pawnbroker's loan is formalized by the drawing up of a pawn ticket.

A pawnbroker charges 8.4% annually on the amount of the loan. In addition to this, the borrower must pay for the safeguarding of the objects given as pledge and bears the cost of their insurance. If the loan is not liquidated within the established period, the pawnbroker has the right after the expiration of the fixed monthly period to give the objects to a trading organization to be sold at actual value but not lower than the established value.

Funds for mutual assistance are organized by professional trade unions and may give loans only to those persons who are members of a given social organization. Loans may be either short-term or long-term; interest is charged for long-term loans.

II. Accounting Relationships

1. *Concept and significance of accounting relationships*. When one party acquires goods or obtains services it pays another party a sum of money, i.e. a monetary settlement takes place between the parties. Cash payments between parties do not result in relationships requiring special legal regulation. The situation is different when socialist organizations order payments through a bank. If the bank participates in the pay-

ments, it enters into relationships with the payer and recipient which are called accounting relationships. *Accounting relationships are relations created between a bank or other accounting institutions and socialist organizations when the latter carry out monetary payments.*

Socialist organizations keep their funds in a bank. There are three banks in the USSR: State Bank of the USSR (*Gosbank SSSR*), the All-Union Bank for Financing Capital Investments of the USSR (*Stroibank SSSR*), and the Bank of Foreign Trade of the USSR (*Vneshtorgbank SSSR*). Banks organize and implement the orders for payment, whose essential feature is that the payments are carried out by transferring funds from the bank account of the paying organization into the account of the credit organization. During the process of monetary settlement, the bank supervises the legality of the payments which are carried out and sees that the socialist organizations use the monetary means for the designated purpose. Payments are made with the agreement of the payer after the products are unloaded, work is completed, or services rendered from the funds in his account or from the credit given to him by the bank.

Articles 391-392 of the Civil Code contain the basic principles for settling accounts. Detailed regulation of accounting relationships are given in special acts of all-union legislation: decrees of the Council of Ministers of the USSR on payments, charters of banks, instructions given out by the Banks of the USSR; and the Rules on Payment orders contained in the Instruction of the State Bank of the USSR No. 2 of 31 May 1979, "On Payment Order in the National Economy."

2. *Contract of payment (current) account*. Socialist organizations open an account in a bank in order to keep and dispose of funds. Organizations based on *khozraschet* open *payment*, and organizations based on a budget which do not carry out economic activity open *current* accounts. The legal relationships between organizations and a bank in regard to accounts are based on a contract of payment (current) account.

Under a contract of payment (current) account, the owner of the account binds himself to keep his funds in the institutions of the bank and to dispose of them by adhering to the rules of the bank, and the bank binds itself to perform accounting/banking services for the owner of the account. The rights and obligations of the parties under a contract of payment and current account do not have significant variations. The contract is concluded when a newly-formed organization makes an application to a bank for opening an account. A copy of the charter (statute) of the organization, photocopies with the signatures of the persons authorized to sign for the account, an impression of the seal, and other important documents are attached to the application. The agreement of the bank manager to open the payment account signifies the acceptance of the proposal of the contract. The contract of a payment account is terminated when the account is closed.

The rights and obligations of the parties in a contract of payment account are defined by the bank rules. The bank is obliged to carry out the instructions of the owner of the account on disbursing and transferring funds from his account. When there are insufficient funds in the account, payments are carried out in installments. The organization-owner of the account may dispose of the funds kept in its account only in accordance with the designated purpose of these funds. The violation of the contract results in *property liability* for the parties. The norms of such liability are contained in the Statute on Penalties for Violating the Rules for the Completion of Accounting Operations,[3] which were confirmed by the Council of Ministers of the USSR. The bank pays a penalty to the owner of the account: 1) for sending out unauthorized or untimely accounting documents, an amount of 25 rubles; 2) for incorrectly entering or delaying in entering the amount due to the owner of the account, 0.5% of this sum for every day of delay through his fault. In turn, the owner of the account pays a penalty to the bank: 1) for delay in informing the bank of an incorrect amount of money due to his account, an

3. *Collection of Decrees of the USSR* 1973 No. 18 item 106.

amount of 0.3% of the amount due for each day of delay; 2) for delay in confirming the balance of the payment or current account, an amount of 15 rubles. Penalties are established for a number of other violations. Liability in the form of compensation for losses arises only in circumstances directly provided by the charter of the bank.

3. *Forms of payment between socialist organizations.* In accordance with the Instruction No. 2 of the *Gosbank* of the USSR, orders for payment for goods and services are carried out: by claims for payment which are accredited in special accounts by payment authorizations including transfers between enterprises of the ministry of transport, by checks, combined claims, and encashment authorizations (orders for undisputed writing off of funds). Payments by payment claims are carried out by encashment with acceptance (agreement) of the payer to pay even without his acceptances. The use of one or another form of payment (acceptance, letter of credit, payment orders, checks) is established by the parties in the contract, except in those circumstances when a specific form of payment is provided for by normative acts. Payments between organizations which are carried out by banking institutions located within one town are called intra-town, in distinction to inter-town, which are carried out by banking institutions located in various population centers.

The most widely used forms of payment are: for inter-town payments, the acceptance for or payments through payment orders if the supply of products or goods by transport organizations does not exceed three 24-hour periods; for intra-town payments, by payment orders or by checks from limited or unlimited purpose accounts. That these forms of payment are recognized as the most prevalent signifies that it is recommended that organizations choose one of them in concluding a contract.

For regular, constant supply, socialist organizations may settle accounts with each other on the basis of planned payments. Under planned payments, settlement is not made after each delivery, but by periodic transfer of funds at periods, and for the amounts agreed to by the parties, for instance, every ten years.

4. *Acceptance form of payments.* Under the acceptance form of payments, an organization delivering products or rendering services writes up a payment claim and submits it to the bank which provides services for it. A *payment claim* [*platezhnoe trebovanie*] is a payment document containing claims against the payer to pay a specific sum. The payment of payment claims is performed by the bank servicing the payer from the account of the payer only upon the latter's consent (agreement). There are two forms of acceptance: tacit acceptance or acceptance by written notification on the part of the payer to pay the claim. As a rule, the tacit form is used. Abstention from refusal of the acceptance of a payment claim for an established period indicates the payer's agreement (acceptance) to pay.

Acceptance payment claims may be *subsequent* or *advance*. In payments by procedure of subsequent acceptance, claims are paid on the day they arrive at the bank of the payer. But the latter has a right to communicate his refusal to accept within three working days after the arrival of a claim at the bank (within ten days in cases provided in law). The transferred amount is restored to the account. In payments by procedure of advance acceptance, a claim is paid on the day following the expiration of the acceptance. The payer has the right to inform the bank of his refusal to accept within three working days for inter-city payments, and within two working days for intra-city payments.

The payer has the right of full or partial refusal of payment of a claim on grounds provided in the instructions of the bank, normative acts in regard to individual contracts, or the contract itself. Thus, in cases of supply, a payer has the right to refuse payment by reason of the delivery of unordered products, delivery of goods to the wrong address, etc.

In instances provided by Instruction No. 2 of the *Gosbank* of the USSR, payments are made with payment orders which are paid without the acceptance of the payer. Claims for the payment of communal services (gas, water), electricity, and heating are paid without acceptance. The non-acceptance form is sometimes used in collecting a fine, for instance, a fine for means of transportation or for the supply of defective goods.

5. *Letters of credit as a form of payment.* Under the letter of credit form of payment, which is used only in inter-city payments, the payer must give instructions in advance to a bank to transfer a specified amount to the banking institution servicing the supplier. A *letter of credit* is an authorization by the bank of the payer to a bank in another city which services the supplier to carry out payments on the account of the supplier for the delivery of goods on conditions provided in the declaration of the payer in regard to the drawing up of the letter of credit. The bank of the supplier carries out the payments in the amount transferred by the bank of the payer after the supplier presents the accounts and transport documents testifying to the delivery of the products and goods. This form of payment guarantees to the supplier prompt payment for delivered goods.

The form of payment by letter of credit may be provided in a contract. The supplier has the right to change to the letter of credit form of payment if the payer is systematically late in making payments or if he refrains from prompt payment for products by payment authorizations or checks.

6. *Payment by payment authorization. Payment authorizations* are used in payments for goods of material value, as well as in carrying out other payments at the initiative of the payer (transfer of a penalty, etc.). In contrast to a payment claim, which is presented to the payer, a payment authorization is a document containing the authorization of the payer to the bank of the deduction (transfer) of a specific amount of money from his account to the account of the recipient of the money. Payment authorizations are used by a bank to make payments irrespective of whether there are any assets in the account of the payer and are paid in installments.

7. *Payments by checks.* A *check* is an instrument of payment which contains the authorization of an organization owning a checking account, to the bank, to transfer a specified amount from its payment account to the account of the recipient of the money. The organization writing the check is the *check giver*, and the recipient of the money, the *check holder*. In order to carry out payments by checks, the check giver endorses the filled out check to the check holder, who gives it to the bank to receive payment. A check may be given to a concrete person or the bearer. Checks are used in intra-city payments, in particular for products received or the transport of goods.

III. Credit Legal Relationships

1. *Concept and significance of credit legal relationships.* Socialist organizations carry out payments from their own financial assets. However, there often arises a need for supplementary sources of payment in the process of economic activity. This may be caused by the fact that payment for supply of goods was not received or by necessary additional expenditures of a seasonal nature.

Funds in addition to those apportioned in the budget are also needed for capital investments. Citizens may also find it necessary to borrow money for the construction of a dwelling or the purchase of livestock. Therefore, socialist organizations and citizens may not always go about their business without borrowing from a bank.

The increased role of credit to stimulate production is characteristic of the contemporary period. The scope of the use of credit has increased, and the periods for granting credit have been extended. The decisions of the XXVI Congress of the CPSU and the decree of the CPSU Central Committee and the Council of Ministers of the USSR of 12 July 1979 on improving the economic mechanism, set forth the task of strengthening the effect of credit in increasing production and also in providing for the development of credit relationships in the area of construction and the use of bank credit for the reconstruction and technical improvement of existing enterprises. In accordance with para. 57 of the joint Party/Government decree of 12 July 1979 which was noted above, banks give credit to be used for the payment on negotiable instruments needed for production during times of temporary need of money by buyers on accounts. The relationships based on the granting of money for a temporary period by the bank of the USSR are known as *credit relationships*. The norms in regard to them are found in Articles 393-394 of the Civil Code, the Fundamental Provisions on Bank Credit which were confirmed by decree of the Council of Ministers of the USSR on 11 November 1982, in other decrees of the Council of Ministers of the USSR, and in relevant instructions of the *Gosbank* of the USSR and of the *Stroibank* of the USSR. These acts strengthen the basic principles of credit: the giving of credit solely by state credit institutions; the purpose of credit; the plan character of credit relationships; the granting of credit on condition of the return of the loan at a specific period; and the reciprocity of credit.

2. *Contract of bank loan*. Credit is provided by a bank on the basis of a contract of bank loan. *Under a contract of bank loan, a bank provides money for a period to the operative management (ownership) of an economic organization for a specific purpose*. The organization receiving the money (loan) binds itself to use it for the designated purpose and to return the loan to the bank at the specific period with interest.

The contract of bank loan is a variant of the contract of loan; however, it has a plan character. The basis of the contract is a credit plan in accordance with which the bank defines the credit ceiling, i.e. the maximum amount of credit which each organization is allowed. The contract of bank loan is reciprocal since the recipient of the credit must pay interest for it. The amount of the interest rates for credit are established by the Council of Ministers of the USSR and depend on the type of loan.

In order to receive a loan, an economic organization presents a claim to its bank. The contract is considered to be concluded from the moment the bank adds the contract loan to the credit account. The basic obligation of the bank is the giving of credit within the credit ceiling. The obligation of the recipient of the credit lies in the use of the bank loans for the designated purpose, to pay interest to the bank for the credit received, and to repay the loan on time.

Depending on the period and the purpose of the credit, a distinction is made between *short-term* and *long-term credits*. Short-term credit is given for a period of twelve months or for longer periods to create circulating assets.

Short-term credit is given to cover the following expenses: to build up a stock of seasonal and non-seasonal commodity-material stock in excess of the norm; for expenses which are related to the manufacture of new products and the acquisition of new types of products; for expenditures on installing new technology and improving the technology of production; to pay for the supply of goods. The contract of temporary credit helps to secure commodity-material stock.

Long-term credit is given for expenditures related to the construction of new enterprises; technical re-equiping, reconstructing or enlarging existing enterprises, and the construction of dwellings and objects at a cultural-everyday nature.

Long-term credit is widely used by dwelling construction cooperatives.

3. *Bank loans to citizens.* Special-purpose credits are given to citizens to help them in building individual dwellings. Credit may be given to repair dwellings and to purchase cows and calves; citizen-members of gardening collectives are given credit to acquire building equipment, to purchase pre-fabricated huts, and to organize the plot. In the latter case, the loan may be up to 3000 rubles to be paid in ten years, starting in the third year after the receipt of the credit.[4] To use the credit, the gardener pays 0.5% interest annually.

The conditions for giving loans and the procedure for their

4. *Collection of Decrees of the USSR* 1981 No. 6 item 37.

repayment are different for different categories of creditors and depend on where a citizen works. If the creditor works in an institution based on a budget, he receives credit independently of a bank; if he works in an organization based on *khozraschet*, the bank transfers the amount of the credit to that organization and the latter pays the received amount to its employee, who is the recipient of the credit.

IV. Deposits in Savings Banks

1. *Deposits of money by citizens.* Citizens may keep money in state worker's savings banks. The operations of savings banks are regulated by the Statute on State Workers' Savings Banks of the USSR, which was confirmed by the Council of Ministers of the USSR.[5] The tasks of the savings bank is to provide people with an opportunity to keep money in a safe manner. Savings banks also help to save money and to use it in the interest of the national economy. The deposits generate contractual relations between the depositor and the savings bank. The contract is reciprocal because interest is given for the deposit.

The Statute on the Savings Banks allows five types of deposits: demand, term, conditional, lottery, and current account. In deposits on demand, the depositor may receive the money at any time, either in the full or partial amount. Deposits for a term must be made for not less than six months. In a conditional deposit, a third party has the right to dispose of the deposit under certain prerequisite conditions which are set up by the depositor (for instance, attainment of legal age). Profits on lottery deposits are paid in the form of winnings. And, finally, checks signed by the depositor are paid on the basis of current accounts. The depositor receives a payment book and a check book for this. This deposit is a variant of the deposit on demand since the money from this account may be paid by check at any time.

Deposits on demand may be made in the name of a specific

5. *Collection of Decrees of the USSR* 1977 No. 21 item 131.

person or a bearer, i.e. any person bearing the savings book. The other types of deposits are made only in the name of a specific person.

2. *Rights and privileges of a depositor.* A depositor is free to dispose of his deposit. He has the right to receive the deposit personally or to authorize its receipt by another person; to examine the entry made in the registration card of the personal account; to transfer the deposit to another savings bank; to commission the savings bank to carry out payments by check from his deposit (to pay for communal services, etc.). The depositor has the right to receive a profit in the form of interest for winnings.

A depositor may will his deposit to one or several parties irrespective of whether or not they are his legal heirs, as well as to the state or specific state, cooperative, or social organizations/juridical persons. The disposition of the deposit in the instance of death is noted in the personal account of the savings bank.

A number of privileges is provided for a citizen-depositor. Deposits may not be taken in execution or subject to arrest other than by procedure established by Article 395 of the Civil Code. The state guarantees the privacy of deposits, their safety, and repayment at the first demand of the depositor. Profits on deposits (interest or winnings) are exempt from taxes and fees.

The depositor himself disposes of the deposit, and the compulsory surrender of the deposit may be carried out only on the grounds of: a judgment or court decision to satisfy a civil claim arising from a criminal case, for instance, recovering the value of stolen goods from the convicted party; a court decision on the recovery of alimony if the person who is obliged to pay alimony has no wages based on work or no property which may be taken in execution; or a court decision on the division of a deposit which is the joint property of spouses.

The confiscation of deposits is possible only by a court judgment which has entered into legal force, or by an order for confiscation of property carried out in accordance with the law (Art. 395 Civil Code).

Chapter 33. Contract of Agency. Contract of Commission Agency

I. Contract of Agency

1. *Concept and significance of the contract of agency.* A contract of agency is used when citizens or socialist organizations are not able to carry out legal acts without the participation of another party. For instance, citizens seek legal advice in entrusting a lawyer to carry out their affairs in court, to formulate rights to inheritance, and to perform other legal acts. The contract of agency is used in the selling of goods on credit when workers and employees entrust a part of their earnings to their enterprise to cancel the credit. Citizens often perform agency contracts for each other while carrying out comradely services (accepting earned wages on the basis of agency, sending mail, etc.).

Articles 396-403 of the Civil Code are devoted to the contract of agency. In addition, the norms of the Civil Code on representation and authorization are also applied to the contract of agency inasmuch as the contract of agency is one of the grounds of representation (Arts. 62-70 Civil Code).

Under the contract of agency, one party (the agent) binds himself to perform specific legal acts in the name and on behalf of another person (the principal) (Art. 396 Civil Code). From this definition, it is clear what the characteristic features of the contract of agency are. The agent binds himself to carry out legal acts for the principal (to conclude a contract, appear in court, receive money or objects, etc.). The agent does not carry out these acts in his own name but in the name and on the account of the principal.

The features listed above make it possible to differentiate the contract of agency from the contract of commission which it closely resembles. The basic difference is that the agent concludes contracts or carries out legal acts in *the name of the principal.* Under a contract of commission, the *commission agent* carries out contracts with third parties in his own name rather than in the name of the client. Thus, the difference between these two contracts is most obvious in the relationships of the parties with third parties (the external part of the relationships constituting the contract).

The norms of the contract of agency do not specify its form, and the general rules regarding the form of transactions are applied. The contract of agency must be formalized in writing if it is concluded with the participation of a socialist organization. If citizens participate, the written form of the contract is obligatory if the contract involves an amount greater than 100 rubles. The fact of agency, on the basis of which the agent enters into legal relationships with third parties, serves as proof of the existence of the contract and its content.

2. *Rights and obligations of the parties*. Under a contract of agency, the agent is *obliged* to carry out personally the task entrusted to him in accordance with the instructions of the principal. If the agent is entrusted with concluding a contract, the instructions of the principal determine the contents of the future contract.

The agent is obliged to inform the client as to how the task entrusted to him is proceeding, and to render him an account and give him everything which was received in relation to the performance of the entrusted task when the entrusted task is completed.

The principal is obliged to receive from the agent without delay everything which was done by him in performing the contract. Since the contract of agency is performed on the account of the client, the latter is obliged to compensate the agent for all the expenses which he incurred in performing the contract.

As a rule, the relationships in a contract of agency are of a *personal-agency* character. Precisely because of this, an exception is made for this contract in regard to the general rule on the impermissability of the unilateral change of a contract or refusal to perform it. According to Article 401 of the Civil Code, a client may change the entrusted task at any time, and the agent may refuse to perform the contract.

Article 396 of the Civil Code provides that the principal is obliged to pay the agent a fee if payment of a fee is provided by law or the contract. Thus, a contract of agency may be either *compensatory* or *gratuitous*. The compensation for the agency must be clearly expressed in the contract.

In case of the improper performance of their obligations under the contract of agency, the parties are liable and must compensate for losses caused to the contracting party (Art. 219 Civil Code).

II. Contract of Commission Agency

1. *Concept and significance of the contract of commission agency*. The contract of commission agency is used by citizens in selling objects, as well as automobiles, through commission stores. The contract of commission agency is also used in relationships between socialist organizations. Collective farms conclude this contract with consumer cooperative organizations in selling agricultural products which are not sold under procurement contracts (on the account of state purchases based on the plan and in excess of the plan). Territorial organs for material-technical supply sell items of material value which are not used by enterprises as well as products for which no market exists on the basis of commission agency. The contract of commission agency is also used in foreign trade. Soviet foreign trade organizations act as agents of Soviet organizations in carrying out transactions for the import of goods.

The norms on the contract of commission agency are found in Article 404-421 of the Civil Code. Here are provided the general rules on the contract of commission agency as well as special norms on the variants of this contract, which are applicable to the contract of commission agency for the sale of agricultural products by collective farms, and to the contract of commission agency which is concluded between commission stores and citizens. These variations of the contract of commission agency are regulated in greater detail by acts of the

Central Union of Consumer Unions of the USSR (*Tsentrosoiuz SSSR*) and the Ministry of Trade of the USSR.[1]

Under the contract of commission agency, one party (the commission agent) binds himself on the direction of another party (the client) to perform for him one or more transactions in his own name for a fee (Art. 404 Civil Code). Like an agent, a commission agent carries out transactions (legal acts) for another party. However, a commission agent acts in his own name while an agent acts in the name of the principal. Thus, the seller in the sale, on commission, of property belonging to a citizen or juridical person is a commission agent.

The obligation of the client to pay a fee to the commission agent, i.e. the compensatory nature of the contract of commission agency, is another feature distinguishing it from the contract of agency which may also be gratuitous.

The contract of commission agency must be in the *written form* irrespective of its participants or the amount of money of the contract (Art. 405 Civil Code). There are various methods of formalizing the contract of commission agency. The contract of commission agency for the sale of agricultural products may be concluded by giving an invoice. Commission stores give receipts upon accepting objects on commission from citizens. Organs of material-technical supply conclude a contract of commission agency or include relevant conditions on the organization of the supply in the contract when they render services in selling unused things of commodity-material value on the basis of commission agency.

2. *Rights and obligations of the parties.* The basic *obligation* of the commission agent is to perform the task entrusted to him in accordance with the instructions of the client and on terms most favorable to the client. If the commission agent performs transactions on terms more favorable than those which were defined in the contract with the client, the entire benefit passes to the client (Art. 408 Civil Code).

Property which passes into the commission agent's keeping

1. *Civil Legislation. Collection of Normative Acts*, Moscow 1974, pp. 913-920.

from the client or which is acquired by him for the client is the client's property. This gives rise to the obligation of the commission agent to take steps to safeguard the property and to secure the safekeeping, transportation, and insurance of the property. The commission agent is liable to the client for the loss, shortage in, or damage to the property of the client which is in his keeping (Art. 412 Civil Code).

Upon completing his commission, the commission agent is obliged to present the client with an account and to hand over everything he received under the commission, and also to give over to the client all rights in relation to a third party which derive from the transaction (Art. 413 Civil Code).

The client is obliged to receive from the commission agent everything done by him in performing his commission, and also to take upon himself all the obligations in relation to third parties which the commission agent assumed in carrying out the commission. The client pays a fee to the commission agent when the commission is completed.

Like the contract of agency, the contract of commission agency has a *personal-principal character*. Therefore, the client has the right to change the commission given by him to the commission agent either entirely or in part before the commission agent concludes a transaction with a third party. However, the commission agent does not have the right to refuse to perform the commission he accepted except when the refusal is based on the impossibility of performing the commission or the violation of his obligations on the part of the client (Art. 419 Civil Code).

The commission agent who performs a transaction (purchase, sale) acquires rights and assumes obligations, although the client was named in the transaction or entered into relationships with a third party during the performance of the transaction. The commission agent is liable to the third party even in non-performed transactions. If the third party does not perform the transaction, the commission agent is not liable to the client (Art. 411 Civil Code) although he is obliged to transfer to the client the money which he receives from the culpable third party.

3. *Obligations under the contract of commission agency for the sale of agricultural products by collective farms.* The Food Supply Program of the USSR for the period until 1990, which was approved by decision of the May 1982 Plenum of the Central Committee of the CPSU, provides for increased commerce of agricultural products which are bought by consumer cooperatives. Therefore, the contract of commission agency for the sale of agricultural products is even more widely used. In accordance with the Conditions and Procedure for Accepting Products from Collective Farms on Commission, which were confirmed by the Central Consumers' Union on 8 June 1967, relationships between collective farms and consumer cooperative organizations are formalized by the signing of a contract, an agreement, or the exchange of letters and telegrams. In addition to the concluded contract, an invoice must be created for the products accepted by every party. The sale price of the products is determined by the parties at the moment of acceptance. In the sale of products beyond the limits of the district of its activity, a consumer cooperative organization has the right to conclude contracts on a subcommission basis with other consumer cooperative organizations. In this case, the commission agent who concludes the contract with them carries out payment with the collective farms. Payments are carried out within three days after the sale of the agricultural products. In the sale of products beyond the limits of the administrative district in which the collective farm is located, the consumer cooperative organization gives to it an advance of 75% of the value of the goods received on commission.

4. *Special features of the contract of commission agency between citizens and commission stores.* The rules of trade in commission stores for the sale of manufactured goods were confirmed by the Ministry of Trade of the USSR on 12 August 1968.

The subject of the contract may be an object which citizens have the right to acquire and to sell. The sale prices must not exceed state retail prices. Commission stores are not allowed to refuse to accept from a citizen on commission objects which are still useful and which do not have more than 50% of wear and tear. Upon accepting objects on commission, the store writes a receipt and a label for the item. The commission fee is established at 7% of the price at which the object is sold. The period for the sale is 45 days. If the object is not sold within that time, the commission agent, with the agreement of the client, has the right to lower the price of the article. The store must pay the client the money for the sold object on the third day after the sale.

Chapter 34. Contract of Deposit for Safe Custody

1. *Concept and significance of the contract of deposit for safe custody.* The purpose of the contract of deposit is to secure safe custody of socialist and personal property. It is widely used in mutual relationships involving citizens and socialist organizations. Citizens deposit articles in check-rooms, in stations, and in ports. Stores provide custodial services for purchased articles. Citizens may give articles of personal and of domestic use into the custody of a pawnbroker, internal loan bonds into the custody of a savings bank, and outerwear to a cloakroom. Often, citizens render custodial services to each other.

As a general rule, relationships involving safe custody are formalized by *contract.* However, the obligation to provide safe custody may also arise under force of *law.* For example, if the recipient of goods which were delivered to him by mistake refuses to pay for them, he is obliged to keep them in safe custody until the organization which owns the goods asks for their return or until they are delivered to another party. Under Article 145 of the Civil Code, an organization is obliged to safeguard a found object which the finder transferred to it, and under Article 147 of the Civil Code, a collective or state farm must safeguard untended or straying cattle given over to it.

The Civil Code devotes Chapter 37 (Arts. 422-433) to the contract of deposit. There are also special norms on deposit. They are included in the Model Statute on Pawnbrokers, the Rules of Deposit of Hand Luggage which are applicable in various types of transport, etc.

Under the contract of deposit, one party (the custodian) binds himself to safeguard the property deposited by another party (the depositor) and to return this property safely (Art. 422 Civil Code).

Deposit is a *manual contract*: it goes into effect from the moment that an object is given over into safe custody. However, in accordance with Article 422 of the Civil Code, socialist organizations may enter into a contract providing for the obligation of the custodian to accept property of the depositor for safekeeping. This contract is *consensual*: it goes into effect from the moment that the parties reach the relevant agreement.

The contract of deposit is *gratuitous*. However, the law on contract may provide for payment for custodial services. This is usually the case when the custodian is a socialist organization whose functions include that of custodian (for example, citizens must pay for the safekeeping of articles at stations or ports, in pawnshops, etc.).

The contract of deposit is *bilateral*: it is not only the custodians but also the depositors who have obligations under this contract (in a compensatory contract of deposit the depositors must pay a fee, and in a gratuitous contract they must compensate the custodian for expenses incurred; in addition, a depositor has the obligation of taking back his property promptly).

The premise of a contract of deposit is that the depositor gives an object to the custodian for temporary possession. This feature distinguishes the contract of deposit from the contract of protection. Under this latter contract, an organization (for example, the in-house security in an institution) binds itself to take measures to protect objects which continue to be in the possession of an enterprise, institution or citizen to whom it belongs.

2. *Parties to the contract. Its form*. Either citizens or *socialist organizations* may be parties to a contract of deposit. Neither the Civil Code nor the special rules contain any norms on the procedure for concluding a contract of deposit. Therefore, the procedure which is provided in the general conditions on the law of obligations, secured in the Civil Code, are applied. As a rule, either party may initiate the conclusion of the contract. However, if it is a plan contract of deposit, the custodian must work out the project.

The contract of deposit may be concluded either *orally* or *in writing*. The written form is necessary for a contract in which a citizen participates and in which the object given for custody is worth more than 100 rubles; in this case it does not matter whether the services for safekeeping are gratuitous or for payment. The violation of this rule gives rise to the consequences provided in Article 46 of the Civil Code: in the case of a dispute, the parties are deprived of the right to refer to the testimony of witnesses in confirmation of the transaction. However, this general rule on the form of the contract has a number of exceptions (Art. 423 Civil Code):

first, if the object is deposited for a short time in the cloakroom of an institution, enterprise, or organization and a ticket or token is received, it is not necessary to have a written contract regardless of the value of the object;

second, if the depositor declares that the custodian returns another article rather than the one he accepted for custody, both the depositor and custodian may refer to the testimony of witnesses in order to confirm or dispute the other's claim;

third, if the object is given into custody under exceptional circumstances (at the time of a fire, flood, etc.), the rules regarding the obligatory written form are generally not in effect. This means that no matter what the value of the object, the depositor is not deprived of the opportunity to refer to the testimony of witnesses in regard to the giving over of the object even in the case of an oral transaction.

3. *Rights and obligations of the parties.* The custodian is *obliged*:

a) to accept the objects given for safe custody from the depositor (only in a contract based on a plan act);

b) to take care of the property given into his custody in a gratuitous contract as if it were his own; in other cases, to take all measures provided in the contract or necessary to safeguard the property (Art. 425 Civil Code);

c) to return the article at the first claim of the depositor. The depositor may present a claim at any time, even before the expiration of the period that the object was given to the custodian.

As a general rule, the custodian may not use the object given to him with the exception of circumstances for which something different is provided in the contract or law. Thus, under Article 374 of the Code of Civil Procedure, a party which received an object for custody from a sheriff may use it if this does not lead to its ruin or lessens its value.

If an object is left until it is to be claimed or without any determined period, the custodian has the right to demand at any time that the depositor take it back. He must only give the depositor sufficient time under the given circumstances to take the object back.

The depositor has the following obligations:

a) to take back his property on time;

b) to pay the fee in compensatory contracts. The amount must be established by lists, schedules, and tariffs, and in their absence, by agreement of the parties;

c) to compensate the custodian for expenses incurred in the safeguarding of the property. This is the case only for necessary expenditures, i.e. expenditures without which the object would have been destroyed or damaged.

4. *Liability of the parties*. In a consensual contract, the custodian is liable for refusing to accept an object for safe custody. In all other instances, the custodian is liable only for the loss, shortage in or damage to the object. As a general rule, the custodian is liable if the loss, shortage in or damage to the property was caused through his fault. But if an organization's functions include providing safe custody (refrigeration organizations, elevators, railroad stations, etc.), it is liable not only for actions which are its fault but also for those which are not. It may be exempt from liability only if the contract was violated as a result of *force majeure* (Art. 427 Civil Code): fire, flood, tornado, etc.

If the object was ruined when the depositor was late in claiming it (did not claim his object on time), the liability of the custodian is decreased: he is responsible for the loss, shortage in or damage to the object in safe custody only if he acted intentionally or was grossly negligent.

If the depositor does not perform his obligation of claiming the object promptly, it may result in yet another unfavorable consequence for him: the custodian-citizen may request the court to allow the enforced sale of the object. If the custodian is a socialist organization, the unclaimed property is sold by a procedure established in its charter (statute). The amount received must be returned to the depositor with a deduction for the service rendered to him by the custodian.

As a general rule, the custodian carries limited liability: if the property is lost or suffers a shortage, the custodian must compensate only for the value of the lost or decreased property, and if the property is damaged, he pays the amount of the decrease in its value. If the object is appraised when it is handed over for deposit, the custodian pays the amount given at the appraisal. He has the right to prove that the actual value of the lost, decreased in quantity or damaged property is lower than the appraisal.

Sometimes it happens that an object changes somewhat while it is in safe custody and it is impossible to use it for its primary function (a violin cracks from the heat, a vase loses one handle, etc.). In these cases, the object is considered to be altered, and the depositor has the right to demand compensation for its value.

Hotels, rest-homes, sanatoriums, hostels, and similar organizations are under stricter liability (Art. 429 Civil Code). They are liable for the safety and preservation not only of the property which was given to them for safe custody, but also of any other property which is in the residences set aside for citizens (for instance, for the theft of a coat in a hotel room). This liability does not extend to money and valuables: there is liability only if they were taken into safe custody.

Article 431 of the Civil Code provides for the responsibility of the depositor: he is obliged to compensate the custodian for losses caused by the nature of the article taken into safe custody where the custodian did not and ought not to have known of its nature (for instance, where a caustic substance leaks from one of the suitcases given for safe custody and damages other objects in the storage room).

5. *The contract of deposit of unidentified things (Art. 432 Civil Code).* Things which share generic traits are kept in the same type of safe custody. For instance, an elevator may store grain and allocate a definite storage area for this purpose to an organization. It may also operate differently: to mix the grain of various organizations in one storage area. The latter case is an example of the so-called irregular (i.e. "incorrect") deposit. In irregular deposit, all the depositors are share co-owners. For instance, if the elevator mixed the grain of ten collective farms, each of which gave twenty tons, each one of them has the right to a share ownership of 200 tons of grain, the share of each collective farm being one-tenth of this. The parties have the right to make provision that the property handed over by them passes into the ownership of the custodian. Then the custodian has the obligation to return to them that number of things of the same type and quality as that which was given over earlier. In the latter instance, a contract of deposit turns into a contract of loan.

Chapter 35. Obligations Arising as the Result of Causing Harm

1. *Concept of obligations which arise as the result of causing harm.* The Soviet state requires of every citizen and socialist organization that it cause no harm to someone else. Those who inflict harm are liable for it, including liability under the norms of civil law. Civil liability consists in compensating for the harm caused. As was pointed out in Chapter 20, this type of liability arises first of all for that party which did not perform, or improperly performed, its contractual obligations.

Civil liability may also arise in matters not connected with the violation of contractual relationships when the parties are not in a contractual relationship. These liabilities have a wide range of application. They arise in cases of damage or theft of another's property, causing bodily harm, etc. Sometimes the party causing harm is subject to both civil and criminal liability. However, civil liability may arise independently of criminal liability.

The purpose of civil liability in this case is to secure compensation for the harm caused to the victim. Besides this, like other legal liabilities, it has the purpose of preventing violations of the law. The obligations for causing harm are used by citizens to realize their right, granted in the USSR Constitution (Art. 57), of judicial protection against attacks on their life and health, their personal freedom and their property. The protection of socialist property is also realized with the help of obligations arising from the causing of harm. The necessity for such protection is provided in Article 10 of the USSR Constitution.

The norms regulating the obligations for causing harm are contained in Chapter 40 of the Civil Code (Arts. 444-471). A number of important norms are contained in the Rules on the Compensation for Harm Caused to Workers and Employees by Mutilation and Other Harm to Health Connected with their Work by Enterprises, Institutions, and Organizations. These rules were confirmed by decree of the State Committee on Labor and the Presidium of the All-Union Central Trade Union

Council on 22 December 1961. The decree of the Plenum of the Supreme Court of the USSR of 23 October 1963, "On Judicial Practice in Suits on Compensation for Harm," and the decree of the Plenum of the Supreme Court of the RSFSR of 7 February 1967, "On Some Problems Arising in Court Practice in Applying the Norms of the RSFSR Civil Code Regulating the Causing of Harm," were devoted to the adoption of these acts. Analogous decrees were issued by the Plenums of the Supreme Courts of the union republics.

2. *General conditions (grounds) for liability for causing harm.* The general grounds for liability for causing harm are defined in Articles 444-445 of the Civil Code. The contents of these articles makes it possible to draw the conclusion that such liability arises if four conditions are present simultaneously: *the occurrence of harm, unlawful behavior of the causer of harm, causal relationship between the unlawful behavior and the harm, and the fault of the causer of harm.*

This obligation arises if the victim suffers *property damage*, that is, harm which may be expressed in money. In cases when clothing is torn, a motorcycle is damaged, a crystal chandelier broken, furniture or dwelling burned, the harm may be directly measured in money: the amount necesary to pay for the repair of the clothing or motorcycle, the value of the chandelier, furniture, or house. If as the result of harm a victim loses an arm, the harm caused may be expressed in the loss of either full or partial amount of wages for work. When a locksmith loses an arm, he may have to work as a janitor. The difference in work wages before and after the injury constitutes the harm caused. In contrast to property damage, non-property damage (for instance, the feelings of illness or mental anguish of a girl whose face has lasting scars as a result of the actions of someone causing harm) is not provided in law. This is because it is impossible to measure mental or physical suffering in money.

Only the party whose action was *unlawful* must compensate for harm. Unlawful behavior may be expressed either in *action* (vacationers did not put out the campfire and as a result part of a forest burned) or *inaction* (the management of an enterprise did not observe the rules of industrial safety which leads to occupational injuries). In the second, someone did not do what ought to have been done according to law.

Sometimes harm may be caused as a result of actions not linked with the violation of the law, that is, *legally*. This happens when a party is authorized by law to cause harm to another (for instance, performing an operation and amputating the arm of a sick person, killing cattle on grounds of a decree of the organs of veterinary supervision because of an epidemic, etc.). Such harm is not subject to compensation. Necessary defense is also an example of lawful action. A person who harms hoodlums or robbers in the course of defending himself against them is not subject to either criminal or civil liability (Art. 13 Criminal Code, Art. 448 Civil Code). Actions taken in instances of urgent need, i.e. to prevent danger to one's own life, to the life of other citizens, or to the interests of the state or society are also considered to be legal (Art. 14 Criminal Code). However, civil legislation also may place liability to compensate for harm on those who act in situations of extreme necessity. Therefore, if a motorcyclist veers sharply and breaks a store window in avoiding a collision with a pedestrian, the court may make him liable to compensate for its value. In addition to this, the law (Art. 449 Civil Code) allows for the possibility of placing the liability to compensate for the harm on the person in whose interest the person causing the harm acted, and discharges from liability a third party as well as the causer of harm. The court makes such a decision after considering the circumstances under which the harm was caused.

Only a harm which is in a *necessary causal relationship with the unlawful behavior of a party* may be compensated. If an automobile is damaged as the result of a collision and later a tool is stolen from its trunk while it is being repaired, the causer of the harm is obliged to compensate for the harm which arose as the result of the damage to the automobile, but not for the result of the theft since this is not directly linked to the actions of the causer of harm.

The obligation to compensate for harm arises if *fault* is an attribute of the actions of the causer of harm: *intention* or *negligence*. Here, the causer's fault is presumed: this means that the causer may be discharged from liability if he can show that the harm was not caused by his fault. There is no compensation for a harm which occurs accidentally.

In cases when harm is caused by the fault of the workers of an organization in the performance of their work (employment) duties, the organization and not the workers is liable to the victim (for instance, if workers who are performing outside repairs to a house ruin the clothing or health of a passerby in dropping building material, the repair organization is liable for the harm caused). The organization is liable not only if permanent, but also if temporary or supernumerary workers are involved. In compensating for harm caused, an organization has the right to file a regressive claim against its workers. The question as to the grounds and amount of regressive liabiltiy is determined by the norms of labor legislation concerning property liability of workers and employees.

The obligation to compensate for harm caused arises if the general conditions noted above are present unless otherwise provided by law. The law distinguishes several circumstances for which special grounds for liability, which do not coincide with the general grounds, are established, as well as a special range of parties which is liable for harm caused. They are: first, liability for harm caused by increased danger; second, liability for harm caused by minor and incapable persons; third, liability for harm, caused by the actions of officials in the area of administrative management.

3. *Liability for harm caused on the grounds of increased danger*. Article 454 of the Civil Code provides that organizations and citizens whose activities involve increased danger to their neighbors (transport organizations, industrial enterprises, automobile owners, etc.) are obliged to compensate for harm caused on the grounds of increased danger unless they can show that the harm arose as the result of force majeure or the intention of the victim. Thus, the party engaged in

activity linked with increased danger is subject to stricter liability: it must compensate for harm not only when it is *at fault* but also when there is *no fault* on its part. The purpose of this norm is to stimulate those parties, who are engaged in activities posing dangers to their neighbors, to be as careful as possible. They must be more careful than others to perfect techniques, to repair machines and machinery promptly, etc.

An activity involving increased danger is one in which objects are utilized whose natural or endowed properties and characteristics cause them to be not entirely controllable, and their use creates an increased probability that harm will be caused. For instance, an activity having the nature of increased risk is one which involves the use of moving objects which attain great speeds and are therefore difficult to brake instantaneously. Such objects include automobiles, trains, motorcycles, ice-boats, machine-tools, threshing machines, etc.

Not only qualitative but also quantitative features are necessary to define such an activity. For example, storing large amounts of flammable material, setting up shooting ranges, and using high-tension currents are activities involving increased danger. However, storing gasoline in a canister or storing a hunting rifle are not considered to be examples of increased danger.

Caring for wild animals is considered an activity of increased danger but not caring for domestic animals. Therefore, the owner of a cow is liable only for actions based on his fault if the cow gores a neighbor and causes bodily harm.

Objects whose use creates increased danger for those in the neighborhood are called sources of *increased danger*. It is a special feature of compensation for harm caused by a source of increased danger that the question of the subject of liability (who must compensate for the harm) and the grounds for liability is decided differently.

According to Article 454 of the Civil Code, the possessor of the source of increased danger is liable to the victim for harm caused. The possessor of the source of increased danger is the party possessing the source on legal grounds: its owner or the organization possessing the object under the right of operative management, as well as the party possessing it under contract of lease or rent, power of attorney, or the right of management and disposition (the transfer of the possession of an automobile belonging to a citizen by right of personal ownership is usually accomplished by power of attorney), etc.

The party to whom the source is given for technical operation only is not liable for harm caused. If B. gives D. the right to drive his automobile, it is B., as the possessor of the source, who is liable to the victim, and not D. who was behind the wheel. Only after B. compensates for the harm to the victim may he demand compensation from D. for that amount. However, if the possessor of the source is liable to the victim irrespective of fault, then the party operating the source is liable to its possessor on general grounds, that is, only if he was at fault. This means that situations may occur under which the possessor, in compensating for the harm to the victim, will not receive anything from the party which operated the source of increased danger.

The party operating the source in the course of a work relationship with the possessor (chauffeur, machine operator, lathe-operator, etc.) is never considered to be its possessor. In these situations, the possessor is considered to be the appropriate organization which must compensate for the harm caused by the worker.

In the examples given above, the possessor of the source himself gives it over to another party. But it can also be otherwise: the source may leave the possession of the possessor without his will. This happens most frequently when automobiles are stolen. If the person who stole someone else's automobile has an accident and causes harm, the question arises: who is liable to the victim, the possessor or the person who stole and drove the vehicle at the time of the accident? The answer is given in the decree of the Plenum of the Supreme Court of the USSR of 23 October 1983: the possessor of the source of increased danger is not liable for harm caused by the use of this source if he can show that it left his possession because of the unlawful actions of third parties through no fault on his part. This means that if the possessor took the key from the lock of the ignition and locked the vehicle, the person stealing the car is liable to the victim according to Article 454 of the Civil Code. If the possessor does not take the necessary precautions and as a result creates conditions conducive to theft, the liability for harm caused may be placed either on the person stealing the vehicle and causing the harm or on the negligent possessor.

As was already noted, the possessor of a source of increased danger is liable for harm caused *regardless of whether it is his fault or not*. He may be discharged from liability only under two circumstances: if the harm occurred as a result of *force majeure* or through the *intention* of the victim. *Force majeure* is an extreme and inescapable event in a given situation. In actuality, the causing of harm by a source of increased danger as a result of *force majeure* occurs relatively seldom. *Force majeure* has significance above all in air transport, which is harmed more than other transport organization by the natural forces of *force majeure*. However, in the interests of the victims, the state has found it necessary to increase the liability of an air transport carrier for the causing of harm even in comparison with all other possessors of sources of increased danger. In accordance with the decrees of the Air Code of the USSR, in the instance that death, injury, or other harm occurs to the health of a passenger as the result of *force majeure*, the carrier may be discharged from liability only if it can show that the occurrence of or increase in harm was caused by the intention or gross negligence of the victim himself.

4. *Liability for harm caused by minors and incapable parties.* The capacity to bear independent property liability for one's actions is an aspect of civil dispositive capacity. Therefore, those who do not have civil dispositive capacity are not responsible for their actions. This applies above all to minors under the age of 15. Liability for harm caused by such minors is placed on their parents (adoptive parents) and guardians who are obliged to provide their education and are

responsible for their constant supervision. In addition to this, Article 450 of the Civil Code provides that if the harm is caused by minors while they are under the supervision of educational, training, or health care institutions (schools, professional-technical schools, hospitals, sanatoriums, etc.), the liability for the compensation for harm is placed on these establishments or institutions.

The liability of the parties mentioned above is based on general grounds; in particular, they must compensate for harm if they cannot prove that the harm was not caused through their fault. The Decree of the Plenum of the Supreme Court of the RSFSR of 7 February 1967 examines the possible fault of parents in not exercising necessary supervision over minors at the moment when the harm was caused or in being irresponsible in regard to their upbringing. The fault of parents (guardians) may manifest itself in the connivance at or the encouragement of mischief or hooligan behavior, lack of supervision of children, lack of attention toward them, etc. The fault of the relevant institutions manifests itself in not carrying out proper supervision over children at the moment that harm was caused. If fault cannot be shown in the actions of parents, guardians, or relevant institutions, they are not considered liable.

Let us consider an example: during recess, pupils started a dangerous game: throwing a needle tied to a pencil against the blackboard. Pupil B. threw the needle into the eyes of his classmate, M. It took a long time for M. to recover, and it was necessary to remove his eye. M.'s father brought a suit against B.'s parents and the school for the recovery of the expenses in connection with the recovery and care of his son: the cost of the application to stay in a sanatorium, expenses for consultants, etc. The People's Court granted his claim, ordering B.'s parents and the school to pay the money. However, a higher court countermanded the decision, deciding that only the school was liable for the harm caused. To affirm this conclusion, the following reasons were given: B. had shown exemplary behavior during the entire time of his schooling, he had never had a claim brought against him, and his parents had shown great concern in regard to his education. The person in charge of the class was at fault in the incident by not exercising necessary supervision over the children. The school reprimanded him for his irresponsibility and for not taking measures to stop the game. The school carries property liability for the faulty behavior of the class supervisor.

Minors between the age of 15 to 18 are considered to have limited dispositive capacity. At the same time, the law establishes that they have civil liability for causing harm (Art. 451 Civil Code). Like parties who are entirely incapable, they must compensate for harm caused on general grounds. However, the lawgivers took into account that these people do not always have the ability to actually compensate for the harm they caused (as a rule, they do not work but are studying and have no property). Therefore, in the interests of the victim it was established: that if the minor who causes harm has no property, earnings, or other income with which to compensate for the harm, the harm must be compensated in the appropriate amount by the faulty parents or guardians who did not raise or supervise their children properly. Thus, the parents (guardians) have only supplementary liability. This liability is limited by time: as soon as the causer of harm himself attains the age of 18 or has the possibility of compensating for the harm, the supplementary liability of the parents is terminated and the obligation to compensate for the harm is transferred entirely onto the causer of harm (Art. 451 Civil Code).

The liability for harm caused by a citizen considered incapable because he is emotionally disturbed or retarded is based on the same principles as are the actions of person below the age of 15: the guardian or organization responsible for his supervision must compensate for the harm. Also, the organization responsible for his supervision is liable for poorly supervising him or refraining from supervision (for instance, if employees in a health-care institution, knowing of the dangerous state of health of an incapable person, refuse to hospitalize him or release him from the hospital too soon).

Special rules are established in regard to parties who cause harm when they are in a state which makes it impossible for them to understand the consequences of their actions or to control them (for example, in a state of violent mental disturbance, narcotic or alcohol poisoning, etc.). These persons are discharged from civil liability. However, the law provides that the party who brought himself into such a state through the use of alcohol, narcotics, etc. is obliged to compensate for the harm caused (Art. 453 Civil Code).

5. *Liability for harm caused by the actions of officials.* This liability of officials is provided in Articles 446 and 447 of the Civil Code. Article 446 defines the procedure for the compensation of harm caused by officials within the scope of their administrative duties. The following situations in which harm may be caused were kept in mind:

first, the official duties of an *official.* Officials are persons who carry out the functions of the representatives of authority, and also perform duties in organizations and enterprises which are related to the performance of organizational-regulatory duties;

second, the actions of the officials were *illegal.* The illegality of actions may appear in their unlawfulness as well as in the fact that they were performed by exceeding authority or without sufficient legal grounds (for instance, a 15-year old pedestrian was fined for violating a traffic regulation while a fine may be imposed only on a person who is at least 16 years old);

third, the unlawful actions of an official performed in the area of *administrative management*, i.e. having authoritative character (for instance, the damage caused by the demolition of a garage on the unlawful order of the chairman of the district executive committee).

The Civil Code establishes various consequences for the causing of harm by officials within the scope of administrative management, depending on whether the victim is a citizen or an organization. The appropriate institutions, organizations, and enterprises whose officials cause harm are liable to citizens on general principles if not otherwise provided by law.

The harm caused by such actions of officials of organizations is compensated by procedure established by special law. Included among such laws at the present time is the Comprehensive Law on the Regulation and Confiscation of Property[1] which provides for the right of an organization to demand from a relevant institution compensation for harm caused by the unlawful confiscation and regulation of property. Under Article 84 of the USSR Merchant Shipping Code, a sea port is liable for a collision caused by the fault of its maritime pilots in performing their official duties. Finally, there is the decree of the All-Union Central Executive Committe [*VTSiK*] of the RSFSR of 16 January 1928, "On the Liability for Losses Caused by the Unlawful Meddling of Organizations of Authority in the Activity of Cooperative Organizations."[2]

The consequences of harm caused to a citizen as a result of unlawful conviction, unlawful institution of criminal proceedings, unlawful measures of suppression to place under arrest, and unlawful administrative penalty in the form of arrest or corrective labor are regulated in a special manner. In all these instances, the harm caused is compensated in full by the state (and not the organization causing the harm) regardless of the fault of the officials of the organizations of inquiry, preliminary investigation, procuracy, or court (Art. 89 para. 2 Principles).

The harm caused is compensated by procedure provided in the Edict of the Presidium of the Supreme Soviet of the USSR of 18 May 1981, "On the Compensation for Harm Caused to a Citizen by the Unlawful Actions of State and Social Organizations, as well as Officials in the Performance of their Official Duties," and the Statute, which was confirmed by the same Edict, "On the Procedure for Compensation for Harm Caused to a Citizen by the Unlawful Actions of Organs of Inquiry, Preliminary Investigation, Procuracy, and the Court."[3]

1. *Collection of Laws and Decrees of the Workers' and Peasants' Government of the RSFSR* 1927 No. 38 item 248.
2. *Collection of Laws and Decrees of the Workers' and Peasants' Government of the RSFSR* 1928 No. 11 item 101.
3. *Gazette of the Supreme Soviet of the USSR* 1981 No. 21 item 741.

Compensation is granted not only for property damage (loss of earnings, pensions, other property); the housing, labor, pension, and other personal and property rights of citizens are also restored.

The norms of the Edict of 18 May 1981 realize the principle which is enshrined in Article 58 of the USSR Constitution under which "citizens of the USSR have the right to compensation for harm caused by the unlawful actions of state and social organization, as well as officials in the course of performing their official duties." These norms promote the further strengthening of the protection of the rights of citizens and strengthen the legality of the actions of institutions and organizations.

6. *Amount of compensation for causing harm. Consideration of the fault of the victim and the property situation of the causer of harm.* As a rule, the harm caused must be reimbursed in kind. If an object is damaged or destroyed, the causer must repair the damaged object or give another object of the same type and quality. If compensation in kind is not sufficient, the causer must compensate for losses. The amount of this compensation for losses is determined by the general norms of the law of obligation (Art. 219 Civil Code). In this case, losses comprise the income lost by the victim, and loss and damage to property, as well as the loss of profits which he would have received had there been no violation of the law (see Chapter 20).

The evaluation of the behavior not only of the causer of harm but also of the victim is very significant in deciding the question of the occurrence of liability and its extent. If the victim helped to create the harm by his intentional actions, then he has no right to demand compensation. This rule is applied in all instances of the causing of harm irrespective of whether the liability of the causer is based on grounds of fault or whether it exists independently of fault (the liability of the owner of a source of increased danger).

In instances when it is the *gross negligence* of the victim which help to cause the harm or to increase its effect, the extent of the compensation may be decreased or compensation may be refused. The *simple negligence* of the victim in general does not influence the liability of the causer of harm.

Thus, the court is often confronted with the complex task of deciding how to distinguish between simple and gross negligence. The Decree of the Plenum of the Supreme Court of the USSR of 23 October 1963 requires that judges examine the actual circumstances of the event. In particular, the Plenum pointed out that if the victim is in an intoxicated state at the moment that the harm occurs, his actions are considered grossly negligent. Let us look at some actual examples from judicial practice. The court decided that there was gross negligence in the action of a motorcyclist who crossed over to the side of the road of approaching traffic with a defective hand brake and was hit by a car; in the actions of a passenger who tried to get onto a train while it was moving, lost his grip and fell under the wheels; in the actions of a pedestrian who crossed the road where this was not allowed and was hit by a car.

7. *Liability for injury to the health or for the death of a citizen.* Compensation for harm caused as a result of injury to the health or the death of a citizen has several particularities. They are manifest in the determination of the amount of the compensation and, in a number of instances, in the determination of the grounds of liability. To a great extent, these particularities are caused by the existence of social insurance in the USSR which covers all workers, employees, and collective farm workers from the first day of employment. Social insurance is paid for at the expense of the state. Enterprises, institutions, and organizations make payments toward social insurance without deducting anything from the earnings of workers, employees, and collective farm workers. In the event of injury to their health, victims receive benefits from the social insurance account during temporary inability to work, or pensions if they become invalids or if the breadwinner dies. But if the amount received as benefits or as a pension is lower than the earnings which the victim had been

receiving, the question arises as to the necessity of compensation for harm in the form of the difference between the prior earnings and the allotted pension (benefits during temporary inability to work).

If the injury to health or the death is the result of the actions of a citizen, the latter is liable to compensate for the harm caused by his guilty actions, and if the harm is caused by a source of increased danger, compensation must be paid regardless of fault. This issue is decided in the same manner when the causer of harm is an organization, and the victim an outsider to the organization (not its worker).

It is another matter when the organization (plant, factory, state farm) causes harm to its worker, that is, to one for whom it contributes payments for social insurance. In this instance, the organization is obliged to compensate for harm only if it was at fault. The rule on the increased liability of the owner of a source of increased danger does not apply in this situation.

The fault of the enterprise usually consists of not observing the rules of industrial safety. An enterprise is also guilty if it allows people to work who have no preliminary knowledge of the rules of safety. The fault of the enterprise may be shown by the acts regarding accidents, decisions under which workers were judged to be guilty of violating the rules of industrial safety, the decree on imposing disciplinary penalties, etc.

Liability for the harm caused to the health of a citizen is shown by compensating the victim for earnings which were lost by him through the loss or decreased ability to work. The decisive factor in determining the amount of compensation is the degree of loss of professional or general ability to work (professional ability to work is the ability to work in a specific profession, and general ability is the ability to work in a job where no special skills are necessary).

If the professional ability to work is only partially lost, general ability to work is not taken into consideration. Only when professional ability is entirely lost is the degree to which the general ability to work is retained taken into consideration. The degree of loss of professional and general ability to work is defined by the Medical-Labor Expert Commission [*VTEK*].

Thus, if the *VTEK* establishes the loss of professional ability to work at 80%, and the general ability at 30%, and if the average monthly earnings prior to the injury were 200 rubles, the amount of lost earnings is established at 160 rubles (80% of 200 rubles).

If the victim lost his professional ability entirely, the compensation is in the amount corresponding to the degree of the retained general ability to work. For instance, a victim received earnings of 200 rubles, lost 100% of his professional ability to work, and 70% of general ability. To determine the amount of his lost earnings, 21 rubles are deducted from 200 rubles (earnings of an unskilled worker are 70 rubles, retaining 30% ability to work).

When the victim is considered to be an invalid, the appropriate pension is given. The compensation for harm is decreased in order to calculate this amount. Therefore, if in the examples above, a victim was granted a pension of 50 rubles, then in the first example, he will receive 110 rubles as compensation for harm (160 minus 50), and in the second example, an amount of 129 rubles (179 minus 50).

Besides earnings, the victim must also be compensated for supplementary expenses which he incurred as a result of the injury to his health. This includes the money spent on special food, prosthetic devices, cures in a sanatorium/resort, special care, acquisition of a motor wheel-chair, etc. As a rule, the necessity for these supplementary expenses must be confirmed by the *VTEK*.

Minors under 15 years of age may also be victims. In this instance, the liability of the causer of harm is limited to compensation for the expenses necessary to restore health. When the minor attains the age of 15 (when he may begin his work activity), his compensation for harm must take into consideration loss of ability to work; his earnings are equated with the earnings of an unskilled worker. Finally, after the minor actually begins his work activity, he has the right to demand an increase in the compensation for harm, based on the amount of compensation for a worker with his skills.

If the victim dies, the right of compensation for harm passes to parties unable to work who were his dependants or who had the right to receive support from him. These parties include children under 16 years of age (if they are students, the age is 18), parents, spouses, or other dependants who have reached pension age (55 for women, 60 for men), and invalids. The spouses and parents of the deceased have a right to compensation regardless of age and capacity to work if they do not work and are occupied with caring for children, grandchildren, brothers, or sisters of the deceased under 18 years of age. All of these persons receive compensation for that part of the deceased person's earnings which they received or would have received when he was alive.

For example, if the earnings of the deceased were 150 rubles and his dependants consisted of a wife and a minor son, the share which his wife and son receive after his death consists of two-thirds of his wages of work or 100 rubles. In determining the amount of compensation, the pension given in case of loss of the provider to his dependants is taken into consideration. In the example given above, if the wife and son of the deceased are granted a pension of 30 rubles each, the amount of compensation totals 40 rubles (100 minus 60) or 20 rubles each. If the dependant wife is able to work, only the minor son has a right to compensation for this amount.

Chapter 36. Obligations Arising as a Result of Saving Socialist Property

1. *Concept of obligations arising as a result of saving socialist property.* The USSR Constitution (Art. 61) places the obligations to safeguard and secure socialist property on citizens of the USSR. It is the duty of a citizen of the USSR to fight against the theft and destruction of state and social property, and to treat the wealth of the people with care.

In performing their constitutional obligations, the Soviet people often risk their property and health in attempting to prevent any harm threatening socialist property. These patriotic deeds do not always end well. Sometimes they lead to a loss of or damage to the property of a citizen, to its mutilation, or even to its destruction. And then the question arises as to the necessity of compensating for the damage to the person who participated in extinguishing a fire, preventing an accident in an enterprise or institution, restraining a dangerous criminal who is robbing or attempting to rob a store, etc. If the person dies, the property interests of the members of his family must be secured.

The state shows its concern for those who suffer injury or other damage to their health while performing their civic duty in protecting socialist property by paying them benefits during the periods when they are unable to work and pensions in the maximum amount. However, benefits and pensions do not always compensate in full for the earnings of the victim. Who must pay the difference between the earnings and the pension (benefits) of the victim and under what conditions?

If one is guided by the general norms on civil liability for causing harm, which were covered in the last chapter, the claims of the victim ought to be refused since the factory or store whose property was saved by the victim may not be considered a causer of harm. And, of course, it is impossible to consider that these organizations are at fault in the occurrence: it is the victim himself who decided to take those actions which resulted in causing him harm.

The questions concerning compensation for harm caused to citizens in saving socialist property are regulated by special norms which are contained in Article 472 of the Civil Code. The primary aim of the norms is to stimulate appropriate actions on the part of Soviet citizens. At the same time, these norms secure compensation for harm to those who suffered it as a result of saving socialist property.

2. *Conditions giving rise to obligations.* According to Article 472 of the Civil Code, an organization whose property was saved by the victim from the danger threatening it must compensate for harm caused to the victim. It is possible to distinguish the following conditions (grounds) which give rise to these obligations:

First, in obligations which arise as the result of saving socialist property, as in other tort obligations, compensation is only for *harm to property*, that is, harm which may be expressed in money. All other harm (for instance, that related to pain and suffering) may not be compensated.

Second, the socialist property which was saved by the victim must have been threatened by a real danger of being ruined or damaged. This danger may be created by the actions of people (for instance, theft, intentional or careless treatment of property, etc.) as well as by natural events (fire, explosion, flood, tornado, etc.). The significance of this condition is that there is no compensation for harm caused by averting an apparent danger, as, for example, if a fire company pretends that there is a fire for instructional purposes. When such measures are undertaken, those in the immediate area must be properly informed. If it can be shown that an organization did not do this, it is liable to a citizen who rushed in to extinguish the fire under the norms on obligations arising as the result of causing harm.

Third, the actions of the victim must have a strictly defined purpose: to save socialist property. The following case was analyzed in judicial practice: T. entered a savings bank with the aim of robbing it and ordered the employees and clientele to keep quiet. D. made a remark to him. T. aimed his pistol at her, and when D. tried to turn it aside, he shot her. D. requested compensation for the harm caused by injury to her health from the savings bank. However, the court decided that Article 472 of the Civil Code could not be applied in this instance. D. tried to defend only herself, and did not know that T. intended to rob the savings bank. D. could demand compensation for harm from T. on the basis of Article 444 of the Civil Code since her claims arose from obligations arising as the result of causing harm, and not from saving socialist property.

In order that there be an obligation to compensate for harm to the victim, it is not necessary that his purpose was entirely achieved. This conclusion may be drawn from the very name of the institute: "Obligations Arising as the Result of Saving (and not rescuing of) Socialist Property." This means that the person who by his actions prevented the destruction, loss, or ruin of property, as well as the person who did not succeed in doing so (the building burned down although he tried to put out the fire), has a right to compensation for harm.

Fourth, it is necessary that there is a causal relationship between the actions of the victim and the harm caused. If this necessary causal relationship does not exist, an organization has no obligation to compensate for harm. Thus, an organization is under no obligation to compensate for harm to a citizen who, while returning from extinguishing a fire, falls into an open sewer along the road. In this instance, the victim must press his claim for compensation for harm against the organization which oversees the sewer system. It must respond to the victim on the basis of the obligation arising as the result of causing harm.

3. *The parties and content of obligations*. In the obligation under consideration, only a *citizen* may be a creditor (victim). If an organization undertakes the actions to save, special rules are applied. These include, for example, Articles 260-272 of the Merchant Shipping Code of the USSR, which are devoted to the procedure of payment of award when one

ship participates in the salvaging of another ship which is in danger. It is not of significance whether or not the victim has dispositive capacity in order that this obligation arise. An organization is obliged to compensate the harm caused to any person.

The right to claim compensation under Article 472 of the Civil Code is granted only to a person whose actions are not linked with the performance of his employment (official) functions. The decree of the Plenum of the Supreme Court of the USSR of 23 October 1963 points out that this norm may not be applied in situations when the victim saved the property of an organization in the cause of performing his official duties. These would include firefighters, rescue workers, mountain-rescue teams, etc. These persons may claim compensation for harm caused only from the organizations with which they are linked by labor relations, and only on grounds established in Article 460 of the Civil Code (it is necessary, for instance, that the organization be at fault in causing the harm).

The organization is the *debtor* in the obligation. This may be a state organization which has operative management of property, or a cooperative or other social organization which owns property under the right of ownership, or an organization which manages property on other legal grounds. Thus, a railroad is obliged to compensate harm to a person who tried to save transported goods. This is related to the fact that the victim acted in the interests of the railroad: if the goods were lost or damaged, the railroad would have had to pay appropriate compensation to the owner of the goods.

From the standpoint of the scope, nature, and amount of compensation for harm, the persons who participate in saving socialist property must not be placed in a situation which is worse than those who are caused harm as the result of wrongful actions. Therefore, in regard to these matter, Article 472 of the Civil Code refers to the appropriate articles of the chapter (of the Civil Code) on "Obligations Resulting from the Causing of Harm" (see Chapter 35).

Chapter 37. Obligations Resulting from the Unjust Acquisition or Retention of Property

1. *Concept of obligations resulting from the unjust acquisition or retention of property.* Citizens and state, cooperative, and other social organizations may acquire property for ownership or operative management only on grounds provided for in law or contract. Everything which surrounds us at home was acquired at some time on the basis of the contracts of purchase and sale, gift or labor contract by money received in the form of earnings or pensions. Besides this, property may also be received through inheritance. A socialist organization may acquire property on the basis of decisions of superior organs, through contracts of supply or work contract, etc.

There also occur situations when property is acquired without legal grounds, as well as situations when the grounds on which property was acquired are no longer in effect. A citizen received a parcel or monetary transfer which was meant for someone else with the same name. The cashier in a store gave a buyer more change than he should have. A combine dispatched metal for several enterprises in Minsk in one truck, but only one bicycle factory in Minsk received the entire amount – both its own as well as that intended for others. A court decision that a citizen may receive part of an inheritance was overturned by a higher court. In these and similar situations, the necessary legal basis for receiving all or part of the property does not exist. Such acquisition is called *unjust*.

A directly opposite situation may exist: a party should spend a specific part of its property, but kept it without having any basis for doing so. For instance, a party discharged someone else's debt by mistake. A factory received top quality goods but was sent a bill for a lower quality. These situations are examples of unjust retention of property.

A party may acquire or retain property as a result of its own *wrongful* and *culpable actions* (for instance, by presenting false documents on length of employment and amount of earnings, an increased amount of pension was received). Similar situations give rise not to an obligation based on the groundless retention or acquisition of property, but rather to an obligation as a result of causing harm (see chapter 35).

Obligations resulting from unjust acquisition or retention of property occur when the party who acquired or retained property without having grounds for doing so did not perform any wrongful actions for this purpose (he did not know that what he received was not owed to him, or that he paid less than he ought to have). Such obligations have the purpose of protecting the interests of those on whose behalf the unjust acquisition or retention of property occurred. Like obligations resulting from the causing of harm, they are meant to protect the right of ownership – both socialist and personal.

A special chapter in the Civil Code (Arts. 473-474) is devoted to obligations resulting from the unjust acquisition and retention of property. Several categories of these obligations are also regulated by special norms. These norms include Articles 430-432 of the Code of Civil Procedure of the RSFSR. These articles concern situations when specific property is given to the plaintiff on the basis of a court decision, and a higher court later revokes the decision of the people's court and denies the claim. Here the plaintiff's grounds for the acquisition of property no longer exist. The special norms also include Article 75 of the Charter of the Railroads of the USSR which regulate the consequences of overpayment or underpayment of transport fees (in the first instance, the carrier acquires the property without grounds, and in the second, it is the supplier who does this).

A mistake by the party which acquires (retains) the property or by the party on whose account the unjust acquisition (retention) of the property occurred also causes unjust acquisition or retention of property. There are also situations when the unjust acquisition or retention is a result of the actions of a third party. For example, because of a mistake by the employees of *Gosbank*, the value of delivered lumber was deducted from the account of a furniture factory, while the actual recipient of the merchandise was a furniture combine located in the same town.

The following three conditions are necessary in order for an obligation for the unjust acquisition or retention of property to arise:

First, the acquisition or retention of the property must occur "at the expense of another." This means, that the property increases or is kept by one party while another party loses it or does not acquire it although he ought to acquire it (these situations include paying a non-existing debt, performing an obligation for a party who is mistaken for a creditor, and the performance of obligations by a person who mistakenly believes himself to be a debtor).

Second, there is no legal basis for the acquisition or retention of the property. It is possible that there never was such a legal basis or that it once existed but later ceased to exist (a person bought a house under a contract of purchase and sale which was later declared invalid by the court).

Third, the party who acquired the property is not at fault. For example, he received the property without grounds while being convinced that the property was intended for him: he paid less than required only because he did not know that a higher payment was required, etc.

2. *Content of claims for the return of property which was acquired (retained) unjustly*. The participants in an obligation which arises as the result of the unjust acquisition and retention of property are: *the debtor*, who acquired and retained the property unjustly, and the *creditor*, at whose expense this was done. The *content* of the obligation consists of the creditor's claim that the property acquired (retained) unjustly be returned. This does not mean any property. In those instances when such property is an individually-defined object, the party at whose expense it was acquired (retained) unjustly may bring suit *in vindicatio* under the rules of Articles 151-157 of the Civil Code. A person who does not receive a package containing a painting by an artist which is sent to him has a right to bring a suit *in vindicatio* for the return of the painting against the person who received the package by mistake. In other instances, the creditor may bring a suit against a debtor on the basis of Article 473 of the Civil Code.

As a general rule, the debtor is obliged to perform his obligation specifically, i.e. to return to the creditor that object which he received (retained). If a store delivers a furniture suite to a citizen who did not order it because of a mix-up in addresses, the citizen is obliged to return the suite to the store. When it is not possible to return an object specifically, its value must be compensated. According to Article 473 of the Civil Code, the value of the property is determined at the time when it was acquired unjustly.

Sometimes a person acquires or retains a property which may reap a profit. In this case, it is necessary to return not only the property but also the profits which the person gained or should have gained from this property from the moment that he knew or should have known of the unjust acquisition of the property. For instance, the rent for the rental of a *dacha* must be returned if the contract under which the *dacha* was rented is later considered to be invalid.

The law (Art. 474 of the Civil Code) provides for instances when the property acquired unjustly must not be returned. This is property which was transferred in performance of an obligation before the term of performance or after the expiration of the period of limitations. In both instances the debtor may not claim the return of the transferred property by alleging that he did not know that the term for performance had not yet arrived or that the period of the limitation of actions had expired.

An organization which voluntarily, and not because of an accounting error or bad faith on the part of the recipient, pays out an excessive amount or on grounds no longer in effect for authors' royalties or royalties for an invention or rationalization proposal, does not have the right to demand that this be returned. Finally, property and any excess amounts which were paid as compensation for harm in connection for injury to health or death may not be subject to return as property which was unjustly acquired if such payments were made in good faith on the part of the recipient.

3. *Grounds for the recovery in favor of the state of property unjustly acquired (retained).* As a general rule, property which is unjustly acquired (retained) is returned to the party at whose expense it was acquired (retained). However, when the property is acquired at the expense of another party not on the basis of a transaction but as the result of other actions carried out contrary to the interests of the socialist state and society – if not subject to confiscation – it is recovered for the favor of the state (Art. 473 para. 3 Civil Code). Such consequences arise, for instance, in relation to property which was acquired as a result of begging, fortune-telling, or other prohibited trade.

Copyright and the Law of Inventions

Chapter 38. Copyright Law

1. *Concept of copyright law*. Copyright law, the object of which consists of spiritual valuables, holds great significance for the life of our state. It secures the transfer of works for the use of society and also protects the author's interests both as a creator and worker.

The basic tasks of Soviet copyright law are expressed in the USSR Constitution. The Soviet state is concerned about the protection, increase, and widespread use of spiritual valuables for the moral and esthetic education of the Soviet people and the raising of their cultural level (Art. 27). The USSR Constitution guarantees freedom of creativity to citizens by securing it with economic, political, and other guarantees, and provides for state protection of the rights of authors (Art. 47). At the same time, the Constitution also provides for the right of citizens to use the achievements of culture (Art. 46). Thus, copyright law has two tasks – to secure the maximum widespread use of spiritual valuables and their systematic increase.

Copyright law is the totality of norms securing for the subjects of copyright law the exclusive right of disposal of works of science, literature, and art, especially as it is linked with the use of these works, the personal and property rights of the authors of these works, as well as the procedure for the realization and protection of these rights.

2. *Objects of property*. According to Article 475 of the Civil Code, the *objects* of copyright are works of science, literature, and art. The Civil Code does not contain an exclusive listing of the objects of copyright and the listing it does contain is meant to give examples. The law distinguishes three traits which make it possible to determine an object of copyright.

First, it must be a work which is the result of the creative activity of the author; the purpose and merit of the works are not significant. *Second*, the creative result, which in itself is intangible, must find outward expression in objective form. What this objective form consists of is not important; it is important that it makes it possible for third parties to perceive the work. The works may be affixed on any type of material (manuscript, film) and may also be objectively expressed through public delivery or performance. *Third*, it is necessary that the objective form of the work not only be perceived by third parties, but that it is also possible to reproduce it, the form of reproduction not being important.

Works are protected regardless of whether they are published (disseminated to the public), i.e. accessible to an undefined range of persons. In order that a work be considered an object of copyright, the important fact is its existence in an objective form; as a rule, no registration or other formality is required.

3. *Subjects of copyright.* The parties who possess copyright are the subjects of copyright. It is possible to distinguish several categories of subjects. They are the authors of the works, their heirs, and other legal successors, including the state.

The author of the work is the person who produced the work through his creative activity. He is the primary, first bearer of the right to the work and possesses the widest range of rights connected with the work. Some rights may belong only to the author. The author of a work may consist of one or several parties. Authorship of several persons (coauthorship) occurs when they produce a work through joint creative activity (Art. 482 Civil Code). In some instances, a single author of a work is a juridical person. The law contains an exhaustive listing of works which may be the object of such a right, and an exhaustive listing of organizations which may be authors having the right to copyright to these objects (Arts. 484-486 Civil Code). Included among such works are scientific and encyclopedic collections, journals and other serial publications, motion picture films, television films, and radio and television programs. The organizations which issue such works are the subjects of these rights. However, the authors of the individual segments of such a work retain the copyright to the results of their own labor.

The heirs of an author may also be *subjects* of copyright law. They receive all the rights of copyright, except the right of authorship and the right to the use of his name. Besides heirs, other legal successors may also be subjects of copyright law. These are parties who obtained some rights to copyright (as a rule, by a contract with the author or his heirs). The state may also have copyright to use a work if it was compulsorily purchased by the state or declared to be the property of the state at the expiration of the copyright term (Arts. 501-502 Civil Code).

4. *Author's rights.* The basic author's rights are defined in Article 479 of the Civil Code. They include: the right to authorship; the right to name the right to the inviolability of the work; the right to publication, reproduction, and dissemination of the work; and the right to remuneration. Some of these rights may belong to any subject of copyright, while others only to the author himself. These legal rights may be divided into *personal* and *property* rights, but this division exists only for the author himself (and partially for his heirs). *Personal* non-property rights include the right to authorship, name, and inviolability of the work, while property rights include the right of remuneration for use of the work. Let us examine the rights under copyright individually, starting with those which may belong to any subject of copyright.

Publication of a work (making it public) consists of making it known to an undefined range of people and may be realized in various ways: issuing it in printed form, public performance, public exhibition. The right to publish is the right to decide whether the work may be made public through one of these methods.

Reproduction is the repeated conveying of the work in an objective form which may be perceived by other parties. This may be the publication of a work (for instance, the publication of books), as well as the reproduction of works which were already published (for instance, the re-issuance of books).

Dissemination may consist of publishing the work or publicly exhibiting a work which had not yet been reproduced (paintings on exhibition); dissemination may be a form of publishing, but works which have already been published may also be disseminated.

Thus, the rights of publishing, reproduction, and dissemination mean the right to dispose of the work. The law allows these rights be transferred to another person including transfer by contract, as for example, a contract for publication.

The bearer of copyright has the right not only to decide on the use of the work in the original, but also whether it may be translated into another language, as well as whether it may be changed from one genre into another in circumstances provided by law, for instance whether a narrative work be changed into a scenario. The translator, scenarist, etc. will possess the copyright for the translation or scenario.

Some rights may belong only to the author of a work. First of all, this includes the right of authorship, the right to consider oneself the author of a work. This may not be alienated or transferred to another person on any basis, and refusal of this right is invalid.

While the right of authorship exists regardless of whether the work is used, the right to name exists only in relation to its use. This signifies the right to determine the name under which use of the work is realized, above all, in the publication of the work. The author has the right to decide whether the work will be used under his own name, under a conditional (but not another person's) name (pseudonym) or without any name (anonymously). The right to name may also not be transferred and belongs only to the author.

The author has yet one more right – the right to the inviolability of the work. This right signifies that no one may: alter or shorten the work, supply prefaces or afterwords, commentaries or other explanations, illustrations, etc. insofar as this may influence how the work is perceived, without the author's consent. The right to the inviolability of the work also includes the author's right to incorporate changes, additions, etc. into his work. The right to protect the inviolability of the work is realized by the heirs of the author (Art. 481 Civil Code). Limitation of actions is not applied in the protection of these property rights of an author (Art. 90 Civil Code).

The basic property right of an author is his right to *remuneration* for the use of his work. Author's remuneration for use is a form of earned wages, but its legal regime differs significantly from that for wages of work. An author's remuneration is regulated by civil, rather than labor law. It is payment for the use of the work. Therefore, the basis of payment of a fee is not the creation of the work but its use. The extension of the boundaries on use (for instance, a second edition of a book) or use for a new purpose (for example, publication of a translation in another language) is grounds for increasing the amount of the remuneration, regardless of the supplementary expenditures on the work of the author.

The system of author's compensation (royalties) varies for different types of works and the methods of their use, but it has some common principal features. A characteristic of Soviet copyright is that author's remuneration is normative, and only if there is not confirmed rates is the remuneration determined by agreement of the parties.

The law provides the periods and procedure for payment of the remuneration. As a rule, the user (publisher, movie studio) pays the author. However, in some circumstances this payment must be carried out by the All-Union Copyright Agency [*VAAP*], which acts as an intermediary between the author and user.

Copyright is valid only within the territory of the USSR. Copyright is recognized abroad only in cases provided for by international law. Copyright is also limited in time. It continues for the lifetime of the author and for 25 years after his death, calculated from January of the year following the year of the author's death. Copyright belonging to an organization in the capacity of author is in effect for an unlimited period. Copyright which is transferred to the state by will is terminated (Art. 552 Civil Code). After the expiration of the copyright, a work may be used freely without any limitation. Only the author's name and the integrity of the work continue to be protected.

5. *Authors' Contracts. The form through which the bearer of copyright realizes his right is by author's contract. Under an author's contract one party, the bearer of copyright, binds himself to transfer (create and transfer) his work for use or to transfer the right to dispose of the work to the other party – the user, and the user binds himself that when the work is transferred he will use it in the manner stipulated in the contract or to perform the obligations provided by the contract on the transfer of copyright, and to pay the bearer of the copyright remuneration for the use of the work or the transfer of the right.*

The law provides two types of authors' contracts: the author's contract for the transfer of a work for use (author's contract for use) and the author's licensing contract, which provides for the transfer of defined author's rights and their disposal to another party. It exists in two variants – the commercial licensing contract (as a rule, this includes a foreign element), and a work licensing contract. The latter is concluded when an author grants to a user-organization the right to permit the translation of a work into another language or its re-working into another genre (for instance, turning a narrative work into a drama or scenario, and vice versa) and its subsequent use in the re-worked form.

The author's contract for use is one type of author's contract which is used in the USSR. It is concluded only by authors and their heirs, but not subsequent legal successors. The other party of this contract is a socialist organization (publisher, movie studio, theater) which uses a work of a specific genre according to its charter. Depending on the type of work and the nature of its use, an author's contract for use may allow for publishing, theatrical performance, or screen play; it may be an order for a painting, or grant use of works of decorative-applied art in industry, etc.

The State Committees of the USSR on Publishing, Printing, and the Book Trade [*Goskomizdat*] and on Cinematography [*Goskino*], as well as other organs which have the legal rights to do so confirm model authors' contracts with the consent of creative unions which define the content of the author's contract. Model contracts establish a list of conditions underlying the agreement with the author and recommend the content of a number of such conditions. Any conditions in a contract concluded by an author which worsen his position in comparison with the law or the model contract are invalid and are changed to conditions established by law of a model contract (Art. 506 Civil Code). This rule is an important guarantee securing the rights and interests of worker-authors.

The basic obligations of an author under a contract for use is to execute the work in accordance with the conditions of the contract and to give it to the user within the established period by the agreed procedure.

The obligations of the user-organization lie in examining the work for a determined period as a result of which its agreement with the conditions of the contract may be established. The user may approve of the work, refuse it, or send it to be re-worked. In the latter two instances, it must send the author a written notification of its reasons; if such a notification is not sent, the author considers that the work is approved. The other obligation of the user is in regard to the use of the object. The model contract established the period by which the user is obliged to realize or begin to use the work, and, as a general rule, he may not exceed two years from the day of approving the work. If this obligation is violated, the user is obliged to pay royalties to the author. A part of the royalty is paid to the author when the work is approved, and the rest is when it is used. Sometimes an advance is paid.

In circumstances provided by law, if the author does not perform or performs his obligations in an improper manner, the royalty paid to him must be returned. But if it can be shown in court that the author did not act in bad faith in carrying out the work required, the author retains the advance received by him in full or in part, as defined by the model contract, but not less than 25% of the amount of the contract.

The law establishes several exceptions to the rule that a work may be used only with the permission of the author (bearer of the copyright) and by a contract with him. In circumstances which are listed and precisely defined by law, the use of published works is permitted without the author's agreement, but with an obligatory indication of his name, and the source of the borrowing which may either be gratuitous (Art. 499 Civil Code) or with payment of royalties (Art. 495 Civil Code). Thus, the use of a work without payment of a royalty is permitted to citizens for reproduction in newspapers, movies, radio and television, etc. The payment of a royalty permits, for instance, public performance of published works.

6. *Protection of authors' rights.* In addition to protecting the author from violations on the part of the user – his counteragent by contract, the law also protects him from violations by third parties. In connection with this, it is necessary to distinguish the methods or protecting the non-property and property rights of authors.

The violation of his personal non-property rights gives the author the right to demand that one of the following be done: a) the restoration of the violated right and the correction of the violation (for instance, if there was a violation of the inviolability of the work, that there be publication in print, or some other method, in regard to the violation); b) prohibiting the dissemination of the work; or c) limiting its range (Art. 499 Civil Code). The latter two methods have particular significance for the right of the author to the publication, reproduction, and dissemination of his work. The right as to the choice of defense of his right belongs to the bearer of the copyright, although he must consider the rights and legal interests of other parties.

When his rights of authorship are violated, the author has the right to demand that his authorship be recognized by judicial procedure. One of the types of disputes on authorship are disputes involving co-authorship, as well as the defense against plagiarisms, which is the appropriation of someone else's work. Disputes as to authorship are decided by the court, if necessary by appointing an expert. If the court so decides, publication is made of who the real author is, or other measures are taken to confirm the author's authorship and name.

The defense of the property rights of an author consists of compensating him for losses, the basic form of which is income not received when the author is not paid his royalties. For the unlawful, non-contractual use of the author's work, he is paid the fee which is due him for the lawful use of the work under a contract.

Chapter 39. Law of Discoveries

The recognition of the right of discoveries by Soviet law is an acknowledgment of the significance which scientific conditions have in the era of the scientific-technical revolution for the development of the economy, of the importance of state confirmation of the reliability of these positions, as well as of the role of moral and material stimuli for increasing the creative activity of discoverers. The norms on discovery are contained in Articles 107-109 of the Principles (Arts. 517-519 Civil Code).

In accordance with para. 10 of the Statute on Discoveries, Inventions, and Rationalization Proposals, which was confirmed by Decree of the Council of Ministers of the USSR on 21 August 1973, a discovery is the establishment of a natural law, property, or phenomenon of the material world which had heretofore been unknown, and which fundamentally changes the level of knowledge. Thus, a discovery is characterized by a number of traits.

A discovery must be a theoretically or factually proven scientific position which is the result of knowledge of objective reality, and not a hypothesis or proposal for changing this reality. A discovery must touch the most basic positions of objective reality, its natural laws, properties, and phenomena: bring about substantive changes in the level of knowledge; and possess universal novelty. A discovery establishes the base for subsequent inventions and has great practical significance.

The legislation on discoveries does not cover the establishment of isolated facts, including facts in regard to geography, archaeology, paleontology, or the discovery of deposits of useful fossils, new types of plants, animals, and microorganisms; nor does it extend to propositions in the areas of the social sciences and material conclusions.

A special procedure is provided for formalizing the rights for discovery, which begins with the presentation of an application to be granted a certificate for a discovery to the State Committee of the USSR on Inventions and Discoveries.

After examining this application, this Committee – in agreement with the Academy of Sciences of the USSR – makes a decision regarding the registration and publication of the discovery. One year after publication, if the decision on the registration was not disputed, the author (co-authors) of the discovery are granted a certificate for the discovery. It certifies the recognition of the scientific status of the discovery, the priority of the discovery, and the author of the discovery. There is no exclusive right to a discovery; any one may use it without having to ask permission and without limitations.

The non-property right of an author of a discovery is the right of authorship which may belong to one or more persons (co-authors). The property right of an author is the right to a remuneration which is paid at the time that the certificate is granted in an amount of up to 5000 rubles for one discovery, regardless of the number of co-authors. The Committee on Discoveries and Inventions determines the amount of the remuneration and its payment, depending on the significance of the discovery. The author of a discovery also has a number of other rights and privileges (the right to name the discovery with his name or to give it a special name, to present the discovery as a thesis in seeking an academic degree, the right to supplementary living accommodations, etc.).

The rights of an author of a discovery are protected by *judicial* and *administrative* procedure. Disputes regarding the authorship of a discovery are examined under judicial procedure, and they may arise only after the application for the discovery is recognized. Other disputes (as to the nature of the application or the registered discovery, the priority date, the remuneration, etc.) are examined by administrative procedure by the State Committee on Discoveries and Inventions of the USSR.

Chapter 40. Law of Inventions

1. *Concept and sources of the law of inventions.* The Soviet law of inventions has two basic, related tasks: to establish conditions for increasing technical achievements based on large-scale creative activity of workers and, at the same time, to promote the use of these achievements in the manner which is most effective for the national economy. Both of these tasks are listed in Article 47 of the USSR Constitution.

The law of inventions embraces several groups of relationships. *First*, the conditions for creating proposals for inventions. *Second*, the formalization of such proposals, including their qualifications, and establishing the legal regimes which are the prerequisites for use. *Third*, the determination of the rights of inventors and rationalizers, particularly their moral and material incentives. *Fourth*, the protection of the rights of inventors and rationalizers. The law of inventions also defines the format for guiding inventive and rationalization activity in the USSR. These relationships are regulated by various branches of law: civil, administrative, labor, and several others. But the basic conditions of the law of inventions are in the area of civil law.

The most important norms of the law of inventions are contained in Articles 110-116 of the Principles and Articles 520-526 of the Civil Code. Detailed regulation of relationships dealing with inventions and rationalizing proposals is provided by the Statute on Discoveries, Inventions, and Rationalization Proposals (from now on to be referred to as Statute), which was confirmed by Decree of the Council of Ministers of the USSR of 21 August 1973, with subsequent additions and changes. There are also government decrees on individual issues of the law of inventions.

In the system of normative acts regarding inventions, the acts issued by departments are very important. First of all, these are acts which are directly listed in government decrees. The State Committee on Inventions and Discoveries issues orders and explanations on the application of the legislation.

2. *Inventions and their protection.* The definition of an invention is given in Article 21 of the Principles: *an invention is a new technical solution of a problem in any field of the national economy, social and cultural construction, or national defense which possesses essential distinctive features and yields a positive effect.* From this definition it follows that an invention must possess the following features: 1) it is a proposal containing the technical solution to a problem; 2) the solution is a new one; 3) it must be substantially different from previously known solutions; and 4) the solution must render a positive effect. The presence of these features must be established by competent organs in accordance with established procedure, in other words, the proposal must be recognized as an invention.

A proposal is an invention if it not only poses a problem but also solves it by providing a practical solution. A solution is technical if it is related to arrangements (including diagrams), methods (technology), or things. It is also the practice to single out such a type of technical solution as the exploitation of known arrangement, methods, or things for a new purpose. Non-technical solutions (for instance, organizational ones, those dealing with methods and system of teaching, etc.), like proposals which are contrary to the interests of society, the principles of humanity, or socialist morality, are not recognized as inventions.

The novelty of a proposal is determined by comparing it with that which is already known. The moment at which the proposal is considered to be created and which determines its novelty is called the priority (the date of the priority) of the invention. As a rule, priority is defined from the day that the claim is received by the Committee. The novelty of an invention is defined on the basis of world-wide data ("world novelty"), which takes into consideration all accessible sources: books, essays, and the results of implementation, as well as public lectures on television, etc.

The positive effect is the new, higher result which society gains from the proposal in comparison with that which is already known (increased labor productivity, efficiency, economization of materials, etc.). There is no positive effect if it is clear that the proposal is useless, or if it is not expedient to use it either at the present time or in the future.

The presence of these conditions is established by a single organ, the State Committee of the USSR on Inventions and Discoveries (from now on to be referred to as the Committee) acting in this instance through its subordinate body, the All-Union Scientific and Research Institute of State Patent Expertise (*VNIIGPE*). The form for recognizing a proposal as an invention is the inventor's certificate of patent. If these documents do not exist, the invention does not exist in the legal sense.

An invention may have two different legal regimes depending on whether an inventor's certificate or a patent were issued for it. If an inventor's certificate was given for the invention, the state has exclusive right to it for 15 years from the day that the claim was filed. The state allows gratuitous use of such inventions to state, cooperative, and social enterprises, organizations, and institutions without special permission.

An inventor's certificate also establishes the authorship of the inventor and secures for him rights and privileges provided by law. These rights and privileges are inalienable, several of them may be transferred only through inheritance, and continue for an unlimited period from the day that the claim was presented. An inventor's certificate also establishes the recognition of the proposal as an invention and the priority of the invention.

In contrast to an inventor's certificate, a patent secures the exclusive right to the invention for 15 years to the holder of the patent – the author of the invention or the party to whom this right was assigned by the author. No one has the right to use the invention without the consent of the holder of the patent; a violator must compensate for losses. Permission to use a patented invention is given by remuneration and is called licensing. A patent may also be transferred entirely (the sale of a patent). The granting of a license or the sale of a patent must be registered with the Committee or risk being invalid. Fees must be paid on a patent, and non-payment within a specified period results in the termination of the validity of the patent prior to its expiration period.

The author of an invention has the right to choose the form of its safeguarding. Social and personal interests cause Soviet inventors to take an inventor's certificate; patents are given almost exclusively to foreign firms. In some instances, only an inventor's certificate, rather than a patent, is issued. It is not possible to give a monopoly over certain objects to individuals because of their importance for the defense of the country, the public health, etc. Therefore, only inventors' certificates are issued for medicine, prophylactic measures, diagnoses for the treatment of people or animals, and secret inventions, as well as for material which is created through chemical procedures. Only an inventor's certificate is given if the invention was made in connection with the author's work in a state, cooperative, or social organization or at its instruction, and also if the author is given financial or other material help by such an organization.

In order that an invention may be evaluated, it is necessary that an application be presented in a form which accords with the requirements of legislation in force. The claim is submitted to the Committee and is passed on to the *VNIIGPE*. Here, first of all, its adherence to the formal requirements are verified ("preliminary expertise"), and then the adherence of the presented proposal with the requirements of inventions themselves are verified ("scientific-technical expertise"). On the basis of the verification, a decision is made on whether to grant an inventor's certificate or patent, to reject the proposal, or to ask the claimant to supply additional material.

The decision to grant a patent or inventor's certificate contains the formula of the invention – the formal definition of the invention, established according to definite rules; on the basis of the formula, the limits of the invention are defined, and the fact of its use is established. A decision to refuse to recognize a proposal as an invention must contain reasons for the refusal.

An objection may be raised against the refusal to recognize a proposal as an invention, as well as in regard to the formula established by the Committee, and a complaint may be brought against the Committee, which is examined by the Supervisory Council on Scientific and Technical Expertise of the Committee, whose decision is final and may not be appealed.

A proposal, which is recognized as an invention, is entered in the State Registry of Inventions of the USSR, which fact is published in the official publications of the Committee, and the author is issued a document called an "inventor's certificate," or the patent holder a "patent."

Inventors' certificates and patents may be disputed and recognized as invalid in whole or in part by administrative procedure for a period of three years from the day of publication of the work as a result of a violation of the requirements for recognizing a proposal as an invention. They may be disputed by judicial proceedings at any time after being granted to the wrong author (co-author) of the invention. Furthermore, a patent may be disputed during the entire period of its validity if it was granted for an invention for which only an inventor's certificate may be given and if the patent holder has no right to the patent.

If an inventor's certificate was given for an invention, the exclusive right of the state to it also extends to the disposal of the invention abroad. A characteristic trait of the legal protection of inventions is the territorial principle of its application. This means that the exclusive right to use an invention exists only in the country where the protective document has been received. Consequently, in order that the right to an invention which is already protected in the USSR is also protected in foreign states, a patent must be obtained in all of these countries which is issued by the national patent authority.

Legislation establishes the procedure for patenting Soviet inventions abroad, as well as the procedure for selling a license abroad. A method for receiving profits from Soviet inventions abroad is the export of objects in which they are applied. For this, it is important that the objects of export do not fall under the patents issued to foreign patent holders in the country of export; this is the assurance of "patent purity." Legislation provides the procedure for securing and verifying the patent purity of exported products.

3. *Rationalization proposals.* Rationalization proposals allow for the adaptation of already known technical solutions to specific conditions in actual production. If the purpose of inventions is to secure the development and extension of scientific and technical development, rationalization proposals must aid in expanding the scope of their practical application. Taken as a whole, they have a significant effect on the national economy.

The definition of a rationalization proposal is given in para. 62 of the Statute: a rationalization proposal is a technical solution which is new and useful for the enterprise, organization, or institution to which it has been submitted, and which envisages a change in the design of articles, the technology of production, or change in the composition of the material. The features of a rationalization proposal resemble those of an invention, but different criteria are established for them.

First, a rationalization proposal, like an invention, must not pose a problem but rather a solution, and it must be of technical character, i.e. directed toward changing arrangement (of products, applied technology), technology, or materials. Organizational proposals to improve the management of organizations and the organization of production are not considered to be rationalization proposals.

Second, a rationalization proposal must be new, but in contrast to an invention, its novelty must be not of worldwide but rather of a local standard, i.e. only within the organization to which the proposal was submitted. The fact that a solution is known in another organization is not a hindrance. The novelty of a rationalization proposal is defined by a listing of sources, and other sources may not be considered. The novelty is defined as of the date of priority – the date when the proposal was submitted.

Third, a rationalization proposal must be the result of the creative activity of its author. While this feature is an inalienable element of the essential (worldwide) novelty of an invention, here it serves to distinguish rationalization proposals, which are of local novelty, from mechanical borrowings.

Fourth, a rationalization proposal must be useful in a way which is different from the positive effect of an invention: the expediency of the use of a proposal is determined in conformity with the conditions which exist at the moment of examining the proposal and must be made in accordance with established plans. In other words, a proposal must be accepted for use.

As an exception to the general rule, under certain conditions a proposal is not recognized as a rationalization proposal even where it conforms to all the listed conditions. Two conditions are necessary for this. First, an author must belong to a definite category of workers, namely engineering/technical workers in scientific-research, design, construction, and technological organizations and analogous subsidiary enterprises. Second, it is excluded if the proposal is related to the official duties of one of these persons.

In order for the legal consequences which the law links to a rationalization proposal to arise, it must be considered qualified to be a rationalization proposal, that is, a competent organ must verify the presence of the necessary features and conditions. The form that such a qualification takes is, as a rule, certification for a rationalization proposal.

In order that a proposal be recognized as a rationalization proposal, a written application must be submitted in conformity with the established requirements. An application is submitted to an organization whose activity is related to the proposal, and if it could be used in several organizations, it is submitted to the proper ministry (department). A decision concerning the application is taken within an established period by the manager of the organization to which the proposal was submitted. If the decision is a positive one, the author is issued a certificate for a rationalization proposal. If the proposal is rejected, the author has a right to bring an appeal which is examined by the head of the organization jointly with the trade union organization where the decision was made, and afterwards by a superior organ. The decision of the head of a ministry or his deputy is final. The certificate which is given for a rationalization proposal may, within established periods, be disputed by administrative procedure for violation of the requirements necessary for rationalization proposals, as well as for issuing a certificate to a person who does not have precedence (in the latter instance, the court makes the final decision).

No one has exclusive right to a rationalization proposal, including the state. This means that any interested party may use a rationalization proposal gratuitously, without need for special permission from any organization.

The rights of an author of a rationalization proposal are defined by its local nature: they apply within the limits of the organization which recognized the proposal as a rationalization proposal. In addition, they apply in organizations to which the proposal is given by a contract for the transfer of scientific and technical achievement. These laws do not apply to proposals analogous in content which were submitted by other persons to other organizations and which were used by them.

4. *Rights of inventors and rationalizers*. The authors of inventions and rationalization proposals have a number of personal non-property and property rights.

The basic personal non-property right is the right of authorship, i.e. the right to be considered the author of a certain proposal. Authorship may belong to either one or more persons (co-authors) if the proposal was created jointly through creative activity; rendering technical assistance is not sufficient for co-authorship.

The basic property right of an author is the right to remuneration, the conditions for payment of this being the qualification of the proposal by the issuance of an inventor's certificate for an invention or a certificate for a rationalization proposal. A procedure is established for the payment of such remuneration (the basis of the payment, amount periods for payment, organizations which are obliged to pay).

The basis of payment for an invention is the use of the invention either in the national economy of the USSR or abroad. In addition to this, an incentive payment is given for the creation of official inventions or for inventions in organizations which are engaged in social activity.

The amount of remuneration is determined by the savings to the economy which result from the invention. It if creates savings, remuneration is 2% of the savings for each calendar year for the first five years of use. Savings are determined by a special method. If an invention does not result in savings, remuneration is determined by the actual value of the invention as measured by special coefficients, taking into account any technical or other positive effect as well as the range of applicability. A third method is used for remuneration for licenses sold abroad – it consists of up to 3% of the amount received for the sale. In all instances, the maximum remuneration for one invention may not exceed 20,000 rubles.

The remuneration is paid every two months from the first day of the calendar year of use, the decision of the transfer of technical documents abroad, or the transfer of money for a sold license. Payment is made by the enterprise or ministry (department) using the invention, or by the Committee.

The basis for the payment of remuneration for a rationalization proposal is its use in the USSR, and only by that organization which issued the certificate or by the organization which received it under a contract for the transfer of scientific and technical achievements. The procedure for determining the amount of payments depends on whether the rationalization proposal creates savings. If it does, the remuneration is determined according to a special scale by the amount of savings for the first year of use. If the scope of use increases in the second year, an addition is made to the payment. Payment for proposals which do not create savings are determined by a system of coefficients similar to those used for inventions. The minimum amount of remuneration is 10 rubles, and the maximum is 5000 rubles.

Authors of inventions and rationalization proposals are also granted a number of other rights and privileges: the right to retain previous norms and wage rates for a period of six months (Art. 106 Labor Code of the RSFSR), tax privileges, the right to assistance in the compiling and submitting of their proposals, the right to participate in the preparation of the prosposal for use, the right to supplementary living space, etc. An entry is made in authors' work books indicating the use of their inventions or rationalization proposals, and when they are given their first author's certificate, authors are given a badge confirming the fact that their invention is being used. Special premiums are paid to persons who have assisted in the creation of the invention or rationalization proposal.

The protection of inventions and rationalization proposals is carried out by judicial and administrative procedure, and for some disputes, judicial examinations precede administrative hearings.

The disputes examined under administrative procedure are mainly disputes relating to the qualification of the proposal, as well as disputes on its use. In order to provide more guarantees to authors, many of such disputes are examined by a specially created organ which is not connected with the activities of the first expertise of the work – the Supervisory Council of Scientific and Technical Expertise of the Committee.

The following basic types of disputes are examined by judicial procedure: disputes concerning authorship (co-authorship); the payment of remuneration and its division between co-authors; the use of an invention protected by patent; precedence to inventions. In a suit on the protection of their rights, authors are relieved from payment of the state fee and court costs.

A special procedure is established for the resolution of disputes on the payment of remuneration. A complaint concerning a violation is first of all given to the organization which made the decision where it is examined, as a rule, with the participation of the trade-union organization. Where there is disagreement with the decision reached or if a decision is not reached within the established period, the complaint may be submitted to a superior organ or a suit may be brought in court. Precedence to a rationalization proposal is examined by an analogous procedure. Disputes concerning the division of remuneration between co-authors are examined in court.

Disputes concerning the labor rights of inventors and rationalizers (retention of norms and wage rates, etc.) are examined through the procedure established for settling labor disputes. The violation of some rights of inventors and rationalizers are prosecuted under criminal procedure.

The Law of Succession

Chapter 41. Law of Succession

1. *Concept of succession.* Article 13 of the Constitution of the USSR establishes that "the personal property of citizens and the right to its inheritance is protected by the state." *Succession is the transfer of the rights and obligations of the deceased to other parties.* The procedure for this transfer is determined by legal norms which in their totality comprise the *law of succession.*

The Civil Code devotes a special chapter (VII), containing Articles 527-561, to the law of succession. The Plenum of the Supreme Court of the USSR issued a Decree on 1 July 1966, "On Judicial Practice in Regard to Cases on Succession." On 26 March 1974, the Plenum of the Supreme Court of the RSFSR adopted the Decree, "On the Application by the Courts of the RSFSR of the Norms of the Civil Code on Succession and the Implementation of the Decree of the Supreme Court of the USSR of 1 July 1966, 'On Judicial Practice in Regard to Succession Cases'."

In order to study the law of succession, it is necessary to understand the meaning of the concepts used in connection therewith, and above all, such terms as "estate," "succession on intestacy," "succession under will," "time of the opening of the succession," "place of the opening of the succession," "estate-leaver," and "heir."

The *estate* is the totality of rights and obligations which are transferred by succession. Only those rights and obligations which citizens may transfer to other parties may be included in the estate. Rights and obligations which are linked with the personality of a citizen (for instance, the right to compensation for harm in connection with injury or the right to receive alimony, as well as the obligation to pay alimony) may not be transferred by succession. It is mainly property rights and obligations that are applied in the procedure for succession. Basically, the matter under consideration involves the right to ownership of articles of everyday or personal use and comfort and of the subsidiary household, a dwelling, and savings from labor. Under succession may also be transferred property rights and obligations arising from legal relationships involving the law of obligations or the law on copyright or inventions (for instance, the right to receive a debt owed to the deceased, including royalties, or corresponding obligations to pay the debt of a deceased creditor). As an exception, in circumstances which are explicitly provided in law, some non-property rights may also be transferred by succession, for instance, the right to secure the inviolability of a work of literature, science, or art (Art. 481 Civil Code) or the right to receive an author's certificate or patent for an invention or certificate for a rationalization proposal (Art. 525 Civil Code).

Succession contains two basic distinctions: *intestate succession* and *succession under will.* The norms of the Civil Code define to whom the rights and obligations of the deceased may be transferred and their scope. Succession based on these norms is called intestate succession. In addition, citizens have the right to dispose of their property in a manner different from that prescribed in law. Disposal for this purpose is called will, and succession based on it is called succession under will.

The *time* and *place* of the *opening of succession* have great significance for the resolution of the problems related to succession. For instance, the time of the opening of the succession determines the heirs and the contents of the property of the deceased; from this moment the periods established by law for receipt of the estate begin to run, etc. The place of the opening of the inheritance has significance when civil legislation of the different republics has differing solutions for the same problems. In these circumstances, in accordance with Article 8 of the Civil Code, the "law of the place of the opening of the succession" is in effect. Also, the notarial office at the place of the opening of the succession (in places where there is no notarial office, the executive committee of the local Soviet of People's Deputies) must take measures to protect the deceased's property (Art. 555 Civil Code).

The time of opening of the succession is the day of the death of the estate-leaver. If the estate is opened in connection with the declaration of a citizen as dead, the day of death is considered to be the day that the court judgment in regard to this is made, and if there are special instructions in the court judgment in regard to this, it is the day of the presumed death of the citizen (Art. 528 Civil Code).

The *place of the opening* of the succession is the last permanent place of residence of the estate-leaver, and if this is not known, then it is the place where the property of the estate, or its main part, is located (Art. 529 Civil Code). Thus, if a Moscow resident dies while on a business trip in Tashkent, it is important to be guided by the norms of the Civil Code of the RSFSR and not of the Civil Code of the Uzbek SSR.

The *estate-leaver* is the person whose rights and obligations are transferred by succession. Only citizens may be estate-leavers.

Heirs are people who have the right to receive an estate. In contrast to estate-leavers, citizens, judicial persons, and even the Soviet state my be heirs. Under succession on intestacy, citizens who are alive at the moment of the death of the estate-leaver, and children born after his death, are called to the succession. Under succession by will, all heirs who are alive at the moment of the death of the estate-leaver or were conceived when he was alive but were born after his death, are called to the succession.

According to law, some citizens do not have the right to succeed (Art. 531 Civil Code). A reason for excluding them from the succession is if they carried out various unlawful actions which are listed in the law. Thus, those who furthered their own chances of being called to the succession through illegal actions directed against the deceased, against one of his heirs, or against the fulfillment of the last wishes of the estate-leaver as expressed in his will may not be successors. For example, a son who murdered his father may not later succeed to his estate. A daughter who falsifies the will to her own advantage may not succeed to the estate after her mother. Of course, these examples concern only those persons who carry out these types of actions intentionally.

A father and mother may not be heirs on intestacy to children in relation to whom they have been deprived of their parental rights, as well as parents and children of full age who maliciously refused to carry out the duties to maintain the estate-leaver which were imposed on them by law (for instance, a father who systematically disappeared in order to avoid paying alimony). All circumstances which could serve as a basis for the exclusion from succession must be confirmed by judicial procedure.

2. *Intestate succession.* Intestate succession occurs in all circumstances except when there is a will. Intestate heirs are divided into two categories. In the first place, heirs belonging to the first category are called to the succession. These include children (including adopted), spouses, and parents (including adoptive) of the deceased. A son or daughter of the deceased born after his death are also included here. A spouse is the person who is registered in marriage to the deceased at the moment of his death. A spouse who has annulled the marriage loses the right to succession. Children are always considered heirs after the mother, but after the father only if paternity is established by legal procedure (see Chapter 44). The question of the succession of the mother and father after their children is decided in the same fashion.

The property is divided evenly among all the heirs of the first category: neither the material situation, age, state of health, length of time of marriage, or any other condition have any bearing on the amount of the share.

If there are no heirs of the first category, of if they all refuse the estate (do not accept it), or if the estate-leaver deprives all of them of their rights as heirs in his will, persons belonging to the second category are called to the succession. The second category of heirs includes brothers and sisters and the grandfather and grandmother of the deceased – both on his father's and mother's side. Brothers and sisters having both the same mother and father of the deceased, as well as his half-brothers and half-sisters (having the same mother and different fathers or the same father and different mothers) are called to the succession.

The Civil Codes of several of the union republics contain other rules as well. For instance, the Civil Codes of the Moldavian and Kazakh SSRs establish three, rather than two, categories of intestate heirs; in Moldavia, nephews and nieces who are unable to work are put into a third category, while in the Kazakh SSR, brothers and sisters of the deceased who are able to work are included in that category (brothers and sisters who are unable to work are placed in the second category). The Civil Code of the Azerbaidzhan SSR includes, in addition, the great-grandfather and great-grandmother among the heirs in the second category.

The Civil Code of the union republics also include as heirs persons unable to work who were dependants of the deceased for not less than one year prior to his death. These may be relatives of the deceased (for instance, grandfather, cousin, niece, etc.), or people who are not related to him. People who are considered unable to work are women of 55 years of age, men of 60, invalids belonging to categories I, II or III, as well as persons under 16, and, if they are students, under 18. In order for a person to be considered a dependant, it is necessary that he be fully supported by the estate-leaver, or that the assistance given to him by the estate-leaver was his basic and permanent means of support.

If there are other heirs, dependants unable to work succeed equally with the heirs of the category which is called to the succession. Thus, if at the moment of the opening of the succession the estate-leaver had a wife, sister, and dependant 10-year old niece who was living with him, two are called to the succession: the wife (heir in the first category) and a niece. But if it turns out that the wife is already dead when the estate-leaver dies, and there is no heir of the first category, the succession is divided evenly between the sister as the heir in the first category and the niece as heir in the second category.

A special procedure is established for the succession of

grandchildren and great-grandchildren of the estate-leaver. They are called heirs by right of representation. They become intestate heirs only if at the moment of the opening of the succession that parent who would have been the heir has died. For instance, Boris Smolin had two sons, Vladimir and Sergei. Sergei died on the front. He had been father of three children. If Sergei had been alive at the moment of the death of his father, Boris Smolin, the succession would have been evenly divided between him and Vladimir. Sergei's children (Boris' grandchildren) would not have been called to the succession. But since Sergei is not alive, he is represented by three children at the moment of the opening of the succession. The share of the succession which would have been received by their father is evenly divided among them. Vladimir Smolin has a right of one-half of the succession, and each of the Sergei's children receives one-sixth.

In intestate succession, domestic furnishings and articles of everyday use are differentiated from the estate in general. These are articles which are used by the members of the family of the estate-leaver in the course of everyday life: furniture, dishes, kitchen utensils, etc. Luxury objects (expensive pictures, decorative crystal vases, etc.) are not under consideration here. Under Article 533 of the Civil Code, household furnishings and articles of everyday use pass to those intestate successor who lived with the estate-leaver for not less that one year prior to his death. It is assumed that they actually used these objects to meet their everyday needs. If an heir of the first category is called to the succession, these objects are given to him in addition to the share to which he is entitled. If an heir of the second category, who is not called to the succession if there are heirs of the first category, lived with the estate-leaver, he has a right to receive only domestic furnishing and articles of everyday use on succession.

3. *Succession by will.* If a citizen wishes to dispose of his property in a fashion other than on intestacy, he may draw up a will. A will must meet the requirements set forth in law both in content and form.

In principle, a testator may dispose of his property as he wishes. He has the right to leave all or part of his property, including domestic furnishing and articles of everyday use, to one or several persons. These may be intestate heirs or persons who do not fall into that category. Property may be left by will to the state or to a specific state, cooperative, or social organization. Sometimes a will is drawn up with the intention of depriving one or more intestate heirs from the right to succession. In this instance, a person deprived of the right to succession in the will is not called to the succession. The testator is also given the right to designate "reserve" heirs in addition to his primary heirs. These are persons to whom the estate is transferred if a primary heir dies prior to the estate-leaver or renounces the estate. This condition of a will is called *substitution* of an heir.

However, the right to devise under a will is not absolute under Soviet law. There is a defined category of intestate heirs whom the estate-leaver may not deprive, by direct or indirect means, of their succession. These include the minor children, as well as the children who have attained majority but are unable to work, of the estate-leaver, as well as the spouse, parents (adoptive parents), and dependants of the deceased. They always have an *obligatory share* of the estate. Regardless of the content of the will, they must receive not less than two-thirds of that which they would have been allotted under intestate succession. When this requirement is violated, the appropriate portion of the will is considered to be invalid.

Thus, for example, Vasilii Kobzev drew up a will under which all his property, consisting of 6000 rubles, was left to his wife. At the moment of his death, his successors included, besides his wife, the children of his first marriage (21-year old Fedor and 19-year old Maria), who were able to work, and his father, Ivan Kobzev, who was 83. Of all these, only Ivan Kobzev, who was unable to work, has a right to an obligatory share. If there had been no will, everything would have been divided equally between the wife, the two children and the father of the deceased as heirs of the first category.

Consequently, Ivan Kobzev would have had a right to his legal share of 1500 rubles. The obligatory share is 1000 rubles (two-thirds of 1500 rubles). Thus, Ivan Kobzev receives 1000 rubles from the estate, and the remaining property of 5000 rubles goes to the wife of the deceased Vasilii Kobzev, for whose benefit the will was drawn up.

It is permissible to draw up a will under which the estate-leaver charges the heir with performance of a specific obligation in favor of one or more parties. For instance, a citizen may leave all his property in his will to his wife with the charge that she pay 70 rubles every month to his parents. This type of disposition is known as a *legacy*. In order to understand the meaning of this, it is necessary to realize that the word "legacy" has not only the usual meaning of "not to agree", but also the meaning "to bequeath, to leave by will." The person in whose favor a legacy was set up is the creditor in an obligation. This debtor is the heir who was charged with the performance of the legacy. It may turn out that the heir did not receive enough property to be able to perform the legacy. This circumstance is taken into account by the law: according to Article 538 of the Civil Code, an heir must perform a legacy only to the extent of the value of the property left to him.

A testator has the right to charge an heir with the performance of actions which are useful to society. For example, an artist may leave paintings and the house belonging to him to a museum with the charge that it organize a permanent exhibit open to visitors.

The legal requirements in regard to the *form* of a will are very strict. As a general rule, a will must be drawn up in writing with a statement of the place and time of execution, and signed by the testator personally. If the testator is not in a condition to sign the will (because he is physically unable to do so, ill or because of some other reason), this may be done by another citizen at his request in the presence of a notary or other person having the right to notarize a will. It is necessary to state in the will why the estate-leaver was not able to sign the will himself.

A will must be certified by a notary. The notarial form is one of the necessary requirements for the authenticity of a will. However, there are specific exceptions to this rule. Under special circumstances, wills which are certified by officials specified in law are considered to be the equivalent of wills certified by notaries. These include:

a) wills of citizens under treatment in hospitals, other institutions for in-patients, sanatoriums, or homes for the aged and invalids. Such will may be certified by head doctors, their deputies in the medical section or duty doctors in those hospitals, treatment institutions, and sanatoriums, as well as by the directors and chief doctors of the homes for the aged and invalids;

b) wills of citizens aboard sea-going ships during their voyages or on ships on internal voyages under the flag of the USSR. These wills must be certified by the captain of the ship;

c) wills of citizens on exploratory, Arctic, and other similar expeditions. These wills must be certified by the leader of such an expedition;

d) wills of military personnel and other persons under treatment in hospitals, sanatoriums, and other military medical institutions. The right to certify such will is given to the commanding officers, their deputies in the medical section, and senior and duty doctors in these military medical institutions;

e) wills of military personnel at locations within military units, groups, institutions, and military training facilities where there are no state notarial offices or other organs which carry out notarial activities, and also the wills of workers and employees, members of their families, and members of the families of military personnel. These wills must be certified by the commanding officers or the proper leaders of these units, groups, institutions, or facilities;

f) wills of persons who are in places of deprivation of freedom. These wills must be certified by the warden of the places of the deprivation of freedom.

Any will may subsequently be revoked or changed by the

testator. As a rule, this occurs in connection with the drawing up of a new will. It is not necessary to refer to the revocation of an earlier will in the new will. The guiding principle is that the new will automatically supercedes an earlier will in any section where there are differences.

4. *Acceptance of the estate.* In order to acquire the estate, the heir must accept it, i.e. indicate in a specified manner his desire to receive the estate. He may do this by giving an appropriate declaration to the notarial office at the place of the opening of the estate. But it is not necessary that he make a declaration to the notarial office: an heir is considered to have accepted the estate if he actually takes possession of the estate (carries out various actions in regard to managing the estate, paying taxes, safeguarding the property, etc.).

The heir must express his agreement to accept the estate through one of these two methods not later than six months after the day of the opening of the succession. If the heir does not accept the estate within this period, he may turn to the court to request an extension of this period. If the reasons for the lapsing of the period are considered important (for instance, that the heir was on a long business trip, very ill, etc.), the court extends the period for the acceptance of the estate. The heir may accept the estate after the expiration of the period set forth in law without resort to a court if the other heirs accepting the estate do not object.

Acceptance of an estate is always without reservation and is unconditional. An heir may not, for instance, accept a house only on the condition that the sister living in it must vacate it. The estate is considered to belong to the heir from the moment of the opening of the estate.

If the heir who is called to the succession on intestacy or under a will dies after the opening of the estate and before he is able to acquire it within the legally established six-month period, the right to acquire the estate passes to his heirs. In this instance, the actions testifying for the acceptance of the estate (for taking possession of the estate or giving a declaration of intent to do so to the notarial office) must be realized within the remaining part of the period for acceptance of the estate by the original heir. If less than three months remain of this period, it is extended to three months.

An heir has the right to renounce the estate within a period of six months. He may renounce the estate unconditionally, in which case his share passes to the heirs on intestacy or under will who are called to the succession. In this instance, the *accrual* of share of the other heirs takes place. However, the heir has the right to renounce the estate in favor of other heirs which he must do in a declaration submitted to a notarial office within the six-month period. An heir may renounce his share of the estate only to heirs on intestacy or under will, and to state, cooperative, or social organizations. After the death of his father, citizen I. gave a declaration to the notarial office renouncing his share of the estate to citizen K., whom he considered to be the wife of the deceased. It turned out that K. had annulled the marriage with the estate-leaver and therefore could not be an intestate heir. Therefore, the declaration of the son of the deceased to the notarial office in favor of the former wife, who was excluded from the succession, is considered invalid.

An heir who accepts an estate is liable for the debts of the estate-leaver. In order to protect the interests of the heir, the law provides that this liability is limited by the actual value of the estate. Specific time limits are also established for the period during which the creditors of the estate-leaver may present claims to his successors. Such claims must always be presented within six months after the opening of the estate, regardless of whether the estate was accepted by the heir or not. In the latter case, the creditors may present their claims to the notarial office at the location of the opening of the estate or bring a suit in court against the estate. If the creditors miss the six-month period, they lose the right to any claim, regardless of whether the period was missed for valid reasons or not.

If the heir wishes, he may request that the notarial office gave him a certificate of the right to the estate. The receipt of such a certificate is a right but not an obligation of the heir. A certificate is usually given not earlier than six months after the opening of the estate, since it is presumed that other heirs may appear who are ready to accept the estate. However, in instances when the notarial office concludes that there are no other heirs, the certificate granting the right to the estate may be given before that period.

5. *Escheat*. The law establishes those circumstances under which the state is a successor. The state is a successor if: 1) the property is willed to the state by the estate-leaver; 2) none of the heirs on intestacy or under will appears at the moment of the opening of the estate; 3) the estate-leaver deprives all heirs in intestacy of the right to succession; 4) none of the heirs accepts the estate (they renounce the estate).

A part of the estate may also be passed to the state. This happens when one of the heirs renounces his share in favor of the state or if only a part of the property is willed and there are no intestate heirs. In this case, the state receives that part of the estate which is not willed. The succession rights of the state are also established by a certificate to the right to the estate, which is given to the appropriate financial organ by established procedure six months after the day of the opening of the estate.

In accepting an estate, the state accepts the rights and obligations of the estate-leaver. Therefore, like all other successors, the state is liable for the debts of the estate-leaver within the limits of the actual value of the estate.

6. *Succession in a collective farm household*. As a general rule, there is no opening of the estate when a member of a collective farm household dies. This is explained by the fact that the property of the collective farm household belongs to all its members by right of joint ownership and the others continue to possess, use, and dispose of the property on these terms after the death of a member of the household. Only when the last member of a collective farm household dies

and the collective farm household in fact ceases to exist, does the property of the household pass by succession under the general rules of succession.

These rules do not apply to the succession of the personal property of a member of a collective farm household. Such property passes to his heirs on intestacy or under will. The question arises of whether an automobile registered in the name of a deceased member of a collective farm household passes by succession or remains in the ownership of the collective farm household. This problem is resolved as follows: if the automobile was acquired by the joint means of the members of the collective farm household, it is considered to belong to the collective farm household and must be re-registered to another member of the household. Such a re-registration takes place with the agreement of all the other members of the household. If the automobile was acquired by the personal means of the deceased member of the collective farm household, it passes to his successors on general principles.

7. *Disposition by a depositor of his deposit on death*. The deposits of a citizen in state savings banks or in the State Bank of the USSR are passed on succession according to special rules. The depositor has the right to instruct the savings bank or other bank to pay the deposit to some other person or to the state in the event of his death. If such instructions are given, the deposit is not included in the estate and the norms of the Civil Code on succession do not apply to it. In particular, the rules on the obligatory share and the liability of the heir for the debts of the estate-leaver do not apply. It is possible to receive such deposit at any moment after the death of the estate-leaver. The deposit is not taken into account in determining the amount of a share in the estate. The deposit is not included in the property of the estate and is not subject to the general rules of succession even when the special instructions regarding its deposition in the event of death are given not to the savings bank, but are in the will.

In instances when the depositor did not arrange for special

disposition regarding the passing of his deposit, the deposit becomes part of the estate and passes on general principles to the heirs on intestacy or under will.

8. *Protection of the estate.* The necessity for the protection of the estate may arise from the moment of the opening of the succession to its acceptance by the heirs. Usually this occurs when there is the threat that the estate may be damaged, ruined, stolen, when it is in need of management, when creditors present claims, etc. Measures for the protection of the estate are taken by a notarial office (or, if there is none, by the executive committee of the local Soviet of People's Deputies). In order to protect the estate, the notarial office takes an inventory of it which it gives for safekeeping to the heir or to another person. Money which is left by the estate-leaver is put into a bank in the account of the notarial office, and gold, silver, and other valuables are placed into a bank for safekeeping. The notarial office must accept the claims of the creditors against the estate and inform the heirs of these claims.

The notarial office protects the estate up to the moment that is is accepted by all the heirs and if the estate is not accepted, for six months after the moment of the opening of the succession. It is necessary to give advance notice of the termination of the protection of the property to the heirs, and if the estate passes to the state, notice must be given to the appropriate financial organ.

Part II

FAMILY LAW

Chapter 42. General Problems of Family Law

I. The Concept of Family Law

1. Before commencing the study of family law, it is necessary to determine precisely what the family is, the nature of its tasks, the significance of the family to life in a socialist society and to the individual, and the demands that should be made upon it.

The family is a necessary cell in human society, without which society itself could not exist. The tasks and functions of the family are ultimately determined by the production relations of a given society. At the same time, law and morality exert a significant influence on the family. The family in turn influences the life of society. Children--the future members of society, the creators of its material and nonmaterial culture--are born and raised in the family. The nature of the relations that are formed in the family determine more than the proper upbringing of children. The family in large measure determines the character of each of its members, the social and labor activism of its members, and thereby influences the successes and achievements of all society.

Concern for the family in the USSR is a concern of the state. Article 53 of the Constitution of the USSR states that the family enjoys the protection of the state and enumerates the types of aid that is rendered to the family in our country.

Relations in the Soviet family are determined by feelings of love, respect, friendship, and mutual aid. The family today not only rears children, but also organizes family life and cares for its elderly members. Therefore, in addition to personal relations, certain property relations also arise between family members. However, these relations in the family are of a subordinate nature that is dependent on personal relations. Therefore, the norms of civil law, which (as shown in Chapter 1) stem from the use of the commodity-monetary form in communist construction, cannot be applied to them.

The exceptionally great importance of the family to our society requires that it be strengthened in every way. The party and the Soviet government have always used legal norms, in addition to other measures, for the creation and further strengthening of the socialist family. The special character of relations which are formed in the family demands that they be regulated by special norms of law. From the first days of the October Revolution, independent legislative acts have been promulgated on marriage and the family; it is these acts that comprise Soviet family law.

Soviet family law proceeds from special, inherent tasks and principles that are in large measure distinct from other branches of law. Family law is therefore an independent branch of law in which family relations, i.e., relations stemming from marriage and affiliation with a family union, are the subject of regulation.

2. *Principles and tasks of Soviet family law* are clearly expressed in Part I of the Principles of Legislation of the USSR and Union Republics on Marriage and the Family. Its basic principles are: voluntary marriage; monogamy; equality of men and women in all family relations; exercise of parental rights in the interest of the children; and the protection of the family, mothers, and children by the state.

As already noted, the further strengthening of the family on the basis of the principles of communist morality is the principal task of family law. Legislation strives to see to it that family relations are based on the voluntary marital union of woman and man, on feelings of love, friendship, and respect of all family members that are free of material considerations. The Principles [*Osnovy*] consider the upbringing of children in close conjunction with their social education in the spirit of dedication to the Homeland, a communist attitude toward labor, and the preparation of children for active participation in the construction of a communist society to be the most important obligation of the Soviet family.

Family law is also aimed at the all-round protection of the interests of mothers and children, at ensuring a happy childhood for every child. Finally, its task is to eliminate once and for all harmful vestiges and customs in family relations and to inculcate a feeling of responsibility to the family.

Family law, like all Soviet law, is based on the full equality of all citizens regardless of origin, social and property status, race or nationality, sex, education, language, attitude toward religion, type and nature of occupation, domicile, or other status. The equality of women and men in this area is of exceptionally great importance since the dependent, subordinate status of women was previously manifested specifically in family relations.

The general principle of separation of church from state also held special significance for family law. Marriage and divorce before the revolution were determined exclusively by the corresponding religious rules that explained and reinforced inequality of the sexes. The regulation of these questions exclusively by the state has made it possible to attain equality for women in the family as well.

3. As already noted, Soviet family law regulates relationships that arise from marriage and from belonging to a family. Legislation on marriage and the family accordingly establishes the procedure and conditions for entering into and terminating a marriage, and regulates personal and property relationships arising in the family between spouses, between parents and children, and between other family members. This legislation also regulates relationships arising in connection with adoption, guardianship, and wardship, and with the acceptance of children for rearing on other grounds. Finally, family law regulates the procedure for recording instruments of civil status (article 2, Principles).

II. Sources of Soviet Family Law

1. *Sources of Soviet family law* are understood to mean normative acts that regulate family relationships.

The Constitution of the USSR, which proclaims that the family is under the protection of the state and which contains the basic provisions of existing legislation on marriage and the family (articles 53, 66), should be mentioned as the primary source of Soviet family law.

The Principles of Legislation of the USSR and Union Republics on Marriage and the Family, which were ratified by the Supreme Soviet of the USSR on June 27, 1968, with amendments introduced by the October 9, 1979 Edict of the Presidium of the Supreme Soviet of the USSR,[1] are the basic

1. *Gazette of the Supreme Soviet of the USSR*, 1968, No. 27, item 241; 1979, No. 42, item 696.

all-union legislative act in this area.

This fundamental act articulates the tasks and principles of the given branch of law (see I above) and contains norms regarding key questions of marriage and the family that require uniform resolution throughout the USSR.

The Principles provide for the promulgation of codes on marriage and the family by the union republics. Guided by the provisions of the Principles, all the union republics have promulgated corresponding codes (the Code on Marriage and the Family) during the period 1968-70. The RSFSR Code on Marriage and the Family, the articles of which are referred to in the present textbook, was adopted by the Supreme Soviet of the RSFSR on July 30, 1969 and took effect on November 1. The codes regulate in detail the relations that arise in family relations with due regard to local and national particulars.

There are differences between the norms of existing codes on certain questions. It is therefore occasionally necessary to know specifically which republic's legislation should be applied in every concrete instance. The Principles of Legislation on Marriage and the Family stipulate that on the territory of each union republic, the legislation of that republic shall be applied. When the validity of an act is questioned (for example, the validity of a marriage), the courts and other bodies apply not the legislation of their republic, but rather the legislation of the republic in which a given act was carried out (see para. 3 and para. 4 of article 7 of Principles).

As indicated in para. 2, article 7 of the Principles, other legislative acts in addition to the Principles and codes may be promulgated by the USSR as well as by the union republics. The Principles and codes also provide for the promulgation of pieces of delegated legislation [*podzakonnye akty*] on certain issues. Examples: the decree of the State Committee for Labor and Social Problems "On Types of Earnings (Income) Subject to Withholding for Support Payments," Instructions on the Procedure for Recording Instruments of Civil Status, etc.

The Plenum of the Supreme Court of the USSR and plenums of the Supreme Courts of the union republics issue decrees containing directives and instructions that are obligatory for courts and other bodies in applying and interpreting existing laws (see, for example, Decree No. 2 of the March 25, 1982 Plenum of the Supreme Court of the USSR "On the Judicial

Application of Legislation in the Examination of Paternity Cases and Petitions for Support for Children and Other Family Members").

All norms of family law derive from the humanistic principles of socialist morality. A requirement for the unswerving observance of existing norms instills in citizens the necessary moral qualities and a feeling of duty and responsibility to the family. At the same time, the state does not use legal norms to interfere in the personal, intimate relations of spouses and other family members. These relations are subject only to moral norms.

The basic principles of Soviet family law, which were discussed above in para. 2 of part I of the present chapter, were expressed in the first decrees of the Soviet state, which were promulgated in December 1917 and signed by V. I. Lenin personally. Lenin had words of high praise for Soviet laws on marriage and the family: "Not a single democratic party in the world, not even in the most advanced bourgeois republic, has done in decades so much as a hundredth part of what we did in this regard in our very first year in power."[2] Soviet legislation on marriage and the family invariably and consistently reflects the Leninist principles and tasks posed by the Program of the CPSU.

III. Period of Limitations

As indicated in Part 1 of "Civil Law" (see Chapter 9), the law establishes a certain period (period of limitations) for the defense of civil rights. There are only certain causes of action that are enumerated in the law to which the period of limitations does not apply.

A different principle is operative in the defense of rights emanating from marriage and family relationships. As a rule, the law does not establish any period [of limitations] whatsoever for the defense of these rights. Every family member is entitled to demand their protection at any time, regardless of the time of origin of a given right. According to article 8 of the Principles, the period of limitations is invoked only in cases explicitly established in the law. The Principles provide that the period of limitationss may be invoked in the following

2. V. I. Lenin, *Complete Collected Works* [Russ. lang. ed.], vol. 39, p. 23.

instances: in cases involving the division of the common joint property of divorced spouses, for which the period of limitations is three years (para. 5, article 12 of Principles; article 21 of the Code of Marriage and the Family), and in cases in which paternity (maternity) is contested, for which the period of limitations is one year (para. 4, article 17 of the Principles; para. 5, article 49 of the Code of Marriage and the Family).

These same cases are covered by the codes of all the union republics. According to the legislation of the Kazakh SSR, the period of limitations also extends to causes of action to declare adoptions invalid. In this republic, interested parties can bring such causes of action only within three years of the rendering of an adoption decision. The period of limitations is not applied if the suit is filed by a procurator or a guardianship or wardship organ (article 113 of the Code on Marriage and the Family of the Kazakh SSR).

Norms governing the period of limitations are applied in family law in accordance with rules (suspensions, stays, etc.) established by civil legislaion, unless otherwise provided by law (articles 11-12 of the Code on Marriage and the Family).

In addition to the periods of limitations indicated above, the law establishes a three-year period of limitations on the recovery of support payments for a past period. Within this period, support payments may be awarded only if the court establishes that measures were taken to recover support payments prior to the filing of the suit, but that they could not be recovered because the obligated persons avoided making these payments (article 95 of the Code on Marriage and the Family).

In cases where support payments have already been awarded by the court, recovery under a court judgment is possible for a period of three years preceding the initiation of execution proceedings with the exception of cases where execution has not been sought owing to a search for the judgment debtor; in the latter case execution is levied for the entire past period regardless of the established three-year period (articles 95-96 of the Code on Marriage and the Family).

Chapter 43. Marriage

I. The Concept and Form of Marriage

1. Since marriage is the basis of the family, in whose further strengthening the state and society are interested, the state cannot regard marriage solely as the private concern of the spouses themselves. The law therefore regulates the most important aspects of the conclusion and termination of marriages as well as mutual relations between spouses.

Unemployment does not exist in the USSR, every Soviet citizen can find work corresponding to his education and interests, and work is remunerated according to the quantity and quality of labor input. Therefore, no one in our society can be forced to marry on the basis of material considerations. In our country, forced marriages or marriages dictated by parents or other persons upon whom the young people are dependent take place as a rare exception where vestiges of the past have not been entirely eliminated. Young people usually marry of their own will and choice on the basis of mutual love and attraction, common interests, and the desire to work together for the common good of their family. Such a family can become a lasting union that is the source of happiness of the spouses themselves and can provide a healthy child-rearing situation.

Thus, *marriage in a socialist society is understood to mean the voluntary union of man and woman, based on love, with the aim of forming a family.*

The only marriage recognized in the USSR is a marriage concluded at state agencies responsible for recording instruments of civil status (para. 2, article 6 of the Code on Marriage and the Family). The fact that a man and a woman have lived together, regardless of the length of time and the nature of the relations between them, cannot create for these persons the rights and obligations that emanate from marriage.

The law attaches special significance to the registration of marriages and obliges ZAGS [registry of acts of civil status] agencies to make marriage registration a festive occasion if the wedding couple so desires (para. 3, article 14 of the Code on Marriage and the Family). Wedding Palaces [*Dvortsy brakosochetaniia*] have been established accordingly; where they have not been organized, ZAGS offices are appropriately furnished. The wedding day, like a vivid holiday, must be memorable for a lifetime; the atmosphere surrounding the wedding must correspond to the significance of this event in the life of a person.

Before the revolution, weddings were concluded in religious form. Religious marriages entered into before the revolution or during the first years of Soviet power, if ZAGS agencies had not yet been organized in the place where the spouses lived, are considered valid and are equated with marriages registered with ZAGS. This rule also applies to marriages concluded during the temporary occupation of part of our country's territory by the fascists, when the spouses were deprived of the opportunity to register their marriage with ZAGS (article 6, Code of Marriage and the Family).

Not until the registration of the marriage do the rights and obligations of spouses originate. This means that the law attaches legal significance to the registration of marriage.

2. Prior to the promulgation of the July 8, 1944 Edict of the Presidium of the Supreme Soviet of the USSR, Soviet legislation recognized the legality of unsolemnized marriages. Therefore, spouses marrying prior to July 8, 1944 were entitled to cite the actual date of their marriage at the time of registration of their marriage. In such a case, the rights and obligations of spouses will be reckoned as of the date indicated by them. In cases where such spouses have not registered their marriage after July 8, 1944, but have continued living together up until the death of one of them, the surviving spouse is entitled to petition the court to recognize the surviving spouse's marriage to the deceased. If the petition is granted, the surviving spouse's entitlement to all rights of a spouse--the right to a pension, the right to succession, the right to part of the property acquired during the marriage, etc.--will be recognized by the court.[3]

II. Conditions for Entering into Marriage and the Procedure for Concluding a Marriage

1. In order to enter into marriage, it is necessary to observe the conditions established by law. These conditions are: *mutual voluntary agreement* to marriage, the *attainment of the age of marriage*, and the absence of legal *obstacles to marriage* (articles 15-16 of the Code on Marriage and the Family).

The necessity of the conscious, considered wish of each of

3. See: Edict of the Presidium of the Supreme Court of the USSR of November 10, 1944, *Gazette of the Supreme Court of the USSR*, 1944, No. 60.

the parties to enter into marriage follows from the very essence of marriage in a socialist society. Only a person who possesses a sufficient degree of spiritual and physical maturity can decide such an exceptionally important question as the entry into marriage and the creation of a family. What is more, medicine has established that early marriage adversely affects the health of spouses, especially the woman, as well as the health of their offspring. This is why the law has defined the *age of marriage*, i.e., the minimum age at which citizens are permitted to marry.

The Principles set the marriage age for both men and women, and specifically set the age of their majority at 18 years. The Constitution of the USSR gives citizens of that age the right to vote; they are also as a rule sufficiently mature to decide the question of starting a family. At the same time, union republics have the right to establish a lower [marriage] age, but no lower than 16 years. Union republics are entitled to lower the age of marriage for everyone getting married and to lower the age of marriage in specific instances.

Legislation of the Ukrainian and Uzbek union republics sets a lower marriage age for women, while the marriage age in other union republiics for men and women alike is 18 years.

The codes of all union republics provide for granting petitions to lower the marriage. RSFSR legislation permits the marriage age to be lowered to 16 years for both spouses in individual, exceptional cases. A district (city) Soviet of People's Deputies may authorize the lowering of the marriage age (article 15 of the Code on Marriage and the Family). This issue is not resolved in the same way in all republics: the legislation of some republics permits the lowering of the marraige age by only one year; in some republics, this reduction is possible only for women.

2. Persons who have reached marriage age and who desire to marry may not do so if there exist *legally established obstacles*. Persons who are already a party to another, undissolved marriage may not marry. Monogamy is one of the basic principles of Soviet family law; polygamy and polyandry are incompatible with our morality.

Close relatives may not marry one another. This prohibition extends to persons related by direct descendant and ascendant lineage (father and daughter, grandfather and granddaughter, etc.) as well as to brothers and sisters. Brothers and sisters may not marry if they have the same father and mother (if they are full-fledged brothers and sisters) or if they have only one parent in common--mother or father (stepbrothers or stepsisters). While farther removed degrees of kinship are not an obstacle to marriage, marriages between people whose relationship is still farther removed (for example, between cousins) are not as a rule common in our country. Marriage between close relatives frequently results in defective progeny and, moreover runs counter to our moral precepts.

Adoptive parents may not marry their adopted children. However, an adopted child may marry a relative of the adoptive parent, regardless of the degree of kinship. For example, an adopted child may marry the daughter of his adoptive parent.

Finally, marriage is prohibited between persons, one or both of whom have been duly declared by way of a procedure established by law (i.e., by a court) to lack dispositive capacity as a result of mental illness or lunacy. A person lacking dispositive capacity is not only incapable of making an intelligent decision to marry, but is also not a fit partner for normal family relations. What is more, mental illness is reflected in the offspring of a mentally ill person (article 16 of the Code on Marriage and the Family).

Our legislation knows no other obstacles to marriage whatsoever. Before the revolution in Russia, and in many countries even today, there exist a number of additional obstacles to marriage--the existence of kinship of a more remote degree or the existence of affinity [*svoistvo*][4]; the necessity of obtaining special permission to marry from parents, from superiors at work, etc.; there are also such disgraceful restrictions as race.

3. Persons desiring to marry submit an application to ZAGS at the place of residence of the bridegroom or bride. They may also submit their application at the domicile of one of their parents. The application must indicate the number of times each applicant has been married and whether the applicants have children. ZAGS officials receiving the application check to see if there are any obstacles to the marriage and to ascertain that the bridegroom and bride are acquainted with the state of each other's health. Persons getting married are also briefed on their rights and obligations as future spouses and are warned about their criminal liability for the falsification of information in their applications (article 150 of the Code on Marriage and the Family). The date for registering the marriage

4. Affinity—relations that arise between the spouse and his or her relatives and the relatives of the other spouse.

is then set with the consent of the bridegroom and bride. This date may not be earlier than one month after the submission of the application.

This one-month period may be shortened or extended to three months by a ZAGS department chief, or in rural areas and settlements by the chairman of an executive committee of a village (settlement) Soviet of People's Deputies (article 14 of the Code on Marriage and the Family). This time gives the bridegroom and bride another opportunity to consider their decision in all earnestness.

It is mandatory that the marriage be registered in the presence of the persons getting married. Relatives, friends, and acquaintances of the bridegroom and bride may be invited to the registration and may, if the latter so desire, serve as witnesses. As already noted, the registration is performed in a festive atmosphere.

III. The Annulment of a Marriage

The court may annul a marriage that is concluded in violation of the established conditions or in defiance of obstacles specified by law, which were discussed in the preceding paragraph. A *sham marriage*, i.e., a marriage that is concluded without the intention of creating a family, but solely for the purpose of obtaining rights resulting from the registration of marriage (for example, with the aim of avoiding assignment to work elsewhere, of obtaining a residence permit, etc.--article 43 of the Code on Marriage and the Family) shall also be annulled.

Only the court may annul a marriage. Until a marriage is annulled by a court, the married persons are considered spouses. But if a court annuls a marriage, it is considered invalid from the time it was concluded and the rights of spouses between the parties to the marriage are considered not to have originated (article 46 of the Code on Marriage and the Family).

The annulment of a marriage on the ground that it was concluded with a minor for whom the marriage age was not lowered according to the established procedure may be demanded only by the minor spouse, by the minor spouse's parents or guardian (ward), by guardianship or wardship agencies, or by the procurator. If by that time, the spouse has reached marriage age, the demand to have the marriage annulled may be made only by that spouse or the procurator. Persons other than those enumerated, including a spouse who is of age, are not entitled to demand that the marriage be annulled on that ground (article 45 of the Code on Marriage and the Family).

A marriage that was concluded with a person below the marriage age may be annulled only if this is required by the interests of the minor spouse.

The demand that a marriage be annulled may be made by the spouses themselves, by other interested persons, i.e., persons whose rights were violated by the conclusion of the marriage, as well as by guardianship and wardship agencies and by the procurator (article 44 of the Code on Marriage and the Family). For example, the heirs of a deceased citizen are entitled to demand the annulment of a marriage concluded by the estate-leaver on the grounds that the decedent had been declared to be lacking dispositive capacity due to mental illness or lunacy; the spouse of a person who remarries without first dissolving an existing marriage is also entitled to make such a demand.

If at the time the court examines the case the previous obstacles no longer exist (for example, if the first spouse has died or if the marriage with the first spouse has been dissolved), the marriage may be annulled, but only from the time the obstacles ceased to exist and not from the moment the marriage was concluded (article 43 of the Code on Marriage and the Family of the RSFSR).

A sham marriage that was originally concluded without the intention of creating a family is recognized as valid if the persons registering the marriage change their intention and in fact establish a family. Thus, when the court examines annulments, it studies not only violations committed in the process of concluding a marriage, but also the circumstances that have transpired by the time the court hears a given case.

As already noted, if a marriage is annulled, it is considered as never having existed and the parties to the marriage acquire no rights whatsoever from the marriage. There are exceptions to this rule. A spouse who was unaware of obstacles to the conclusion of a marriage is entitled to keep the name chosen by that spouse upon concluding the marriage. The unwitting, deceived spouse of a bigamist is not only entitled to retain the married name, but is also entitled to demand maintenance payments if unable to work and needy, and to demand the division of property acquired in this marriage on the same grounds as a spouse in a lawful marriage (article 46 of the Code on Marriage and the Family). However, the deceived spouse derives no other rights from this marriage (for example, the right to a pension following the death of the spouse).

The annulment of a marriage is not reflected in the rights of children born or conceived in a marriage that is subsequently annulled. Children enjoy the same rights as those born from a lawful marriage (para. 6, article 46 of the Code on Marriage and the Family).

IV. Personal and Property Rights and Obligations of Spouses

1. As a result of the conclusion of a marriage, a number of *rights* and *obligations*, both *personal* and *property*, accrue to the persons entering into the marriage--the spouses. Although the basic relations of the parties to a marriage are personal, these relations usually cannot be regulated by legal norms; the mutual relations of spouses in this area are defined by moral norms. After all, the law cannot be used to forcibly compel anyone to love or respect another. Therefore, the legal norms pertaining to the interpersonal relations of spouses establish only the rules that ensure the full *equality of both spouses* and the free exercise of the rights extended to them as citizens of the USSR.

Accordingly, the law provides that questions pertaining to child-rearing and other questions in family life shall be *resolved by the spouses jointly*. Neither may force his or her will on the other. Everything must be resolved by mutual consent. At the same time, the law emphasizes that each spouse is *free to choose type of employment, occupation and place of residence* even though it goes without saying that in normal family relations these questions are decided by both spouses (article 19 of the Code on Marriage and the Family). Legislation provides a number of measures ensuring the possibility of spouses to live together. When young specialists are assigned to jobs, consideration is always given to where the husband (wife) of the assignee lives; family relocation grants are a part of transfers to work in other regions, etc. However, one spouse cannot be forced to accompany the other spouse to a new destination. In this regard, it should be recalled that according to the legislation of tsarist Russia, the choice of domicile belonged exclusively to the husband and the wife was obliged to follow him; if she refused, she could be forcibly transported to him under arrest.

When they marry, spouses may, if they so desire, choose the

surname of one of them as their *common surname.* It may be the husband's surname or the wife's surname. Spouses may also retain their premarital surnames. The legislation of some union republics (Belorussian SSR, Georgian SSR, and others) also provides that spouses may use a common combined [hyphenated] surname (for example, Ivanov-Sokol'skii, Petrov-Vel'skii, etc.). If one or both spouses already have double names, the legislation of most republics does not permit them to combine their names.

2. The exercise of property rights, unlike personal rights, can always be compulsory, in connection with which they can be regulated in more detail by legal norms. The property rights of spouses consist of rights to property acquired during marriage and rights to maintenance (alimony).

Entry into marriage does not create for the spouses any kind of right to property that belonged to the other spouse prior to that time. Such property remains the personal, separate property of each of them. Conversely, all property acquired by the spouses during their marriage is their *common joint property* (article 20 of the Code on Marriage and the Family). The very fact that they are married is the ground for common joint ownership of property acquired during their marriage. Therefore, spouses are not entitled to agree that certain property will not be part of their common property or that some rules other than those established by law will be applied to its division, management, etc. The legally established regime of common property acquired during marriage is thus compulsory.

3. Common joint ownership attaches to acquired property. "Acquired" property refers to property amassed as a result of the personal activity of a given person. In the USSR, property is primarily acquired as a result of remuneration received for labor in social production, but may also be acquired as a result of labor on one's individual household (on a private plot, as a result of individual labor activity), and, finally, as a result of income permitted by the law (interest on deposits, winnings from the obligation bond lottery).

Property that has been acquired not as a result of the activity of the spouses themselves (i.e., that was not earned by them), but has been acquired gratuitously from other persons (as a gift or an inheritance) does not become the common property of the spouses, but is the personal property of the spouse who is the recipient of the gift or inheritance.

Property amassed during marriage in the form of items of personal use is declared to be the personal property of the spouse that uses them. However, if they are luxury items (expensive jewelry, for example), they are declared to be the common joint property of the spouses (article 22 of the Code on Marriage and the Family).

Property belonging to each of the spouses prior to their marriage remains their personal property. However, if in the course of the marriage this property increases significantly in value due to investments from the spouses' common property, the court may declare it to be common joint property (for example, if a house underwent capital repairs and reconstruction during the marriage).

Thus, the personal property of each spouse consists of: property belonging to the spouse before marriage; property received during the marriage through gifts or inheritance; and items of personal use. Other property is the common joint property of the spouses.

4. The common joint property of spouses differs in large measure from common shared property under civil legislation. As already noted, the basis of common joint property is marriage.

Common joint property accrues only to spouses. Therefore, joint property does not accrue to persons who, although living together and maintaining a common household, have not registered their marriage. The common joint property of spouses ceases when the marriage is terminated.

Common joint property is unshared. Prior to the termination of common property relations, each spouse has a right to all property comprising their common joint property, but not to any part of it. The spouses' shares are determined only with the division and termination of common property.

The law speaks of the common joint property of the spouses. Accordingly, it can be concluded that only items belonging to the spouses by right of ownership will become joint property. However, judicial practice rightfully gives a broad interpretation to the common joint property of spouses, which also includes among the common property of spouses promissory [*obiazatel'stvennye*] obligations resulting from their joint (or individual) use of common property. Examples of common promissory obligations are a claim on Gosstrakh organs to pay insurance benefits on a burned-down house that belonged to the spouses by right of personal ownership, claims on savings banks to pay out deposits made from the spouses' common resources, etc.

As long as common joint property exists, each spouse has the equal right to possess, use, and dispose of all common property, and not only some part of it (article 20 of the Code on Marriage and the Family). Thus, each of them is entitled to alienate (sell, make a present of, etc.) any of the spouses' common property. It is assumed that each spouse acts with the consent of the other. Therefore one spouse is not entitled to demand that a transaction entered into by the other spouse be declared invalid on the grounds that the former did not consent to the transaction. Such a claim can be satisfied only if it can be proven that the party entering into the transaction with the given spouse was aware that the other spouse was opposed to that transaction.

Only when certain items of common joint property of spouses are alienated, is it necessary to obtain the consent of the other spouse beforehand (in the RSFSR, this consent is required for the sale of a dwelling; in certain other republics, it also extends to the sale of other property). If consent is not obtained beforehand even though it is necessary for the given transaction, it may be declared invalid by a court.

5. Common joint property may be divided at the demand of one or both spouses during the marriage as well as upon its termination. If the division is made in the course of the marriage, common joint ownership of property acquired at the time of division is terminated and consequently becomes the personal property of each of the spouses. However, the division cannot change the legally established regime of the spouses' common property, and therefore all property acquired after the division, but before the termination of the marriage, will be regarded as the common joint property of the spouses.

If the division is made in the course of the marriage, all property existing at the time of the division is divided. If, however, the division is made after the marriage has been terminated as a result of the death of one of the spouses or the dissolution of their marriage, the property that exists at the time of termination of the marriage is divided. All those items that comprise the personal, separate property of each of the spouses are naturally excluded from the existing property.

Because husband and wife enjoy equal rights, their property

is as a rule divided into two equal shares. The fact that one spouse has earned more than the other and that various things have been acquired on the basis of that spouse's earnings is irrevelant. Even when one spouse has no earnings (income) whatsoever because that spouse is raising the children or running the household or is not working for other valid reasons, that spouse's rights to common property remain equal and that spouse must be adjudged on a par with the other spouse.

In life there may occasionally be circumstances in which the division [of property] into equal shares would be unjust. The law empowers the court to deviate from the equality principle. Nevertheless, such deviation is possible only in cases provided in the law (article 21 of the Code on Marriage and the Family). These cases are as follows:

The court may award a larger part of the property to the spouse with whom the children will reside after the dissolution of the marriage if this is in the interests of the children. For example, children must be provided with housing and hence the spouse with whom the children will reside may be awarded a larger part of a dwelling or a larger share in a housing-construction cooperative.

A larger part of the property may also be awarded if this is required by the spouse's interests that merit attention, for example, if the spouse is an invalid and requires certain things. It was noted above that a spouse has equal right to property even if that spouse did not work. However, if that spouse did not work and avoided participating in socially useful work in production or in the family, i.e., was a sponger, the share of that spouse could be reduced. The share is also reduced when it is found that one of the spouses has squandered property contrary to the interests of the family (for example, when, as a result of a passion for alcohol, the spouse fails to use earnings for family needs, sells items of common property for the pursuit of his/her own interests, etc.).

After the shares of each of the spouses have been determined, in accordance with these shares, an allocation is made of the specific items that should be awarded to each of them. In the process, the interests of each spouse and the interests of their children are taken into account. For example, professional tools (a musician's instrument, a scientist's library, etc.) are awarded to the person who uses them. If the value of such items is greater that the share that is due the given spouse, the other spouse is awarded cash compensation. If a certain item is essential to a child's education (for example, a piano if the child is attending a music school), it is awarded to the parent with whom the child will be living.

In some cases, the property includes items that cannot be divided (an automobile, for example). Such things are awarded to the spouse who, in the court's opinion, has the greater need. In such a case, the other spouse is awarded a larger part of the other objects or cash compensation.

6. Spouses may have various kinds of obligations, for example, they may have borrowed money, purchased something they have not paid for, etc. If the debtor does not meet his (her) obligation on schedule, a court may enforce collection of the amount due and impose a penalty on the debtor's property. But on what property can a penalty be imposed if the debtor is married and if, in addition to things that are his (her) personal property, there are also things that are the common property of the debtor and his (her) wife (husband)? The law has established special rules in this regard. Relative to the obligations of one of the spouses, a penalty may be imposed not only on the given spouse's personal property but also against that spouse's share of the common joint property of the spouses, which would be due the given spouse if the property were divided. When a penalty is imposed for this purpose, it is frequently necessary to divide the spouses' property.

If damages caused by the criminal actions of one of the spouses are compensated, a penalty may also be imposed on items that are the common property of spouses vis-à-vis whom the judgment will state that these specific items were purchased with funds obtained by criminal means, e.g., that an automobile that is the common property of spouses was purchased with stolen money (article 13 of the Code on Marriage and the Family).

The codes of some union republics provide that where that which was obtained by the debtor under an obligation was used in the interests of the entire family, then, even though the obligation was assumed by only one of the spouses, both spouses are answerable with their common property (for example, article 24 of the Code on Marriage and the Family of the Moldavian SSR). Thus, if furniture purchased on credit is used by the entire family, the spouses will be answerable with all common property; however, a penalty will not be imposed on the personal property of the spouse that did not assume the obligation. While this is not explicitly stated in the Code on Marriage and the Family of the RSFSR, judicial practice proceeds from the same premises. They follow from the sense of those legal norms that establish the common property of the spouses.

7. Common joint property relations of spouses cease with the termination of the marriage. Therefore, each spouse loses the right to dispose of all common property. Both spouses are entitled to dispose of only their respective shares, and these shares are not defined before the property is divided. Consequently, after the termination of the marriage, neither spouse is entitled to sell, give away, or otherwise dispose of any items of common property without the consent of the former spouse.

Spouses themselves usually divide their property after the dissolution of their marriage. If they are unable to agree on the division, the division is made by the court. The period of limitations for petitioning the court to divide property is three years (article 21, Code on Marriage and the Family). This period is calculated from the time when the spouse knew or should have known that his or her right was being violated. Since, as we have seen, neither spouse is entitled to further dispose of common property, this period begins, for example, from the moment that one of them sells certain property without the consent of the other or begins posing obstacles to the other's use of this property, etc. Only the Code on Marriage and the Family of the Armenian SSR provides that the period of limitations for demanding the division of property shall commence with the termination of the marriage. In other republics, this period commences with the termination of the marriage only if the rights of a spouse have already been violated by that time and if the spouse knew or should have known about the violation.

8. A husband and wife usually keep a common household in which they invest their common resources. When one of them is in a position to invest more than the other, no one thinks that one spouse is maintaining the other. But when relations deteriorate, even if matters have not yet reached the divorce stage, husband and wife frequently cease to maintain a common household and sometimes even separate. If both of them are able to work and are consequently able to support themselves through their work, they are not entitled to make any demands on one another. Even though the law provides that spouses are

obligated to provide financial support for one another, this obligation remains only as their moral duty if the grounds provided in the law for not only moral, but also legal obligation--maintenance obligation (see I, Chapter 45 concerning the maintenance obligation concept)--is absent.

The right to receive maintenance accrues to one of the spouses vis-à-vis the other if that spouse is unable to work, i.e., is an invalid (in any group), has reached pension age (55 years for women, 60 years for men), and needs help. A spouse who has absolutely no independent source of subsistence (earnings, pension, etc.), as well as a spouse who is unable to satisfy his/her needs through his/her own resources are considered to be needy. What is more, in order for the maintenance obligation to accrue, it is necessary that the spouse against whom the claim is filed have sufficient funds to render the necessary aid to his/her wife/husband. If this spouse, for example, is an invalid himself/herself and receives a small pension, he/she is not obligated to maintain his/her spouse.

Thus, the grounds for origination of the maintenance obligation between spouses are the *inability to work* and the *neediness* of the person entitled to receive the same and the availability of sufficient funds on the part of the obligated person.

During pregnancy and for one year after giving birth, a wife has a right to receive maintenance from her husband. For the sake of the health of mother and child, it essential that the woman have improved living conditions during this period. Thus, during this period the right to maintenance originates regardless of an inability to work. Two other grounds are necessary for the origination of this obligation. However, it should be noted that the greater needs of the mother during her pregnancy and the time when she is nursing the child are always taken into account in determining need in these cases (article 25 of the Code on Marriage and the Family). The right to receive maintenance even after the dissolution of the marriage is preserved if the spouse's inability to work originated before the dissolution of the marriage or no later than one year after that time. If the inability to work originated one year after the dissolution of the marriage, the right to maintenance does not accrue. There is one exception to this rule: if the spouses were married for a long time and one spouse became unable to work as a result of attaining pension age, the court is entitled to award maintenance if the spouse reached this age within five years after the termination of the marriage. The discussion here is of a spouse (usually the wife) who has not worked the time required to qualify for a pension because she devoted her efforts to her family. It is just that the responsibility for maintaining her in old age be placed on the person to whom she gave her strength and her health, i.e., her spouse (article 26 of the Code on Marriage and the Family).

The court may refuse to award maintenance or else may limit maintenance payments to a certain period if the spouses have been married only a short time and also if the spouse demanding the payment of maintenance has been guilty of misconduct (article 27 of the Code on Marriage and the Family) (for example, if he/she drank excessively during the marriage or became unable to work as the result of his/her commission of a crime, etc.).

If the spouses themselves have not agreed on the amount of aid, the court determines the amount of maintenance based on the family and material status of both spouses (article 28 of the Code on Marriage and the Family). The fact that the spouses have agreed on a certain sum does not mean that the needy spouse may not petition the court to set the amount of the maintenance payments if in that spouse's opinion this amount has not been correctly determined. In the event of change in the material or family status of the spouses, each of them is entitled to demand an increase (or decrease) in previously awarded maintenance payments.

If one of the grounds for receiving maintenance payments is no longer present, i.e., if the spouse receiving the maintenance payments is once again able to work or if he no longer needs the maintenance payments (if, for example, he has been granted a pension that is sufficient to meet his needs) or if the material status of the person making the maintenance payments has changed to the extent that he can no longer assist the other (for example, if he became an invalid and now receives a small pension), the maintenance obligation ceases, and the spouse loses the right to demand the payment of maintenance. The right to receive maintenance is also lost if the recipient remarries. In such a case, the new spouse will bear the responsibility for the given spouse's maintenance.

If maintenance is paid involuntarily, on the basis of a court order, only a new court order can relieve the payer of maintenance of these payments at the latter's request (article 29 of the Code on Marriage and the Family).

V. Termination of Marriage

1. When spouses marry, they hope that their union will last a lifetime. Happy are the ones who are able to walk hand in hand for all their life, sharing joys and sorrows together. Therefore, as a rule, a marriage continues until one of the spouses dies. The death of a spouse terminates a marriage. Since the consequences of the declaration of a citizen's death (see Chapter 7) are the equivalent of death, a marriage is also considered terminated if the death of one of the spouses is announced (para. 1, article 30 of the Code on Marriage and the Family). In such a case, the other spouse is entitled to remarry. If the spouse who has been declared dead is in actuality alive and the court judgment declaring him dead is rescinded upon his return, the marriage is considered to be automatically restored. However, if the spouse of that person has remarried in the meantime, the previous marriage is not restored (para. 1, article 42 of the Code on Marriage and the Family).

2. It sometimes happens that marriage does not bring happiness and the spouses find it impossible to continue living together. This may be the result of rushing into marriage, but this is not the only or principal reason for the breakup of the family. Sometimes the undignified behavior of one of the spouses or other circumstances make it impossible for the spouses to continue living together and result in the termination of the marriage. Not only is it not in the interests of the spouses to compel them to live together at all costs, divorce is sometimes necessary for the proper upbringing of the children. Therefore, the law provides for the termination of marriage while both spouses are still alive through its dissolution (divorce) (para. 2, article 30 of the Code on Marriage and the Family). Marriage is the basis of the family; its termination entails the breakup of the family. Therefore, the state cannot view divorce, like marriage, entirely as the private affair of the spouses themselves. Accordingly, the procedure and grounds for dissolving a marriage are established by law.

3. As a rule, a marriage is dissolved by the court. Spouses may jointly or singly request the dissolution of their marriage. The petition is filed with the district (city) court nearest the domicile of the defendant, i.e., the spouse against whom the

petition to dissolve the marriage is filed. If the spouse filing the petition (the spouse-plaintiff) has minor children or is unable for health reasons to travel to the place where the defendant lives, the petition may be filed in the place where the plaintiff lives (article 118, Code of Civil Procedure).

4. The only ground for dissolving a marriage is the disintegration of the family. The law states that a marriage is dissolved if the court finds that the further joint life of the spouses and the preservation of the family have become impossible (para. 3, article 33 of the Code on Marriage and the Family). In order to arrive at such a conclusion, it is necessary to make a detailed examination of all circumstances in the family's life, to discover how the spouses related to one another, to the family, to the children, and what specifically generated the desire to dissolve the marriage. Witnesses should be summoned and documents submitted if necessary.

The court has an obligation to take the necessary measures to reconcile the spouses. After all, a divorce petition is sometimes filed under the influence of a single act by one of the spouses, while in actuality it is in the interests of both spouses to preserve the family. The court may postpone examination of the case and give the spouses time (a maximum of six months) to effect a reconciliation (article 33 of the Code on Marriage and the Family). Thus divorce lawsuits must be very carefully prepared.

5. If the court concludes that the family has disintegrated and the marriage should be dissolved, in the necessary instances, it takes measures to protect the interests of children and the spouse who is unable to work.

The court decides which of the parents the children will live with after the dissolution of the marriage only if the spouses themselves are unable to reach agreement on this point (article 34 of the Code on Marriage and the Family). However, regardless of whether such a dispute exists, the court explains to the parents that they both have an obligation to care for their children in the future and that therefore the parent living separately has the right and obligation to take part in the upbringing of their children and that parent with whom the children will be living may not prevent the other parent from doing so. In the event of a dispute, the court must determine which parent must pay child support payments and in what amount.

At the request of a spouse who is unable to work and who

has a right to maintenance payments, the court determines the amount of maintenance the other spouse should pay (article 35 of the Code on Marriage and the Family).

At the same time that it dissolves a marriage, the court must divide the spouses' common joint property if requested to do so by one or both spouses (article 36 of the Code on Marriage and the Family). If such division affects to any degree the rights of third parties (for example, the division of a share in a housing construction cooperative), the division dispute must be made the subject of a separate proceeding. In cases involving the interests of third persons, they must be summoned to participate in the case, whereas divorce cases are of a personal, intimate nature and there can be no place for the participation of third parties therein.

At the request of the spouse who adopted the surname of the other spouse at the time of marriage, ZAGS will restore to him/her his or her premarital surname at the same time the dissolution of the divorce is registered (article 41 ofthe Code on Marriage and the Family). The premarital surname may be restored only at the request of the person who hcanged his or her surname; the other spouse is not entitled to demand such a change.

When the court hands down its judgment, it also determines the state fee that must be paid to obtain divorce papers. The fee ranges between 50 and 200 rubles and may be collected from one or both spouses. In the latter instance, the court determines the sum that each of them is obligated to pay (article 37 of the Code on Marriage and the Family). A marriage is considered terminated only after the divorce has been registered with ZAGS (article 40 of the Code on Marriage and the Family).

6. A marriage may be dissolved by registering the divorce with ZAGS. Such a procedure is applied in the following cases.

Spouses who do not have minor children in common may, with their joint consent, dissolve their marriage at ZAGS. The marriage is dissolved three months after the spouses file their divorce petition with ZAGS (article 38 of the Code on Marriage and the Family).

However, if a dispute arises between spouses regarding the children, the division of property, or support payments for a needy spouse who is unable to work, the divorce proceedings go before a court (article 39 of the Code on Marriage and the Family).

Chapter 44. Parents and Children

I. Grounds for the Development of Parental Rights and Obligations

1. The very fact that a child is born to its parents is the ground for the development of parental rights and obligations. Article 16 of the Principles (article 47 of the Code on Marriage and the Family) provides that the mutual rights and obligations of parents and children are based on the parentage [*proiskhozhdenie*] of the children, which is certified according to a procedure established by law. Therefore, in order to confirm parental rights and obligations, the parentage of the children must be appropriately certified. Parentage is certified by an entry in ZAGS records. A birth certificate attesting to this entry is issued.

Since parental rights originate the instant the child is born, parents are entitled to care for and rear their child even before its birth is registered with ZAGS. However if the point at issue is the defense of parental rights or such exercise of these rights as requires official confirmation of the fact that the given persons are the child's real parents, it is necessary to submit a birth certificate identifying the given persons as the child's parents. This demand applies in equal measure to the child's father and mother. Thus the mother has to submit this document in order obtain a residence permit for her child, a maternity grant, etc. Moreover, if for some reason the child is in someone else's custody, his mother cannot demand his return unless she can submit the ZAGS certificate proving that she is the child's real mother.

2. The name of the child's mother is recorded in ZAGS records in accordance with the certificate issued by the medical institution in which birth took place. Therefore, as a rule, there are no difficulties whatsoever in registering the mother as the parent of the child to whom she gave birth. Only in extremely rare cases is it necessary to establish maternity: for example, if the mother abandoned the child and the child was registered without any indication of the mother's identity; if an erroneous entry was made; or if another woman's name was registered because the mother abused her child. In such exceptional cases, maternity must be established by the court. If another woman is recorded as the mother, it should be simultaneously de-

manded that this registration be declared invalid.

Maternity may be demonstrated by any proof permitted by law--the testimony of witnesses, various kinds of written proof, expert testimony, etc.

3. The registration of the child's father is a more complex matter. If the parents are married to one another, there are usually no difficulties whatsoever. The legal presumption is that the mother's spouse is the father of the child. Therefore, when the marriage is registered, the mother's spouse is registered as the father of her child at the request of either of them (para. 2, article 47 of the Code on Marriage and the Family).

Parents are registered at their request. In those rare cases where the mother's husband is not the actual father of her child, but is nevertheless registered as the child's father, the registration may be contested in a judicial proceeding. Only persons who are registered as the father or mother may petition the court to declare this registration invalid. The registration may be contested only within one year of the time when the parent registered as a parent knew (or should have known) of the registration. Such a short period [of limitation] is established in the interests of the child because this question must not be allowed to remain vague for an extended period of time; what is more, the contesting of paternity (maternity) over a long period of time causes considerable difficulties, one of which is the period of limitation. Therefore, if the court finds valid reasons why a petition was not filed within the period of limitation, it is entitled to restore the period of limitation and to examine a petition submitted after its expiration.

4. If a child's parents are not married to one another, the parents can establish paternity by filing a joint declaration with ZAGS. This declaration is the basis for registering the establishment of paternity in accordance with which the necessary information about the father is entered in the child's birth certificate.

The declaration must be filed by both parents. The following rules exist for cases where the declaration cannot be filed jointly due to the death of one of the parents: if the mother dies, if she is declared to be lacking dispositive capacity, if she is deprived of her parental rights, or if her whereabouts cannot be ascertained, paternity can be established on the basis of a declaration by the father alone (para. 3, article 49 of the Code on Marriage and the Family).

If the father dies, however, registration must be based on the establishment of paternity by the court. Since the defendant (i.e., the father) is not living, such cases are usually examined as a special proceeding instituted to verify facts that are of legal significance. The decedent's parentage of the child will be such a fact in these cases.

Paternity is usually established under such a voluntary procedure in the case of children born out of wedlock. However, in some cases such a declaration is not submitted. Most often the child's father refuses to submit such a declaration because he either doubts that he is the father or because he wishes to avoid the fulfillment of parental obligations. There are cases where the mother does not agree to submit a joint declaration in the belief that she can raise her child by herself and because she does not wish to be connected with a person to whom she is not married.

In the absence of a joint declaration by the parents, paternity may be established by the court. Such a declaration may be filed with the court by the child's parents (equally by the mother or father), by the child's guardian (ward), or by the child himself when he reaches majority. According to the legislation of the RSFSR and other republics, such a declaration can also be filed by the person on whom the child is dependent (para. 1, article 48 of the Code on Marriage and the Family).

In order to satisfy the suit, it must be proven that the child is truly the offspring of the persons whose paternity the plaintiff wishes to establish. However, the court is not entitled to establish paternity unless at least one of the circumstances enumerated in the law is present. These circumstances are: joint residence and maintenance of a common household by the child's parents during the mother's pregnancy; joint rearing or joint maintenance of the child; and, finally, positive proof confirming that the father acknowledged his paternity before and after the birth of the child (para. 2, article 48 of the Code on Marriage and the Family). If none of these circumstances can be established, the suit cannot be granted even though there is no doubt that the child is the offspring of the defendant. On the other hand, the establishment of one or even all of these circumstances does not relieve the court of the necessity of establishing the true origin of the child. Just as the husband of the mother may prove not to be the father of her child, it may also happen that the man with whom the mother lived, who helped to rear and maintain the child, and who acknowledged this child as his own, will not be the actual father.

The court occasionally must resort to forensic medicine to determine the true origin of the child in paternity cases. If, for example, blood tests show that the defendant could not be the father of a given child, the suit is dismissed even though the fact of joint residence during pregnancy is established, etc.

The same rules are correspondingly applied where the father has died and the paternity suit is examined in a special proceeding that is instituted for the purpose of verifying the legal facts.

5. The Principles took effect on October 1, 1968. Therefore, everything that has been said above regarding the establishment of paternity applies to children born after that date. The July 8, 1944 Edict of Presidium of the Supreme Soviet of the USSR, which was previously in effect, did not permit the determination of the paternity of children born out of wedlock. Not only was the mother deprived of the right to petition the court to establish paternity, the father, too, was not entitled to request that he be recognized as the father of a child born out of wedlock. Such recognition was granted in only one case: when [the father] subsequently married the child's mother. The birth certificates of children born out of wedlock did not list information concerning the father; the appropriate column was either left blank or crossed out. How was the question of children born during the effective period of this edict resolved?

The law on the ratification of the Principles states that the paternity of children born of parents not married to one another prior to the effective period of the Principles can be established on the basis of a mutual declaration of the mother and father. If, by that time, the child had reached majority, in order to establish paternity, it was necessary to obtain his consent (article 3 of the June 27, 1968 Law "On the Ratification of the Principles of Legislation of the USSR and Union Republics on Marriage and the Family"). Every citizen who is of age is entitled to reject such recognition.

These facts can be established by the court where the father has in fact died without filing such a petition if, during his lifetime, he acknowledged the child as his own and if the child was his dependent. Such cases are examined by the court in a special proceeding. Unlike these cases, if a child was born after October 1, 1968, the court ascertains not the child's origin (paternity), but the fact that the decedent acknowledged the child as his son (daughter) and the fact that the child was his dependent. The child's mother, guardian (ward), and the child himself, upon attaining majority, is entitled to file a petition for the recognition of this fact. When in connection with the establishment of this fact, a civil law dispute arises (for example, a dispute concerning the deceased person's estate which is claimed by other persons), the case is decided as a lawsuit.

If a joint declaration cannot be sumitted as a result of the death of the mother, the declaration that she is lacking in dispositive capacity, because she has been deprived of her parental rights, or because it is impossible to determine her whereabouts, paternity can be established on the basis of the father's declaration.

When the father does not desire to make such a declaration or when the father dies without recognizing and maintaining the child, the determination of paternity of children born before October 1, 1968 is not permitted.

If paternity is established by ZAGS or by decision of the court, the child's rights vis-à-vis the father are entirely equal to the rights of children born of wedlock. Rights and obligations between child and parents are considered to have originated at the moment of the child's birth and not the time of establishment of paternity (maternity).

II. Personal Rights and Obligations of Parents and Children

1. The child is given the same surname as his parents and a patronymic based on his father's first name. If the child's parents have different surnames, the child is given the surname they agree upon; if such agreement is impossible, the surname is assigned on the basis of a recommendation by the guardianship and wardship agency (article 51 of the Code on Marriage and the Family). The parents themselves choose their child's name. ZAGS may not refuse the name chosen by the parents, but at the time of registration may advise the parents not to give the child a strange-sounding name or a name that might prove embarrassing. In the Lithuanian SSR, the child may be given a double name.

As a rule, the first name, patronymic, and surname given a child are not changed and a citizen bears them for an entire lifetime. When citizens reach their age of majority, they have the right to request that their first name, patronymic, or surname be changed.[5] These requests are occasioned by various factors: citizens sometimes have strange sounding names, they may not wish to continue to bear their fathers' surname if he has abandoned the family and not given it any assistance, etc. Requests for surname changes must be addressed to the local ZAGS office.

The surnames of minors may be changed if after the dissolution or annulment of their parents' marriage, the child lives with a parent whose surname is different from that of the child (article 51 of the Code on Marriage and the Family). In this case, the custodial parent is entitled to request that his/her surname be assigned to the child. Since the assignment of one or another surname does not influence the parents' rights and obligations, this question is resolved by guardianship and wardship agencies exclusively on the basis of the child's interests. Therefore, the objections of the other parent need not necessarily be taken into account.

2. Parents are endowed with parental rights. Parental rights are understood to mean not the entire aggregate of the rights and obligations of parents, but rather their right to raise their own child. In the USSR, every child's right to an upbringing is secured. Bearing in mind the natural love of parents for their child, the obligation to rear the child is placed first of all on his parents; they are vested with the parental rights required to carry out this obligation. Thus, parental rights are simultaneously the right and obligation to rear the child.

Parents have an obligation to rear their children, to care for their physical development and education, to prepare them for socially useful labor, and to raise them to be worthy members of socialist society.

Parents have an obligation to protect the personal and property rights and interests of their minor children. They act as informal guardians (wards) of their children. In the defense and exercise of the civil rights belonging to the children, parents are restricted to the same degree as guardians (wards) assigned to the children (see Chapter 46, III). Parents may not exercise their rights in a way that contradicts the children's interests

5. See: Edict of the Presidium of the Supreme Court of the USSR of March 26, 1971, "On Changing the Surnames, First Names, and Patronymics of Citizens of the USSR," *Gazette of the Supreme Soviet of the USSR*, 1971, No. 13, item 146.

(articles 52-53 of the Code on Marriage and the Family).

The state does its utmost to assist parents in their important task of rearing children and protects the rights of parents. Since parents personally have the right to rear their children, they are entitled to demand the return of their child from the custody of any person that is not based on law or a court order. In the event of refusal to return the child, the parent is entitled to bring suit in court. However, notwithstanding the preferential right of the parent, the court is entitled to reject the parents' demand if it concludes that the transfer of the child to the parents is not in the child's interests (article 58 of the Code on Marriage and the Family). The following example can be cited: following the mother's death, the child was left to her parents to be reared under favorable conditions. The child became accustomed to them and loves his grandfather and grandmother. The father had not taken any interest previously in the child's life and was essentially a stranger to him. Under these circumstances, it would not be advisable to transfer to child to his father's care. Such a transfer would traumatize the child and the exercise of parental rights, as noted above, is not permitted if it contradicts the children's interests.

3. The mother and father have equal child-rearing rights and obligations. The law provides that if parents have any disputes on child-rearing questions (for example, on the organization of the child's recreation, on whether to send the child to a special school, etc.), these disputes are resolved by guardianship and wardship agencies with the participation of the parents themselves. However, when the parents live together, such appeals are not encountered. In a normal family, the parents themselves always ultimately come to a common decision even if they at first disagree. The situation is different if the parents have dissolved their marriage. Even when there is no formal divorce, but the parents have separated, they frequently quarrel over the children.

When parents separate,the first question that has to be resolved is who shall be the custodial parent--the father or the mother. If the parents themselves do not agree, such disputes are resolved by the court (article 55 of the Code on Marriage and the Family). While in a dispute with third parties who have custody of a child, the parents have preferential rights and their demands are rejected only in exceptional instances when the transfer of the child to the parents does not correspond to the interests of the latter, in disputes between parents each parent has equal rights. Neither the father nor the mother has any advantage in the given instance. Therefore, disputes are resolved exclusively in the interests of the child. The child is given into the care of the parent who by virtue of his/her personal qualities as well as the attachment of the child for him/her is able to provide more favorable conditions for the child's upbringing and development.

These disputes are examined by the court with the mandatory participation of guardianship and wardship agencies. Prior to examining the case on its merits, the latter have the obligation to make a detailed examination of the living conditions of each of the parents, to ascertain their personal qualities and their child-rearing ability, the attitude of the child toward the parents, the child's attachment for and desire to live with one or the other parent, and also all other circumstances that can influence the solution of the given question. After completing their examination, the guardianship and wardship agencies arrive at a conclusion as to which of the parents, in their opinion, should be given the child. The conclusion of these agencies, even though not binding on the court, is of great importance. This conclusion is rendered as a result of the detailed clarification of all circumstances in the life of a given family by public educators who have the necessary pedagogical experience and a knowledge of child psychology. If the court, after examining the case, nevertheless comes to another conclusion, its judgment must describe in detail the reasons why the court did not find it possible to agree with the given conclusion of the guardianship and wardship agencies.

The child's wish to live with the father or the mother is taken into account by the court together with other circumstances but in itself is not of decisive importance. Since the participation of the child in the courtroom deliberation can traumatize the child and make the further interrelations with parents and the upbringing process more difficult, the child's wishes are ascertained by guardianship and wardship agencies in the process of their investigation. When the court deems it necessary to interrogate the child in person, such interrogation is conducted outside the courtroom and always in the presence of a teacher [*pedagog*].

4. The dissolution of the marriage between parents does not influence the scope of their parental rights. Even after the divorce, both parents are equally entitled and obligated to parti-

cipate in the child's upbringing and are responsible for his actions. It is very important that the divorce not result in the child being made a semi-orphan who is deprived of the love, care, and attention of one of his parents. If the parents approach this matter intelligently, they themselves will find ways for separated parents to keep in contact with and participate in the upbringing of their children. This will significantly attenuate the adverse impact of divorce on children. In a number of cases, the parents' hostility and insensitivity toward each other prevent them from reaching agreement and as a result inflict great, sometimes irreparable harm on their children.

The law emphasizes that the separated parent has the right to see his children and the obligation to participate in their upbringing and that the custodial parent may not prevent the former from doing so. If the parents are unable to reach agreement, the procedure for visiting and participating in the upbringing of children is determined by guardianship and wardship agencies with due regard for the parents' opinions. If parents do not adhere to a decision of a guardianship and wardship agency, the latter is entitled to request the court to compel the parents to perform their obligations without depriving them of their other rights (article 56 of the Code on Marriage and the Family). The legislation of certain union republics (the Ukrainian SSR and Uzbek SSR, for example) also gives the interested parent the right to address such a demand to the court.

The [custodial] parent sometimes refuses to obey a decision of a guardianship and wardship agency on the grounds that contact withthe separated [non-custodial] parent interferes in the normal upbringing of the child and has a negative impact on him. If this contention is true, the guardianship and wardship agencies may deprive the parent of visiting rights for a certain period of time. However, in the absence of such a decree, one parent may not deny the other parent's visiting rights or the right to participate in the child's upbringing.

5. Children, in turn, have the obligation to care for their parents. This, their moral obligation, is confirmed by article 66 of the Constitution of the USSR. All children should care for their parents and help them in every possible way, if their age permits them to do so, regardless of whether they have reached their age of majority. Children and adolescents can help their parents by looking after their younger brothers and sisters,

caring for elderly family members, tidying up, shopping, etc. From the earliest age, children must realize that everyone possesses not only rights but obligations as well. Parents give their all to their children and the children in turn must give their parents love and respect, carry out their just demands, be diligent in their studies, and behave themselves properly.

Grownup children must do their utmost to ensure their parents' tranquility in their old age. In addition to their obligation to maintain their needy parents who are unable to work (see Chapter 45), they must care for their parents, help them to satisfy their everyday needs, etc. (article 77 of the Code on Marriage and the Family).

When they themselves become parents, they must teach their children to respect their grandfather and grandmother. In particular, they do not have the right to prevent the grandfather and grandmother from seeing their grandchildren. Contact with older relatives instills children with respect for the older generation and a feeling of duty to their family. These meetings give much to the old people, who usually feel a deep attachment for their grandchildren. After divorce, it sometimes happens that the mother (or father) will prevent the parents of the ex-spouse from seeing their grandchildren. In these cases, the guardianship and wardship agencies, at the request of the grandfather (grandmother) oblige the parents (or one of them) to make it possible for them to visit their grandchildren and establish visitation rules (article 57 of the Code on Marriage and the Family).

III. Deprivations of Parental Rights

1. Parental rights are at the same time parental obligations. Therefore, the failure of parents to perform their obligations deprives them of certain rights (for example, parents may be refused the custody of the child if they have not taken proper care of the child in the past), and under certain circumstances, such behavior leads to the deprivation of all parental rights.

One or both parents may be deprived of their parental rights if it is found that they are not performing their child-rearing obligations, are abusing their parental rights, are cruel to their children, or exert a harmful influence on the children through their amoral, antisocial behavior. Chronic alcoholics and drug addicts are also deprived of their parental rights (article 59 of the Code on Marriage and the Family).

The deprivation of parental rights is an extreme measure that is applied to unfit parents when efforts to induce them to relate properly to their duties as parents are unavailing. Parents are stripped of their parental rights if they are clearly guilty of misbehavior. They may not be deprived of their parental rights if their failure to perform their parental obligations is due to illness (for example, mental illness) or other circumstances beyond the parents' control.

2. Only a court order can deprive parents of their parental rights. A petition for the deprivation of parental rights may be filed by a guardianship and wardship agency, other state or social organizations, a guardian or ward of a minor, or the procurator. This petition may also be filed by juvenile affairs commissions operating under the auspices of the executive committees of district, city, area, regional, and territorial Soviets of People's Deputies.

If the action for the deprivation of parental rights is filed against only one of the parents, the court determines the whereabouts of the other parent and without fail summons that parent to participate in the case. The participation of the procurator is mandatory. If petition is not filed by a guardianship and wardship agency, the latter must be summoned to participate in the case. In such a case, the guardianship and wardship agency must conduct a thorough investigation of the family prior to the hearing and present its conclusions.

A parent deprived of parental rights loses his right to the personal upbringing of his child. The child is taken away from that parent and the guardianship and wardship agencies decide the child's future fate--whether to enroll the child in a childcare institution, place the child in the care of a guardian (ward), or put the child up for adoption (article 61 of the Code on Marriage and the Family). The parent does not participate in the resolution of these questions. Not only is the parent's consent to the disposition of the child not required, there is not even any need to ascertain the parent's opinion. At the parents' request, guardianship and wardship agencies may allow them to visit the child if, in their opinion, such a visit will not have a harmful influence on the child (article 62 of the Code on Marriage and the Family). A refusal of permission to visit may not be appealed in court.

3. The deprivation of parental rights also entails the ter-

mination of all other rights based on kinship (article 60 of the Code on Marriage and the Family). The parents not only lose rights that arise from family relationships (for example, the right to receive maintenance payments from children vis-à-vis whom they have lost their parental rights), but also all rights in other areas of legislation based on kinship with a given son (daughter). Thus, a parent who is deprived of parental rights is not entitled to inherit or receive a pension after the death of his children, etc.

At the same time that the court strips a parent of his parental rights, it may also order the eviction of the parent if that parent occupies the same premises as the child vis-à-vis whom the parent is deprived of his rights and the court deems their further joint residence impossible on the grounds that the parent violates the standards of socialist conduct.

Parents, however, are not relieved of their obligation to maintain those of their children vis-à-vis whom they have been deprived of their parental rights. At the same time that courts render a decision depriving parents of their parental rights, they must order the given parents to pay child support maintenance. This is the only case where maintenance payments may be ordered at the initiative of the court itself regardless of whether such a claim has been made.

If the child remains with the other parent or with the guardian (ward) appointed to him or if the child is already in a childcare institution, maintenance payments will be made to these persons/parties. If the child is placed in the care of guardianship and wardship agencies, maintenance payments are made to the person or institutions to whom the child is transferred. If a parent who is deprived of his parental rights was previously awarded child support payments from the other parent or received a pension or grant for the given child, the payment of all these sums is terminated.

4. As already mentioned in section I above, the parent is deprived of his parental rights when he is found to be derelict in the performance of parental obligations. If the poor child-rearing conditions are due to objective factors beyond the parent's control, the application of such an extreme measure as the deprivation of parental rights would be unwarranted. Nevertheless, it is sometimes impossible to leave the child with the given parent since the child's physical or moral development may be thereby imperiled. For example, in some cases a child cannot

be left with a mentally ill parent even though that parent has not been declared to be lacking in dispositive capacity.

In such cases, the law provides for the removal of the child without the parent being deprived of his parental rights (article 64 of the Code on Marriage and the Family). In this case as well, the child can be taken from his parents only by court order.

If the court orders the removal of the child, the guardianship and wardship agencies place the child in a childcare institution or appoint a guardian (ward) in the family in which the child is placed. Under these conditions, the parent is also deprived of the possibility of personally rearing his child. However, since this parent is not deprived of his parental rights, he is entitled to have contact with his child. The parent's opinion is taken into account in deciding questions pertaining to the child's upbringing (for example, the question of where to place the child). The child may not be put up for adoption without the parent's consent (with the exception of cases where the parent is declared to be lacking in dispositive capacity as a result of mental illness or lunacy).

5. The possibility is not excluded that parents who are deprived of their parental rights will eventually change their behavior and it will be possible to entrust them with the upbringing of their children. The very fact that the court deprives the parents of their parental rights frequently has such an edifying impact on the parents that they change their behavior. In the case of parents, who, while behaving improperly toward their children, nevertheless loved them, the separation from their children evokes in them the desire to strive to have their children returned to them. Therefore, even though parents are deprived of their parental rights for an indefinite period, the law provides the possibility of restoring these rights (article 63 of the Code on Marriage and the Family).

The reinstatement of parental rights is permitted only by the court at the request of the parent deprived of these rights or the procurator. Before the parent can be restored to his rights, it is necessary first of all to be certain that the parent's behavior has dramatically changed and that the child can be safely returned to him. However, this is not enough. The law permits the restoration of parental rights when this is in the interests of the children. It is not always feasible to return the child to his parents even when they have modified their behavior. In the elapsed time, the child may have grown away from his parents to the extent that they are essentially strangers to him, that his return to them would not create the conditions necessary for his upbringing. The child might have bitter memories of his life with his parents (for example, he may remember them as always being drunk, their cruel treatment of him, etc.) and his return to them might severely traumatize the child. In such cases, the request that they be reinstated in their parental rights may be denied despite changes in the parents' behavior. When the child has been given out for adoption, parental rights are not restored.

A court order is also required to return a child who has been taken from his parents without the deprivation of the latter of their parental rights. Return must be requested by the parent from whom the child was taken by court order or by the procurator. If the reasons that necessitated the child's removal no longer exist, the court may order the child's return; the interests of the child must also be taken into account here.

Chapter 45. Maintenance Obligations

I. The Concept and Content of Maintenance Obligations

All family members as a rule care for and help one another. Each family member helps the other out of a feeling of love, respect, and the desire to help dear ones without being aware that he is at the same time also performing his legal obligation. But the law imposes on certain members of the family, regardless of their wishes, the obligation of providing family members who are minors, unable to work, or needy with the means of subsistence. Maintenance obligations are established for this purpose.

A maintenance obligation is understood to mean the legally established obligation of certain family members to maintain other, needy members of their family. Maintenance obligations are always based on one or another family relationship (marriage, kinship, or relations equated with kinship).

The grounds for the development of a maintenance obligation are always established by the law. The specific legal facts that comprise the grounds for the development of maintenance obligations will be shown in the process of examining their individual types.

The law also indicates the persons who are obligated to pay maintenance and the persons who are entitled to receive it. Maintenance obligations are of a strictly *personal nature.* Therefore, the death of the person obligated to make the payments or the person entitled to receive them always terminates this obligation.

The content of a maintenance obligation (duration, amount) is determined by law and cannot be changed by agreement between the parties. Repudiation of the right to receive maintenance under a maintenance obligation is devoid of legal significance: the repudiater is still entitled to demand payment of sums due to him under the law.

Parents and children have a constitutional obligation to support one another (article 66 of the Constitution). Therefore, the deliberate evasion of this obligation is a criminal offense.

II. Reciprocal Maintenance Obligations of Parents and Children

1. The very fact of the child's birth is the ground for parents' maintenance obligations toward their minor children. Parents have an obligation to maintain their children from the time they are born until they attain their age of majority. The fact that the child may have his own funds that could be used for his support (for example, inherited money) or that an adolescent has gone to work and is earning enough money to support himself cannot be a ground for relieving parents of their obligations to maintain him. As will be subsequently shown (see para. 3), the existence of [the child's] own earnings is grounds for reducing the size of the maintenance award, but does not mean total exemption from the payment of maintenance.

Parents may be exempted from child maintenance payments to children at the age of 16 if they marry after receiving the appropriate permission to marry at an earlier age. As already stated, in these cases citizens who marry before reaching the age of 18 years are declared to have full dispositive capacity. They are therefore able to support themselves by their own labor and in the event they are unable to work, are entitled to receive maintenance from their spouse.

The obligation to pay maintenance to grown children arises only when the children are unable to work and are in need of assistance, i.e., they are unable to pay their own way. Thus, the grounds for maintenance obligations in the given instance, in addition to birth, is the children's inability to work and their neediness (article 67 of the Code on Marriage and the Family).

2. The grounds for payment of maintenance by grown children to their parents, in addition to the fact of birth, are the parents' inability to work and neediness. Children are not entitled to request that they be exempted from maintaining their parents on the ground that it is difficult for them to help their parents because their earnings are too small. No matter how little the children earn, it is their duty to allocate part of their earnings to their parents. Children are not obligated to support parents who have been deprived of their parental rights. The court is also entitled to exempt children from the obligation to maintain their parents if it is found that the parents, even though not deprived of their parental rights, were derelict in the performance of their child-rearing and main-

tenance obligations (articles 77-78 of the Code on Marriage and the Family).

3. The size of maintenance payments for minor children is explicitly stated in the law. It is established as a certain share of the parents' earnings (income): one-fourth for one child, one-third for two, and one-half for two or more children (para. 1, article 68 of the Code on Marriage and the Family).

The types of earnings (income) from which maintenance may be withheld are enumerated in the January 22, 1969 decree of the USSR State Committee on Labor and Social Questions. In accordance with this decree, maintenance may be withheld from all types of earnings and additional remuneration from both basic and secondary work subject only to insurance deductions. Maintenance payments are also withheld from pensions and state social insurance grants. At the same time, an exemption is made for sums that are of a compensatory nature (severance pay, travel allowance, etc.). The decree presents a detailed list of payments that should be subject to withholding as well as those that should be exempt. There is special discussion of the procedure for calculating sums received by various categories of workers (collective farmers, service personnel, creative workers, etc.). Maintenance payments are collected from after-tax earnings (income).

Maintenance payments may not be collected in a share greater than the amount specified by law. A smaller share may be paid only in cases explicitly provided in the law. Such a reduction is permitted if: (a) other of the maintenance payer's minor children residing with him would be in a worse financial position than the child receiving the maintenance payment if the amount specified in the law were paid; (b) the payer is a Group I or II invalid; (c) the child for whom the maintenance is paid is working and has sufficient earnings; (d) the child is fully maintained by the state or by a social organization (para. 2 and para. 3 of article 68 of the Code on Marriage and the Family).

If one of the enumerated circumstances is established, the court is entitled to reduce the amount of the maintenance payment. However, the court is not obligated to lower it in all cases. Based on the specific circumstances of the case, [the request to] reduce the size of a maintenance payment may be refused if, for example, the mother receiving the payment is an invalid or receives a small pension.

The new maintenance payment is also a share, but a small share [of earnings and income], for example, one-fifth or one-sixth instead of one-fourth, etc.

In some cases, it is difficult to collect maintenance as a share of earnings. The maintenance payer may have irregular earnings, some of which may be in kind, etc. At the request of a judgment creditor, under these conditions, the court is entitled to determine the amount of maintenance that must be paid in hard cash every month. The size of this sum is determined on the basis of the projected earnings (income) of the payer and the size of the corresponding shares. Unless the receiver of the maintenance payments files such an action, the court is not entitled to grant the payer's request to set the size of the maintenance payment in a fixed sum.

Maintenance is also paid in a fixed sum when the children do not live together, but instead a certain number of children live with each parent. Under such conditions, maintenance is paid by the more affluent to the less affluent parent. The size of the payment is a fixed share, depending on the financial position and family status of both parents, established by law (article 71 of the Code on Marriage and the Family).

When a maintenance payer has additional income from a private plot, a certain amount of money can be collected from this income in addition to the account collected from earnings (income) (article 72 of the Code on Marriage and the Family).

If the children for whom the maintenance is paid are in a childcare institution or are fully maintained by the state or by a social organization, the court may reduce the amount of the maintenance payment or exempt the payer from making such payments altogether. When it is established that a mother or other person receiving maintenance payments is compelled to spend money on a child in a childcare institution, the discussion can center only on the reduction [of these expenditures]. If such expenditures are not made, the payer is exempted from making maintenance payments to the given person. However, it does not follow from this that he has no maintenance obligation whatsoever. When a child is placed in a childcare institution, each of the parents (both the maintenance payer and the previous recipient of maintenance) may be charged the cost of keeping the child in the childcare institution in the same amount as the maintenance payments. Based on the financial plight of the parents, the court may exempt them entirely or partially from these payments (article 69 of the Code on Marriage and the Family). There are no charges when the law provides for the gratuitous maintenance of the child in an appropriate institution (for example, children of a single mother, sick children placed in special childcare institutions).

As already noted, the court may not order the payment of maintenance higher than the share set by law. However, when exceptional circumstances require additional outlays (for example, as a result of the child's severe illness or injury, etc.), a parent who pays maintenance may be directed to participate in the payment of such costs during the time for which they are necessary. The court determines his share of the costs (article 73 of the Code on Marriage and the Family).

The amount of maintenance payments for grown, needy children who are unable to work and the size of payments for the maintenance of parents are in each case established in a fixed, monthly sum. The size of this sum is determined according to the family and financial status of parents and children. The fact that there are still other persons to whom payments must be made [*alimentnoobiaziannye litsa*] (for example, the spouse of a person who is in need of maintenance) is taken into account in determining the size of the maintenance payment, but cannot be the ground for exemption from the payment of maintenance. If the parents have several children, the share of each child in the maintenance of their parents is determined by the court depending on the children's material and financial status. The court takes into account all children, notwithstanding the fact that the parents may, for some reason, claim support from only one (or certain) children (articles 76-77 of the Code on Marriage and the Family).

Grown children have the obligation not only to maintain their parents, but also to care for them. Naturally, the court cannot force children to care for their parents if the children are too thoughtless and callous to perform their moral duty to their parents. However, in such a case, the court can increase the amount of maintenance collected from the children, so that these additional sums could be used to pay others to take care of their parents. This obligation of children to care of their parents is also taken into account in the resolution of a number of other cases. For example, in the process of resolving a dispute over the division of a house or the exchange of living accommodations, the court takes measures to prevent the infringement of the interests of parents, the care for whom is the responsibility of their children.

3. Paragraph 2 indicated the instances in which it is possible to reduce the amount of maintenance payments for minor children. If the maintenance payments were awarded in the amount established by law, but the circumstances conferring the right to demand a reduction in the size of the payments came later, the payer is entitled to petition the court to reduce the size of his payments. On the other hand, if the maintenance was awarded in a lesser amount and the circumstances responsible therefore no longer exist (for example, the child living with the payer reaches his majority), the person to whom the maintenance was awarded is entitled to demand that it be increased accordingly (article 75 of the Code on Marriage and the Family).

In the event of a change in the material or family status of the parents or their grown children, each of them is entitled to petition the court to raise or lower, accordingly, the size of the payment previously awarded by the court for the maintenance of parents or children who are unable to work (article 79 of the Code on Marriage and the Family).

III. Maintenance Obligations of Other Family Members

1. Parents, children, and spouses, whose maintenance obligations we have examined, have primary responsibility for maintenance payments [*alimentnoobiazannyi pervoi ocheredi*]. This means that they have a maintenance obligation regardless of whether there are other persons who also have the obligation to maintain the given member of the family.

Maintenance obligations of other family members than those enumerated arise only if there is no possibility of obtaining maintenance from persons who have primary responsibility for such payments. Family members who bear secondary responsibility for payments [*alimentnoobiazannyi vtoroi ocheredi*] are also indicated in the law. The Principles of Legislation on Marriage and the Family list grandfathers, grandmothers, brothers, sisters, stepfathers, and stepmothers as family members with an obligation to maintain minor children. Grandchildren, stepsons, and stepdaughters have an obligation to maintain needy grown-ups who are unable to work. At the same time, union republics have the right to establish other grounds for the development of rights and obligations relating to the reciprocal maintenance of relatives and other persons (article 21 Principles). These powers are widely used by union republics in connection with which there are differences in both the range of persons who have the secondary obligation to maintain members of their family as well as the grounds for the origin of this obligation under the legislation of individual union republics. The subsequent discussion is based on legislation of the RSFSR.

2. Grandfathers and grandmothers have an obligation to maintain minor grandchildren in need of assistance if the grandchildren are not able to obtain maintenance from their parents. Unlike the obligations of parents, who must maintain their children regardless of whether the children have sufficient funds, grandfathers and grandmothers are obligated to maintain only their needy grandchildren. Thus, if grandchildren are awarded a sufficient pension on the death of their parents, if they can be maintained by the property inherited by them, etc., grandfathers (grandmothers) have no maintenance obligation toward them. What is more, under the law, parents always have the obligation to allot part of their earnings to their children; it does not matter whether the part of the earnings that is left after the maintenance payment is sufficient for the needs of the parent himself. The obligation of grandfathers (grandmothers) arises only if they have sufficient funds with which to help their grandchildren. If the court decides that the incomes of the grandfather (grandmother) do not permit them to help, it dismisses the suit for maintenance.

The same obligation arises on the part of the grandfather (grandmother) vis-à-vis their grown grandchildren if they are unable to obtain maintenance from their parents or spouse. If the law establishes a need for maintenance, it can be collected from the paternal as well as the maternal grandfather and grandmother.

Grandchildren in turn have an obligation to maintain grandfathers and grandmothers who are in need of help and unable to work. This obligation also arises only if the grandchildren have sufficient funds (article 84 of the Code on Marriage and the Family). We recall that children have a maintenance obligation vis-à-vis their parents regardless of the children's financial status.

Since grandchildren have primary responsibility for payments, their obligation arises only if a grandfather or grandmother is unable to obtain maintenance from their children or spouse.

3. Stepfathers and stepmothers have an obligation to maintain their minor stepsons and stepdaughters. They have this obligation only toward stepsons and stepdaughers who are being reared or maintained by the stepfather (stepmother). If the children have not lived with the stepfather (stepmother) and if the latter have not reared them or rendered them any kind of assistance, the stepfather (stepmother) has no obligation to maintain them. Thus, a stepfather (stepmother) does not have a maintenance obligation merely because he or she has married the child's parent; for such an obligation to arise, they must voluntarily assume the obligation to maintain or rear the children of their spouse. When children are able to obtain sufficient funds from their parents, the obligation of the stepfather (stepmother) does not arise. At the same time, in order for this obligation to arise, it is not obligatory that the stepmother (stepfather) have sufficient funds. When they assume the obligation to rear or maintain the children of their spouse, they thereby place themselves in the position of the child's parents (article 80 of the Code on Marriage and the Family).

The Code on Marriage and the Family of the RFSFSR does not call upon the stepfather (stepmother) to maintain grown stepsons (stepdaughters) even if the latter are unable to work and are in need of help. Moreover, the Code on Marriage and the Family of a number of republics (for example, the Kazakh SSR) also establishes this obligation with respect to grown stepsons (stepdaughters).

Stepsons (stepdaughters) in turn have an obligation to maintain their needy stepfathers (stepmothers) who are unable to work if these persons have participated in their upbringing or maintenance. The collection of payments in this case is not due to the impossibility of obtaining maintenance from the stepfather's (stepmother's) own children. However, the court is entitled to relieve stepsons (stepdaughters) of this obligation if the stepfather (stepmother) reared or maintained them less than five years or did not rear them properly (article 81 of the Code on Marriage and the Family).

4. Brothers and sisters as well as stepbrothers and stepsisters have an obligation to maintain their minor brothers and sisters who are in need of help. Like the obligation of the grandfather

(grandmother), the maintenance obligation of brothers and sisters arises only with respect to minor brothers and sisters who are in need of help and when those obligated to render aid have sufficient funds to do so. Brothers (sisters) have the same obligation toward their grown brothers and sisters who are unable to work and in need of help. As with those who bear secondary responsibility for payments, their maintenance obligation arises when their minor brothers and sisters are unable to obtain maintenance from parents and when grown brothers and sisters are unable to obtain maintenance not only from their parents, but from spouses or children as well (article 82 of the Code on Marriage and the Family).

5. The maintenance obligation also arises on the part of persons who, having taken children into their home on a permanent basis with the intention of rearing and maintaining them, repudiate the children's continued dependence on them. No one can be forced to assume the obligation of rearing a child, but once assumed, it cannot be arbitrarily repudiated. This obligation arises on the part of those actually rearing the children when the children are unable to obtain maintenance from their parents. They [those actually rearing the children] bear the same obligation toward these children even after the latter have attained their majority if they are unable to work and in need of help (article 85 of the Code on Marriage and the Family).

Such an obligation does not arise if they took the child under their care only temporarily (for example, while he was ill or while his parents were away). Nor does this obligation arise on the part of persons who in the capacity of guardians (wards) rear children, even though they have voluntarily maintained their charges.

Those being raised [*vospitanniki*], in turn, have an obligation to maintain the persons who actually reared them if these persons are unable to work, need help, and are unable to obtain maintenance from their children or spouses (article 86 of the Code on Marriage and the Family).

The legislation of some union republics does not establish a maintenance obligation for *de facto* upbringers and those being raised [*vospitateli i vospitanniki*] (Code on Marriage and the Family of the Lithuanian SSR).

6. The amount of maintenance is determined in each individual instance by the court on the basis of the material and family status of the person receiving the maintenance as well as

the person who pays it. The maintenance payment is set in a fixed monthly cash sum.

If a person has the right to obtain maintenance from several people with secondary responsibility for payments (for example, from a brother or grandfather), the share of each of them in the maintenance of a needy family member is determined with due regard to their material and family status. Even when the demand is made on only one of them (let us say, the grandfather), the court in determining the amount of the maintenance payment considers the fact that the needy person can obtain a certain sum from another person--in the cited example, from a brother (article 87 of the Code on Marriage and the Family).

The size of the maintenance payment can also be established on the basis of agreement if the funds are paid voluntarily and if the person in need has not applied to the court. However, such an agreement is not binding on the court if the case is subsequently examined by the court. The court may set the maintenance payment in a different amount. In the event the family or financial status of the payer or payee change, either of them may petition the court to increase or reduce the size of the maintenance payment (article 88 of the Code on Marriage and the Family).

IV. The Procedure for Collecting and Paying Maintenance

1. In the majority of cases, maintenance is paid voluntarily. When family relations are normal, family members maintain a common household; children and other members of the family who are in need and unable to work receive everything they need and hence the question of collecting maintenance does not arise. When family members entitled to maintenance live apart from those who under the law are obligated to maintain them, the latter usually voluntarily pay the sums specified by the law. These sums may be paid by them in person during visits to their child or another family member, may be sent by mail, etc.

For the convenience of the payer and payee, maintenance may be paid through the administration at the payer's workplace or at the place where he receives his pension or stipend. In order to arrange payment in this way, the payer notifies the administration of the specific sum to be transferred. In the case of payments for the maintenance of minor children, the amount of payment is not specified by the payer but is determined by the administration based on the payer's total earnings and the shares established by law for the maintenance of minor children. If the maintenance payments are made to grown children or other family members, the payer declares the size of the payment to be transferred. Upon receiving such a declaration, enterprises (institutions, organizations) must within three days after payday remit or transfer the withheld sums to the person indicated in the declaration (articles 89-90 of the Code on Marriage and the Family).

2. If the maintenance is not paid voluntarily or if the recipient for some reason prefers to receive the money due him on the basis of a court order, maintenance is collected by way of judicial proceedings.

In such a case, the maintenance recipient enjoys a number of procedural benefits. They may file a claim to collect maintenance with the court in the place where the defendant (the person obligated to make the payments) has his domicile or the court where they reside (article 116 of the Code of Civil Procedure of the RSFSR). No fees are paid on these declarations (article 80 of the Code of Civil Procedure of the RSFSR). Fees and other costs are paid by the judgment debtor if the plea is satisfied. The court has an obligation to render all manner of assistance to the applicant for maintenance (to request the necessary documents, information, etc.). If the defendant's whereabouts are unknown, the court orders that a search be made for him by agencies of the ministry of internal affairs.

Maintenance is awarded from the moment the appropriate cause of action is filed with the court. Maintenance may be collected retroactively only if the court finds that the claimant's efforts to collect maintenance were unsuccessful due to the defendant's refusal to pay. Where maintenance was previously paid on the grounds of a declaration filed by the payer, the amount due is recovered as of the time when payments were terminated on the basis of that declaration. In no case shall retroactive maintenance payments exceed a period of three years (article 95 of the Code on Marriage and the Family).

After the court grants a suit for maintenance, the claimant is issued a court order that is transferred to the administration of the payer's workplace.

If the claimant has for some reason not transferred the court order for execution, maintenance for the period between its issuance and the time when this order is submitted may be collected for a maximum of three years. Cases where maintenance was not collected because the payer, who was trying to avoid paying it, could not be found, are an exception. Under these conditions, maintenance is recovered for the entire past period.

3. Indebtedness for past time is recovered on the basis of the earnings (income) the payer received at that time. If information on earnings has not been submitted, indebtedness is determined on the basis of the defendant's earnings (income) at the time of recovery. The child's attainment of his age of majority does not relieve the payer of the obligation to pay off the indebtedness that has accrued to date.

Relief from or a reduction of the amount are possible only if the court establishes (1) that the nonpayment of maintenance was the result of the payer's illness or for other valid reasons and (2) that his present material and family status does not permit him to pay off his indebtedness (article 96 of the Code on Marriage and the Family).

4. If the whereabouts of the debtor are unknown, a search for him is conducted by internal affairs agencies. Criminal proceedings are instituted against parents who are guilty of deliberately evading payments for the maintenance of their children or against children who deliberately evade payments for the maintenance of their parents. Parents who deliberately evade paying maintenance awarded by the court may be sentenced to the deprivation of freedom for up to one year, to exile for a maximum period of three years, or to correctional labor for a period up to one year (article 122 of the Criminal Code of the RSFSR). Children who deliberately evade the payment of court-awarded maintenance may be sentenced to correctional labor for a period up to one year, may be held up to public censure, or may may be the subject of social pressure (article 123 of the Criminal Code of the RSFSR).

An appropriate entry is made in the passports [identity cards] of persons who are sentenced for the deliberate evasion of the payment of maintenance in order to prevent such evasion in the future. When a person whose passport contains such an entry starts a new job, the administration is obligated to carry out the appropriate withholdings even before the court order is presented (article 94 of the Code on Marriage and the Family).

A bailiff of the court checks to see to it that the administration at the payer's place of work is properly collecting maintenance payments.

Chapter 46. Securing the Interests of Children Who Have Lost the Support of Their Parents

I. Forms of Security

Our country does everything necessary to give all children a happy childhood. The state helps parents to raise a happy, healthy child. But how can one bring happiness to those who have suffered the misfortune of losing their parents? The death of near ones is more than mental anguish--it is also the loss of those who cared for the child. [Children] must not feel that they are orphans. They must not only be provided for in a material sense. They must also be given everything that is necessary for their spiritual/mental development. They must be surrounded with care and attention. The word orphan used to evoke the grimmest associations: homelessness, poverty, lack of care. In our society, this word has lost its former meaning.

In the USSR there are enough childcare institutions to take care of all children who lose their parents. But their proper upbringing requires the combination of social and family influences. Children must attend childcare institutions, but it is better that they live in a family setting because human feelings and emotions develop specifically in the family collective. Regardless of the quality of conditions in childcare institutions, the child is always drawn to the family. Therefore, if possible, children who have lost their parents are placed in the families of workers.

It was shown in Chapter 44 that special norms articulating the mutual rights and obligations of children and parents have been instituted for the purpose of protecting the interests of children and parents. It is all the more necessary to establish rules for people who have assumed the obligation of rearing in their family children who are not their own. In this regard, the law provides the following family law institutions for the upbringing of these children in families of workers: adoption, guardianship, and wardship. In some republics, there is yet another form--foster care.

II. Adoption

1. As a result of adoption, the adopter assumes vis-à-vis a child who is not his son (daughter) all the obligations that the

law imposes on parents and acquires all the rights that are enjoyed by a parent.

In order to secure fully the children's interests, the law provides certain conditions that must be observed in the adoption process. The adoption process itself must be sanctioned by an agency of state power.

Adoption is carried out in accordance with a decision of the executive committee of a district, city, or city-district Soviet of People's Deputies. Executive committees of village and settlement Soviets of People's Deputies are not entitled to perform adoptions (article 98 of the Code on Marriage and the Family).

Adoption requests must be addressed to guardianship and wardship agencies at the place of residence of the adopter or adoptee (see para. 1, III of the present chapter). Pending the decision of the executive committee, the executive committee's department of public education conducts a thorough check to determine whether the necessary conditions for adoption are being observed and whether the adopter's family is truly suitable for the child. In addition, the necessary documents are assembled, the adopter and the child undergo a medical examination, etc. The examination findings and all necessary documents are submitted to the executive committee.

Rights and obligations arise the moment the executive committee decides to authorize the adoption. This decision must be registered with ZAGS. However, if this registration does not take place for some reason, this circumstance does not influence the origination of rights and obligations. Registration in the given instance is of certifying [*udostoveritel 'nyi*] rather than a right-forming [*pravoobrazuiushchii*] nature; its purpose is to facilitate the protection of the citizen's rights. For registration purposes, the law obliges the guardianship and wardship agency to report its decision on adoptions within one month after reaching this decision. Therefore if the adopter himself does not register the adoption, the registration will be carried out on the basis of a copy of the adoption certification received from the guardianship and wardship agency.

As a result of adoption, the rights and obligations of relatives by birth [*rodstvenniki po proiskhozhdeniiu*] arise not only between the adopter and adoptee, but also between the adoptee and all relatives of the adopter (for example, the father of the adopter will have the rights and obligations of the grandfather of the adoptee), while all progeny of the adoptee will be connected by appropriate rights and obligations with the adopter and his relatives (the son of the adoptee will be considered the grandson of the adopter, etc.). All rights and obligations between the child and his parents and other relatives by birth are terminated simultaneously with adoption (article 108 of the Code on Marriage and the Family).

2. Only a person who can properly perform his child-rearing obligation should be an adopter. Minors and persons who are declared by the court to be entirely or partially lacking in dispositive capacity are not entitled to adopt children. Persons who have been deprived of their parental rights are also barred from adopting children (article 99 of the Code on Marriage and the Family). Their inability to rear their own child properly is all the more reason for not entrusting them with rearing a child who is a stranger to them.

The legislation of some republics also contains other restrictions. For example, the Code on Marriage and the Family of the Moldavian SSR provides that persons who have previously adopted a child but where the adoption was rescinded due to their failure to fulfill their obligations properly may not adopt children. While there are no such additional restrictions in the RSFSR, this does not mean that anyone will be allowed to adopt a child. Adoption is permitted only when the guardianship and wardship agency is convinced that the good upbringing will be offered by the adopter's family. The denial of permission to adopt can only be appealed in administrative proceedings to the executive committee of a higher Soviet of People's Deputies.

3. The rearing of the child in a family setting is the aim of adoption. Therefore only minors may be adopted. A child upon reaching the age of 10 years must consent to being adopted. In the given instance, the child's wishes are not only taken into account, but are decisive. If the child does not give his consent, the adoption may not take place.

4. Since parents lose their rights as a result of adoption, it is necessary to obtain their consent to adoption in order to protect these rights when the parents are still living. It is not necessary to obtain the parents' consent if they have been deprived of their parental rights or have been duly declared to be lacking in dispositive capacity or to be missing (article 100 of the Code on Marriage and the Family).

As an exception, it is possible to adopt a child without his parents' consent in cases when they, although not deprived of their parental rights, are derelict in rearing their child. Therefore, if it is found that parents have not lived with their child for more than a year and, notwithstanding warnings to them from the guardianship and wardship agency, are not participating in his upbringing, and show no parental attention and care (are not in contact with the child, are not interested in his development), the child may be adopted without their consent. If the parents are derelict in the fulfillment of their obligations, there are no grounds for defending their rights (article 101 of the Code on Marriage and the Family).

Parents may give their consent to the adoption of their child by certain persons known to them or to adoption in general, leaving the choice of adopters to guardianship and wardship agencies. It is advisable to obtain consent in such form since it makes it possible to preserve the secret of adoption more completely. Therefore when a child enters a childcare institution, the administration of that institution will find out whether the child's parents plan to raise the child themselves subsequently or whether they will agree to give him into the care of another family and will give their consent to the adoption of their child in the future (article 102 of the Code on Marriage and the Family).

The consent of parents must be expressed in written form. Until the adoption is concluded, the parents may at any time repudiate the consent they have given and raise their child themselves.

If the child is in the care of a guardian, in order to carry out the adoption, it is necessary to have the consent of the guardian or the consent of the administration of the childcare institution in which the child is enrolled. The consent of these persons is necessary because, having a good knowledge of their charge, they can cite grounds that would show that adoption in the given instance will not be in the interest of the child. However, if the guardianship and wardship agencies conclude that adoption is entirely in the interests of the child, they may dismiss the guardian and carry out the adoption contrary to his opinion and the opinion of the administration of the childcare institution.

5. If adoption is performed by only one of the spouses, the consent of the other is required. Such consent is not required if the spouse of the adopter is lacking in dispositive capacity or if

the spouses have in fact terminated their relations, have not lived together for more than a year, or the whereabouts of the adopter's spouse is unknown (article 104 of the Code on Marriage and the Family).

6. Adoptions should be kept secret in the interest of both the child and the adopters. After all, the child is placed in the position as a child born of his adoptive parents and if he learns that his parents are not his birth parents, this knowledge may traumatize him and complicate relations in the family. The child will misinterpret the adopters' explanations and will believe that their treatment of him would be different if they had given birth to him. Therefore, the law provides a number of measures aimed at keeping adoptions secret.

At the adopter's request, when the decision is rendered on the adoption of the child, the adopter's surname and a patronymic based on the adopter's first name may be given to the child. The child's first name may also be changed. The adopters may wish the child to bear a name that is customary in their family, a name that corresponds to their nationality, etc. The adopters may also request that they be recorded as the child's real parents rather than as adoptive parents in the birth register and on the birth certificate (articles 105-106 of the Code on Marriage and the Family).

What is more, at the adopters' request, the child's place of birth may be changed in the documents; the child's date of birth may also be changed, but not by more than six months. The legislation of some republics does not allow changing the date of birth.

Without the consent of the adopters and, in the event of their death, without the consent of guardianship and wardship agencies, it is forbidden to communicate any information or to release excerpts from the books of ZAGS agencies revealing that the adopters are not the adopted child's real parents (article 110 of the Code on Marriage and the Family).

Criminal proceedings are instituted against persons who divulge the secret of adoption against the will of the adopter. They may be sentenced to a maximum of one year of correctional labor, or fined up to 50 rubles, or held up to public censure (article 124 of the Criminal Code of the RSFSR).

7. An adoption may be declared invalid if the following conditions were not observed: if the adopter did not have the right to adopt (see para. 2 above); if the adoption was based on

forged documents (for example, the forged consent of parents or a forged declaration in the name of the adopters, etc.); if the adoption was a sham, i.e., if the adopter did not raise the child and had no intention of doing so, but carried out the adoption for some other reason (for example, so that the child would receive a pension after the adopter's death, a larger apartment, etc.).

Only the court may declare an adoption invalid. Any interested person, i.e., person whose rights were infringed by the adoption, may bring an action to have an adoption declared invalid.

If an adoption is declared invalid, no rights and obligations are considered to have resulted from the adoption, and the rights and obligations of the child vis-à-vis his relatives by birth are restored. When the court declares an adoption invalid, it simultaneously decides whether to return the child to its parents or, if that is not in the child's interests, turns the child over to guardianship and wardship agencies, which determine his future fate (articles 111-112 of the Code on Marriage and the Family).

8. A violation of terms of adoption other than those enumerated in para. 6 cannot be grounds for declaring it invalid. In some cases, if demanded by the child's interests, such violations may become grounds for revoking an adoption. An adoption may also be revoked when, even though no terms of adoption have been violated, the situation resulting from the adoption is not in the child's interests.

An action to revoke an adoption on the grounds that the parents' consent was not obtained may be made only by the parent whose consent was not obtained in violation of the law.

The action to revoke the adoption is brought against the adopters. But in some cases, the parents do not know specifically who adopted their child. In order not to divulge the secret of adoption prematurely, the action in the given instance is submitted to the guardianship and wardship agency that rendered the adoption decision. The letter informs the adopters of the claim. The adopters may participate in the case, but may also entrust the defense of their interests to guardianship and wardship agencies and take no part therein (article 114 of the Code on Marriage and the Family).

RSFSR legislation does not provide for the possibility of any other persons bringing an action to revoke an adoption even though their rights may have in some measure been affected by the adoption. Such an action can be brought only by guardianship and wardship agencies or by the procurator. State and social organizations as well as individual citizens who believe that the adoption should be revoked are entitled only to inform guardianship and wardship agencies or the procurator, who decide whether the appropriate suit shall be brought in court (article 115 of the Code on Marriage and the Family). Nor is the adopter himself entitled to demand that the adoption be revoked. Just as parents are not entitled to demand that they be stripped of their parental rights, the adopter is also not entitled to demand that he be relieved of the child-rearing obligations that he voluntarily assumed. Only the Code on Marriage and the Family of the Estonian SSR and Lithuanian SSR entitles the adopter to demand the revocation of an adoption, but only if the child turns out to be mentally ill or weak-minded--a fact that was not and could not have been known at the time of adoption. Since such children cannot be reared in the family and are placed in special childcare institutions, under these conditions guardianship and wardship agencies in other republics always raise the question of revoking the adoption themselves.

9. If an adoption is revoked, reciprocal rights between the adopter and adoptee and the adopter's relatives are terminated at the moment the revocation decision takes effect. The court may, however, obligate the former adopter to make child maintenance payments. Thus, if the adoption was revoked because the adopter failed to perform his obligations properly and the child is in need of funds for his maintenance, it would be unjust to exempt the adopter from his obligation to pay maintenance.

Revocation of the adoption is accompanied by the restoration of reciprocal rights between the child and his birth parents and relatives. The court also determines whether the child should be returned to his parents or whether he should be transferred to guardianship or wardship agencies so that they may arrange his future. The question of whether the child will retain the first name, patronymic, and surname given to him upon adoption is also decided (article 117 of the Code on Marriage and the Family). The child may have become accustomed to that name and in some instances it might be inadvisable to change it. This question is decided in the child's interest and his wishes are taken into account if he has already reached the age of 10 years.

When the judgment to revoke an adoption takes effect or when an adoption is declared invalid, the court is obligated to send a copy of the judgment to the executive committee of the district (city) Soviet of People's Deputies that rendered the adoption decision and to the ZAGS office that registered the adoption (article 118 of the Code on Marriage and the Family).

III. Guardianship, Wardship, Foster Care

1. Executive committees of district, city, city-district, settlement, and village Soviets of People's Deputies are guardianship and wardship agencies. Guardianship functions in respect to minors are performed by departments of public education; in respect to persons declared by the court to be partially or entirely lacking in dispositive capacity--by departments of public health; and in respect to persons who have dispositive capacity, but require help due to the state of their health--by departments of social services (article 120 of the Code on Marriage and the Family).

The goal of guardianship and wardship is the rearing of children and providing the necessary care and attention to grown persons in need thereof due to the state of their health. Guardianship and wardship agencies also pursue the goal of protecting the property interests of wards and defending their lawful rights and interests (article 119 of the Code on Marriage and the Family).

A number of measures have been provided which envisage that guardianship and wardship agencies must establish in a timely manner guardianship (wardship) over citizens in need thereof. All institutions and persons knowing of minors left without parental care have an obligation to immediately inform guardianship and wardship agencies of the actual whereabouts of the child (article 122 of the Code on Marriage and the Family). Such an obligation accordingly rests not only with the personnel of childcare institutions and schools in respect to the children enrolled in these institutions, and with ZAGS personnel who have learned upon the registration of death that the deceased is survived by children who have no one to care for them, but also with all other organizations and citizens (neighbors, parents' acquaintances, etc.) who in one way or another have become aware of children who are uncared for. Article 66 of the Constitution assigns all citizens the responsibility to concern themselves with the upbringing of children; no one has the right to be indifferent to the fate of children who have lost their parents, or to fail to take measures to place the child in the proper surroundings.

In accordance with the civil procedure legislation, the participation of guardianship and wardship agencies in court hearings pertaining to the declaration of citizens to be partially or entirely lacking in dispositive capacity is mandatory. Therefore, these agencies are usually aware that such a decision has been rendered. In addition, the court has an obligation to report this decision to guardianship and wardship agencies within three days after the decision enters into effect so that they can establish guardianship (wardship) over a citizen who has been declared to be partially or totally lacking in dispositive capacity (article 124 of the Code on Marriage and the Family).

Wardship over persons with dispositive capacity who require the help of a warder may be established only at their personal request.

Guardianship (wardship) is established by the executive committee of a local Soviet of People's Deputies at the place of residence of the person over whom guardianship (wardship) is established or at the place of residence of the guardian (warder). The guardian (warder) must be appointed no later than one month after the guardianship and wardship agencies become aware of the need for such an appointment.

Guardianship is established over minors below 15 years of age; wardship--over adolescents between 15 and 18 years old. Guardianship is also established over minors who are declared by the court to be lacking in dispositive capacity. Wardship is established over persons with limited dispositive capacity as well as over persons with full dispositive capacity who require assistance due to the state of their health.

2. A guardian or warder may be appointed only with his consent. Minors, persons stripped of parental rights, or persons declared to be partially or totally lacking in dispositive capacity may not be guardians (warders), just as they may not be adopters.

Guardians are chosen primarily on the basis of their personal qualities, their child-rearing ability, or their ability to care for an adult ward. Guardians (warders) take minors into their families and therefore it is important to ascertain the situation in the guardian's (warder's) family, and to determine the type of relationships that may develop between them and the children who will be their wards. The wishes of the ward are also taken into account as far as possible. However, it is not mandatory that the ward's consent be obtained in the appointment of guardians (warders) for minors entirely (or partially) lacking dispositive capacity. Consent is necessary only for appointing a warder for a person with dispositive capacity, since in the given instance the warder may be appointed only at the behest of the ward (article 126 of the Code on Marriage and the Family).

If children are placed in childcare institutions that take full charge of their upbringing (children's homes, etc.), the administration of the corresponding institution has the responsibility for performing the obligations of guardian (warder). In precisely the same way, the obligation of exercising guardianship and wardship over adult citizens in the corresponding institutions (boarding facilities for the mentally ill, homes for the aged, etc.) is performed by the administration of these institutions. Only if the administration of the corresponding institution is unable to safeguard the property interests of the ward may a general guardian [*opekun nad imushchestvom*] be appointed (articles 127-28 of the Code on Marriage and the Family).

3. Guardianship (wardship) of minors is a form of child rearing in a family. Therefore, the guardian (warder) has the obligation to live with his charge and to perform all the obligations pertaining to his upbringing and the protection of his rights and interests that should be performed by the child's birth parents. Only in individual cases, when the ward reaches 16 years of age, may the guardianship and wardship agencies authorize him and his warder to live separately from one another, if this will not adversely reflected in the upbringing and protection of the minor's rights and interests (for example, if the ward attends school in another town, etc.).

Guardians (warders) have an obligation to report any change in their whereabouts to guardianship or wardship agencies (article 129 of the Code on Marriage and the Family). This is essential so that the latter can oversee their activity. Like parents, guardians (warders) may bring an action for the return of their ward from any persons detaining the child without lawful grounds (article 130 of the Code on Marriage and the Family).

Guardians (warders) of adults have an obligation to concern themselves with the maintenance of their charges, to create the necessary living conditions for them, to provide for their care and treatment, and to protect their rights and interests. In respect to those who are mentally ill, they also have an obligation to see to it that they are under constant medical supervision. If they recover, the guardians (warders) have an obligation to petition the court to declare them to have dispositive capacity. The enumerated obligations do not apply to curators of persons who are declared by the courts to have limited dispositive capacity as a result of having abused alcoholic beverages or narcotic substances.

4. Guardians of minors younger than 15 years of age and grown persons lacking dispositive capacity are their legal representatives and perform all necessary transactions in their name and in their interests. Warders of adolescents between 15 and 18 years of age give their consent to the execution of transactions that the latter may not execute for themselves.

Warders of persons with limited dispositive capacity consent to their wards' receiving payments due them (wages, pensions, maintenance payments, etc.) and to their disposition over these sums of money and other property.

Warders of persons with dispositive capacity, who need assistance due to the state of their health, assist their wards in exercising their rights and protect them against abuses by third parties (articles 132-33 of the Code on Marriage and the Family).

Guardians and warders do not have the obligation to use their own funds to maintain their wards: sums that are paid to their wards in the form of pensions, grants, maintenance payments, etc., are placed at the disposal of the guardian (curator) and are expended by him on the maintenance of his ward. If the guardian (warder) does not have sufficient funds, guardianship and wardship agencies may award a grant (article 134 of the Code on Marriage and the Family).

5. Guardianship and wardship agencies oversee guardians (warders) at the wards' place of residence. Their actions may be appealed to these agencies by any person, including the ward himself (article 136 of the Code on Marriage and the Family).

Guardians and warders may be relieved of their obligations at their personal request for valid reasons. They are also relieved when there is no longer any need for guardianship or wardship--the child is returned to his parents, put up for adoption, assigned to a childcare institution; a mentally ill person recovers or is hospitalized, etc. The warder of a person with dispositive capacity is also relieved at the demand of his ward (article 137 of the Code on Marriage and the Family).

If the guardian (warder) does not properly perform his obligations, he may be relieved thereof (article 138 of the Code on Marriage and the Family). If the guardian has used his guardianship for selfish purposes or has left his ward without supervision and the necessary assistance, criminal proceedings may be instituted against him and he may be punished by deprivation of freedom up to two years or by correctional labor for a period up to one year (article 124 of the Criminal Code of the RSFSR).

Guardianship of a minor ceases when he reaches 15 years, and the guardian becomes the minor's warder without a special decision. Wardship is terminated when the ward reaches 18 years or marries with the consent of the executive committee.

6. The Latvian and Uzbek union republics also provide for the transfer of children to the families of workers on the basis of a contract for foster care.

Chapter 47. Acts of Civil Status

1. It is mandatory that the basic events in a person's life be recorded with agencies for registering acts of civil status. The following events must be registered: birth, marriage, adoption, establishment of paternity, divorce, death, change of first name, patronymic, and surname (article 140 of the Code on Marriage and the Family).

The origin, change or termination of rights and obligations in various branches of law are associated with all of the indicated events. Therefore the prompt registration of these events and the receipt of a document incontestably certifying that a given event truly took place at a certain time and in a certain place hold great significance for the protection of the rights of citizens.

The registration of the enumerated events is also in the interests of the state, which must have precise knowledge of the movement of the population. This information is necessary for the correct planning of the nation's total economic life.

2. Departments (offices) for registering acts of civil status attached to the executive committees of district, city, city-district Soviets of People's Deputies (ZAGS) are the agencies that register acts of civil status in cities and district centers. Registration in rural areas is performed by executive committees of village and settlement Soviets of People's Deputies. Soviet citizens outside the USSR may register acts of civil status at USSR consular facilities (article 141 of the Code on Marriage and the Family).

3. An entry in registers of acts of civil status is proof that the corresponding event took place and may be contested only in a judicial proceeding. In the absence of a dispute between interested persons, incidental errors and inaccuracies are corrected by ZAGS agencies. The refusal of a ZAGS agency to make the necessary corrections may be appealed in court (article 145 of the Code on Marriage and the Family).

As a rule, an entry in the registers of ZAGS agencies is of a certifying nature. However, in certain cases specially indicated in the law, legal and binding significance is attached to this entry, i.e., the corresponding rights and obligations do not arise before it is made. Such significance is attached to entries pertaining to marriage, divorce, and changes of first name, patronymic, and surname.

Russian-English Glossary

administrativnyi administrative
  administrativnoe pravo administrative law
  administrativnoe pravootnoshenie administrative law relationship
  administrativnoe upravlenie administrative management
  administrativnyi akt administrative act
akkreditiv letter of credit
akt act
  administrativnyi akt administrative act
  akt grazhdanskogo sostoianiia instrument of civil status
  akt kodifikatsionnogo kharaktera act of codified nature
  nepravomernyi akt unlawful act
  normativnyi akt normative act
  planovyi akt na postavku plan act of supply
  podzakonnyi akt substatutory enactment
  pravovoi akt legal act
aktsept acceptance
  aktseptnaia forma raschetov acceptance form of payments
  molchalivyi aktsept tacit acceptance
alimenty alimony, maintenance
  alimentnoe obiazatel'stvo maintenance obligation
  alimentnoobiazannoe litso a person to whom maintenance payments must be paid
  alimentnoobiazannyi pervoi (vtoroi) ocheredi person with primary (secondary) obligation to make maintenance payments
  platel'shchik alimentov maintenance payer
analogiia analogy
  analogiia prava i zakona analogy of law and legislation
  analogiia zakona analogy of law
arbitrazh arbitrazh
arendovat' to lease
avans advance
avariia accident; damage
  obshchaia avariia general damages
avtor author, composer
  avtorskii dogovor author's contract
  avtorskoe pravo copyright
  avtorskoe svidetel'stvo inventor's certificate

balans balance
  samostoiatel'nyi balans independent balance
beskhoziainyi ownerless
bessrochno indefinitely
  bessrochnost' permanence

bezdolevoi unshared
bezvestno otsutstvuiushchii missing person
bezvozmezdnaia gratuitous
bilet ticket
  zalogovyi bilet pawnticket
blago good
  lichnoe neimushchestvennoe blago personal non-property valuable
brak marriage
  brachnyi marital
  brachnyi soiuz marital union
  brachnyi vozrast age of marriage
  deistvitel'nyi brak lawful marriage
  dvorets brakosochetaniia wedding palace
  fakticheskii brak unsolemnized marriage
  fiktivnyi brak sham marriage
  sostoiat' v brake to be married
  zakliuchenie braka conclusion of a marriage
bronirovat' to reserve
  bronirovat' pomeshchenie to reserve an accommodation
bukval'nyi literal
  bukval'noe tolkovanie literal interpretation
byt way of life
  bytovoe obsluzhivanie consumer services
  bytovye nuzhdy everyday needs

chek check
  chekodatel' check giver
  chekoderzhatel' check holder

darenie gift
  daritel' giver
deesposobnost' dispositive capacity
deiatel'nost' activity
  khoziaistvennaia deiatel'nost' economic activity
deistvie action
  konkliudentnoe deistvie implied action
  nepravomernoe deistvie unlawful action
  pravomernoe deistvie lawful action
  vinovnoe deistvie culpable action
deistvitel'nost' validity
  deistvitel'naia sdelka valid transaction
  deistvitel'nost' akta validity of an act
  deistvitel'nost' braka validity of a marriage
  deistvitel'nyi brak lawful marriage
deistvuiushchii active, functioning
  deistvuiushchee zakonodatel'stvo existing legislation, legislation in force
  deistvuiushchii kodeks existing code
delikt tort
delo case
  grazhdanskoe delo civil case
  delo o rastorzhenii braka divorce lawsuit
demeredzh demurrage
den'gi money
  denezhnaia kompensatsiia cash compensation
  denezhnye vklady deposits of money

denezhnyi raschet cash payment
deti children
deti, rozhdennye v brake children born in a marriage
deti, zachatye v brake children conceived in a marriage
detskoe uchrezhdenie childcare institution
dispasha average statement
dispasher average adjuster
dispozitivnost' dispositivity
dispozitivnaia norma dispositive norm
dobrovol'nyi voluntary
dobrovol'noe obshchestvo volunteer society
dobrovol'noe strakhovanie voluntary insurance
dobrovol'nost' vstupleniia v brak voluntary marriage
dobrovol'nyi poriadok voluntary procedure
dogovor contract
arendnyi dogovor lease
avtorskii dogovor author's contract
dogovor arendy imushchestva contract of property lease
dogovor bankovskoi ssudy contract of bank loan
dogovor bezvozmezdnogo pol'zovaniia imushchestvom contract of the gratuitous use of property
dogovor buksirovki contract of towage
dogovor bytovogo prokata contract of domestic hire
dogovor dareniia contract of gift
dogovor imushchestvennogo naima property-hire contract
dogovor khraneniia contract of deposit for safe custody
dogovor khraneniia s obezlicheniem veshchei contract of deposit of unidentified things
dogovor komissii contract of commission agency
dogovor kontraktatsii procurement contract
dogovor kupli-prodazhi purchase-sale contract
dogovor kupli-prodazhi zhilykh domov contract of purchase and sale of dwellings
dogovor meny contract of barter
dogovor morskoi perevozki contract of carriage by sea
dogovor naima sluzhebnogo zhilogo pomeshcheniia contract of lease of employment living accommodations
dogovor naima zhilogo pomeshcheniia contract for the rental of living accommodations
dogovor o konkurse contract of public competition
dogovor o zaloge contract of pledge
dogovor perevozki carriage contract
dogovor perevozki gruza contract for carriage of goods
dogovor perevozki passazhirov i bagazha contract for the carriage of passengers and baggage
dogovor perevozki v mestnom soobshchenii contract of carriage for local carriage
dogovor podriada work contract
dogovor podriada na kapital'noe stroitel'stvo work contract for capital construction
dogovor porucheniia contract of agency
dogovor postavki contract of delivery; contract of supply
dogovor pozhiznennogo soderzhaniia contract of maintenance for life
dogovor prodazhi doma s usloviem pozhiznennogo soderzhaniia prodavtsa con-

tract for the sale of a house with the condition that the seller is maintained for life
dogovor prodazhi v kredit contract of credit sale
dogovor prokata imushchestva contract of property rent
dogovor roznichnoi kupli-prodazhi contract of retail sale
dogovor strakhovaniia contract of insurance
dogovor subpodriada subcontractor's contract
dogovor taim-chartera contract of time-charter
dogovor v pol'zu tret'ego litsa contract in favor of a third party
dogovor za pol'zovanie zhiloi ploshchadi v domakh, prinadlezhashchikh grazhdanam na prave lichnoi sobstvennosti contract for use of living space in houses belonging to citizens by right of personal ownership
dogovor zaima contract of loan
dogovor, zakliuchaemyi po usmotreniiu storon contract concluded at the discretion of parties
dogovor zaloga doma contract for the pledge of a house
dogovor zhilishchnogo naima contract for the rental of housing
dogovornaia neustoika contractual liquidated damages
dogovornaia tsena contractual price
dogovornoe obiazatel'stvo contractual obligation
dogovornoe otnoshenie contractual relationship
dvustoronnii dogovor bilateral contract
konsensual'nyi dogovor consensual contract
mezhdunarodnyi dogovor international treaty
planovyi dogovor plan contract
planovyi dogovor postavki plan contract of supply
real'nyi dogovor executed contract
srochnyi trudovoi dogovor employment contract for a fixed period
tipovoi dogovor model contract
trudovoi dogovor labor contract
vozmezdnyi dogovor compensatory contract
vozobnovit' dogovor to renew a contract
vzaimnyi dogovor mutual contract

dokazatel'stvo proof
pis'mennoe dokazatel'stvo written proof

dokumentatsiia documentation
proektno-smetnaia dokumentatsiia design and estimate documentation

dolevoi shared
bezdolevoi unshared
dolevoi sosobstvennik share co-owner

dolg debt
dolgosrochnyi kredit long-term credit
dolgovaia raspiska promissory note
dolzhnik debtor
dopolnitel'nyi dolzhnik supplementary debtor
dolzhnik po obiazatel'stvu debtor under an obligation
pogashat' dolg to liquidate a debt
solidarnye dolzhniki joint and several debtors

dolzhnostnoe litso official

domovladenie house and lot

dopolnenie addition
dopolnitel'naia zhilaia ploshchad' supplementary living space
dopolnitel'noe obiazatel'stvo auxiliary obligation
dopolnitel'nye uslugi supplementary services
dopolnitel'nyi dolzhnik supplementary debtor

dostizhenie achievement, attainment
  dostizhenie brachnogo vozrasta attainment of the age of marriage
doverennost′ letter of authority; power of attorney
  po doverennosti per procuration
    doveritel′ principal
  lichno-doveritel′nyi personal agency
dushevnyi mental, emotional
  dushevnaia bolezn′ mental illness
dvorets public building; palace
  dvorets brakosochetaniia wedding palace
dvustoronnii two-way, bilateral
  dvustoronniaia restitutsiia bilateral restitution
  dvustoronnii dogovor bilateral contract

edinobrachie monogamy
ekonomicheskii economic
  ekonomicheskie otnosheniia economic relationships
  ekonomicheskie tsennosti things of economic value
  ekonomicheskii oborot economic activity
ekspertiza expert testimony
  ekspertiza krovi blood test
  sudebno-meditsinskaia ekspertiza forensic medicine
ekspluatatsionnyi operational; exploitative
  ekspluatatsionnye raskhody operating expenses
ekvivalentnyi equivalent
  ekvivalentno-vozmezdnyi kharakter equivalent-reciprocal character
etalon standard

fakt fact
  iuridicheskii fakt juridical fact
  fakticheskii brak unsolemnized marriage
  fakticheskii sostav factual composition
familiia surname
fiktivnyi sham
  fiktivnyi brak sham marriage
filial subsidiary; branch
finansovyi financial
  finansovoe pravo financial law
fondoderzhatel′ fundholder
formulirovka formulation
  slovesnaia formulirovka verbal formulation
frakhtovatel′ charterer
funktsiia function

garantiia guarantee
  garantiinyi srok guarantee period
gosudarstvo state
  gosudarstvennaia planovaia distsiplina state plan discipline
  gosudarstvennoe strakhovanie state insurance
  gosudarstvennoe upravlenie state administration
  gosudarstvennyi organ state organ
  gosudarstvennyi zhilishchnyi fond state housing fund
grammaticheskoe tolkovanie grammatical interpretation
grazhdanin citizen

grazhdanskaia otvetstvennost′ civil liability
grazhdanskaia pravosposobnost′ civil legal capacity
grazhdanskii kodeks civil code
grazhdanskii oborot civil intercourse, civil activity
grazhdansko-pravovaia sdelka civil law transaction
grazhdansko-pravovoe regulirovanie civil law regulation
grazhdansko-pravovoe zakonodatel′stvo civil law legislation
grazhdansko-pravovoi metod regulirovaniia civil law method of regulation
grazhdansko-pravovoi spor civil law dispute
grazhdansko-pravovoi akt civil law act
grazhdanskoe pravo civil law
grazhdanskoe pravootnoshenie civil law relationship
grazhdanskoe sudoproizvodstvo civil procedure
gruppa group
gruppovoi nariad group warrant
gruz cargo; freight; goods
gruzootpravitel′ shipper
gruzovladelets owner of goods
gruzovaia nakladnaia shipping invoice

iavka appearance
iavnyi nedostatok obvious defect
immunitet immunity
imperativnyi imperative; mandatory
imperativnaia norma imperative norm
imperativnye normy zakona imperative norms of law
imushchestvo property
beskhoziainoe imushchestvo ownerless property
imushchestvo, nazhitoe vo vremia braka property amassed during marriage
obosoblennoe imushchestvo separate property
zalozhennoe imushchestvo pledged property
imushchestvennoe polozhenie property situation, property status
imushchestvennoe strakhovanie property insurance
imushchestvennye otnosheniia property relationships
ingosstrakh foreign state insurance
instruktivnyi instructive; directive
instruktivnye ukazaniia instructive orders
invalid invalid
invalid (I, II, III) gruppy group (I, II, III) invalid
isk suit, action
vstrechnyi isk counterclaim
iskovaia davnost′ period of limitation
iskovoe zaiavlenie complaint
iskliuchitel′nyi exclusionary
ispolnenie performance, fulfillment, execution
ispolnenie obiazannostei performance of obligations
ispolnenie obiazannosti v nature specific performance of an obligation
neispolnenie obiazatel′stva failure to perform an obligation
nenadlezhashchee ispolnenie obiazatel′stva improper performance of an obligation
nevozmozhnost′ ispolneniia impossibility of performance
nevypolnenie non-fulfillment
obespechenie ispolneniia obiazatel′stv securing performance of obligations
prinuditel′noe ispolnenie enforced execution

ispolniat' obiazannosti to fulfill obligations
ispolnitel'nyi list court judgment
istechenie expiration
istechenie sroka deistviia dogovora expiration of the period of validity of a contract
istets plaintiff
istochnik source
istochnik povyshennoi opasnosti source of increased danger
iuridicheskii juridical, legal
iuridicheskii fakt juridical fact
iuridicheskii obychai legal custom
iuridicheskoe litso juridical person
iurist lawyer
izdanie promulgation
izhdiventsy dependents
izmenenie change
izobretenie invention
izobretatel'skoe pravo law of inventions

kabotazh cabotage
bol'shoi kabotazh grand cabotage
malyi kabotazh petit cabotage
kapital'nyi capital
kapital'noe stroitel'stvo capital construction
kapital'nyi remont imushchestva capital repair of property
kassa cashier's office
gosudarstvennaia trudovaia sberegatel'naia kassa state workers' savings bank
kassa obshchestvennoi vzaimopomoshchi mutual assistance benefit fund
kharakter zaniatii nature of occupation
khoziaistvennyi economic
khoziaistvennaia deiatel'nost' economic activity
khoziaistvennaia organizatsiia economic organization
khoziaistvennoe stroitel'stvo economic development
khoziaistvennoe zakonodatel'stvo economic legislation
khoziaistvennyi raschet economic accountability; self-financing
khozraschet economic accountability; self-financing
khranenie deposit for safe custody
khranitel' custodian
khuligan hooligan
khuliganskie deistviia hooligan behavior
kniga book; register
kniga zapisei rozhdenii birth register
kodeks code
grazhdanskii kodeks civil code
Kodeks Torgovogo moreplavaniia SSSR Merchant Shipping Code of the USSR
transportnyi kodeks transport code
zhilishchnyi kodeks housing code
kolkhoz collective farm
kolkhoznoe pravo collective farm law
komandirovka business trip
komandirovanie za granitsu official travel abroad
komissioner commission agent
komissionnyi magazin commission store

komitent principal
kommunisticheskaia moral' communist morality
kompensatsiia compensation
  denezhnaia kompensatsiia cash compensation
kompetentsiia competence
  kompetentnyi organ competent organ
konfiskatsiia confiscation
  konfiskatsiia imushchestva confiscation of property
konosament bill of lading
konsensual'nyi consensual
  konsensual'naia sdelka consensual transaction
  konsensual'nyi dogovor consensual contract
konstitutsiia constitution
  konstitutsionnoe pravo constitutional law
kormilets breadwinner
kredit credit
  kratkosrochnyi kredit short-term credit
  kreditnoe pravootnoshenie credit relationship
  kreditor creditor
krushenie wreck, accident
kustarnyi home-produced
  kustarno-remeslennye promysly crafts and trades
kvartira apartment
  sluzhebnaia kvartira service apartment

lichnyi personal
  lichnaia razdel'naia sobstvennost' personal, separate property
  lichnaia sobstvennost' personal property
  lichno-doveritel'nyi personal agency [adj.]
  lichnoe neimushchestvennoe blago personal non-property valuable
  lichnoe strakhovanie personal insurance
likvidatsiia liquidation
  likvidatsiia iuridicheskogo litsa liquidation of a legal person
liniia line, lineage
  niskhodiashchaia liniia descendant lineage
  voskhodiashchaia liniia ascendant lineage
lishenie deprivation
  lishenie svobody deprivation of freedom
list document
  ispolnitel'nyi list court judgment
litso person
  litso bez grazhdanstva stateless person
  alimentnoobiazannoe litso a person to whom maintenance payments must be paid
  dolzhnostnoe litso official
  iuridicheskoe litso juridical person
  obiazannoe litso obligated person
  odinokoe litso single person
  upravomochennoe litso person empowered by law
  litso, v pol'zu kotorogo zakliucheno strakhovanie insurance beneficiary
lombard pawnshop
l'gota privilege
  l'goty subsidized services
  l'gotnyi mesiats month of grace

material'nyi material
  material'no-tekhnicheskoe snabzhenie material-technical supply
  material'noe obespechenie material support
  material'nye predmety material objects
  material'nye tsennosti goods of material value
materinstvo maternity
mera measure
  mera otvetstvennosti measure of liability
  mery k primireniiu suprugov measures to reconcile spouses
  mery prinuditel'nogo vozdeistviia means of enforcement
mesto place
  mesto nakhozhdeniia iuridicheskogo litsa situs of a legal person
  mesto otkrytiia nasledstva place of opening of an inheritance (or succession)
  mesto zhitel'stva place of residence
  mestoprebyvanie whereabouts
metod method
  metod vlasti method of authority
mezhdunarodnyi international
  mezhdunarodnyi dogovor international treaty
mnozhestvennost' lits plurality of persons
molchalivyi silent; tacit
  molchalivyi aktsept tacit acceptance
molchanie acquiescence

nachislenie added charge
  nachislenie peni extra charge of penalty
nadlezhashchii appropriate, proper
  nadlezhashchee kachestvo proper quality
naimodatel' lessor
nakladnaia invoice
  gruzovaia nakladnaia shipping invoice
nakoplenie savings
nariad warrant
  gruppovoi nariad group warrant
narushenie violation
  narushenie dogovornoi distsipliny violation of contractual discipline
  narushitel' violator
naslednik heir
  nasledniki pervoi ocheredi heirs of the first category
  nasledniki po zakonu heirs ab intestato
  nasledniki vtoroi ocheredi heirs of the second category
  nasledovanie inheritance
  nasledovanie po zakonu succession on intestacy
  nasledovanie po zaveshchaniiu succession under a will
  nasledovatel' estate-leaver
  nasledstvennaia dolia share of the succession
  nasledstvennoe imushchestvo succession
  nasledstvennoe pravo law of succession, the right to an inheritance
  nasledstvo estate
natsional'naia prinadlezhnost' nationality
nedeesposobnyi not having dispositive capacity
nedeistvitel'nost' invalidity
  nedeistvitel'nost' braka annulment of a marriage
  nedeistvitel'nyi invalid

nedelimyi indivisible
  nedelimye predmety items that cannot be divided
nedostatok defect
  skrytyi nedostatok hidden defect
  nedostatochnaia osvedomlennost' lack of sufficient knowledge
nedostoinyi unworthy
  nedostoinoe povedenie undignified behavior
  nedostoinyi roditel' unfit parent
neimushchestvennye otnosheniia non-property relationships
neispolnenie non-fulfillment, non-performance
  neispolnenie obiazatel'stva failure to perform an obligation
nenadlezhashchii improper
  nenadlezhashchee ispolnenie obiazatel'stva improper performance of an obligation
  nenadlezhashchee litso improper party
  nenadlezhashchee vremia improper time
neobiazatel'nyi non-obligatory
  neobiazatel'noe tolkovanie non-obligatory interpretation
neobkhodimyi necessary, essential
  neobkhodimyy uchastnik necessary participant
neosnovatel'nyi unfounded, unjust
  neosnovatel'noe preobretenie imushchestva unjust acquisition of property
  neosnovatel'noe sberezhenie imushchestva unjust retention of property
neostorozhnost' negligence
  grubaia neostorozhnost' gross negligence
  prostaia neostorozhnost' simple negligence
neperedavaemyi not assignable
nepravil'nyi incorrect, false
nepravomernyi unlawful
  nepravomernoe deistvie unlawful action
  nepravomernyi akt unlawful act
  nepravomernyi postupok unlawful conduct
nepreodolimaia sila force majeure
nesdacha produktsii non-supply of products
nesobludenie non-observance
nesovershennoletnie minors
nespravedlivo unfair
nesvoevremennyi inopportune, late
  nesvoevremennaia oplata late payment
netrudosposobnost' inability to work
  netrudosposobnyi unable to work
neustoika liquidation of damages; forfeit
  al'ternativnaia neustoika alternative liquidated damages
  dogovornaia neustoika contractual liquidated damages
  iskliuchitel'naia neustoika exclusionary liquidated damages
  shtrafnaia neustoika penalty liquidated damages
  neustoika za prosrochku platezha kvartplaty liquidated damages for late payment of rent
  zachetnaia neustoika calculated liquidated damages
  zakonnaia neustoika lawful liquidated damages
nevostrebovannyi unclaimed
  nevostrebovannye nakhodki unclaimed finds
nevozmozhnost' ispolneniia impossibility of performance
nevypolnenie non-fulfillment

nezhilye pomeshcheniia non-residential accommodations
nichtozhnyi insignificant, worthless
  nichtozhnaia sdelka void transaction
  nichtozhnaia (polnostiu deistvitel'naia) sdelka void (totally invalid) transaction
niskhodiashchaia liniia descendant lineage
norma norm
  norma zhiloi ploshchadi norm for living space
  normy grazhdanskogo prava the norms of civil law
  imperativnaia norma imperative norm
  normativnyi akt normative act
nositel' carrier
  nositel' sub"ektivnogo prava bearer of a subjective right
notarial'nyi notarial
nravstvennost' morality
nuzhdaemost' neediness

ob"ekt avtorskogo prava object of copyright
obespechenie ensuring
  obespechenie ispolneniia obiazatel'stv securing performance of obligations
  obespechitel'naia funktsiia function of securing
obiazannost' obligation
  obiazannost' vozmestit' prichinennyi vred obligation to compensate for harm caused
  obiazannosti nanimatelia obligations of the lessee
  obiazannaia storona liable party
  obiazannoe litso obligated person, person under obligation
obiazatel'no obligatory, mandatory
  obiazatel'naia dolia v nasledstve obligatory share of the estate
  obiazatel'noe strakhovanie obligatory insurance
  obiazatel'noe tolkovanie obligatory interpretation
  obiazatel'noe velenie obligatory order
obiazatel'stvo obligation
  alimentnoe obiazatel'stvo maintenance obligation
  dogovornoe obiazatel'stvo contractual obligation
  dopolnitel'noe obiazatel'stvo auxiliary obligation
  glavnoe obiazatel'stvo primary obligation
  obiazatel'stvo s uchastiem grazhdan obligation involving citizens
  obiazatel'stvo, voznikaiushchee vsledstvie pricheneniia vreda obligation arising as the result of causing harm
  vstrechnoe obiazatel'stvo mutual obligation
  obiazatel'stvennoe pravo law of obligations
  obiazatel'stvennye trebovaniia promissory obligations
obladatel' possessor
  obladatel' prava operativnogo upravleniia possessor of right of operative management
obman fraud
obmen exchange
  obmen zhilogo pomeshcheniia exchange of a living accommodation
obnaruzhenie discovery
obnovlenie zakonodatel'stva revision of legislation
oborot turn, circulation
  ekonomicheskii oborot economic activity
obosoblennoe isolated, separate

obosoblennoe imushchestvo separate property
obrashchenie appeal
obrashchenie vzyskaniia na zalozhennuiu veshch' levy of execution against pledged article
obratnaia sila retroactive force
obrazets model
obshchaia general, common
obshchaia dolevaia sobstvennost' common shared property
obshchaia sobstvennost' common property
obshchaia sovmestnaia sobstvennost' common joint ownership, common joint property
obshchee khoziaistvo common household
Obshchee polozhenie o ministerstvakh SSSR General Statute on Ministries of the USSR
obshchestvo society
obshchestvennyi social
obshchestvennaia organizatsiia social organization
obshchestvennoe otnoshenie social relationship
obshchestvennyi zhilishchnyi fond social housing fund
obshchezhitie communal life; hostel
obsluzhivanie service
bytovoe obsluzhivanie consumer services
obstoiatel'stvo circumstance
obstoiatel'stva iskliuchaiushchie otvetstvennost' dolzhnika circumstances excluding liability of the debtor
obychai custom
odinokoe litso single person
odnostoronniaia sdelka unilateral transaction
oformliat' to register
oformliat' dokumenty to formalize documents
ogovorka clause, reservation, stipulation
ogovorennoe uslovie stipulated condition
ogranichenie delimitation
ogranichenie otvetstvennosti limitation of liability
ogranichitel'noe tolkovanie restrictive interpretation
okhrana protection
okhrana nasledstvennogo imushchestva protection of an estate
okhrana obshchestvennykh interesov protection of social interests
okhrana prav potrebitelei safeguarding the rights of consumers
okhrannoe svidetel'stvo securing warrant
okrug area
opeka guardianship
opekun guardian
opekun nad imushchestvom general guardian
operativnoe upravlenie operative management
oplata payment
nesvoevremennaia oplata late payment
oplata produktsii payment for goods
oporochivaemoe litso defamed person
oproverzhenie retraction
oproverzhenie nepravil'nykh svedenii retraction of false information
oproverzhenie porochashchikh svedenii retraction of defamatory information
opublikovanie publication
order order

order edinoi formy single-form order
order na zaselenie kvartiry order for residence in an apartment
orderoderzhatel' holder of an order
organ organ
organ upravleniia administrative organ
kompetentnyi organ competent organ
osnovaniia grounds
osnovaniia vozniknoveniia grounds for development
osnovnoi fundamental, basic
osnovnoi zakon fundamental law
osnovnye polozheniia general provisions; basic positions
osnovy principles
Osnovy Grazhdanskogo Sudoproizvodstva SSSR i Soiuznykh Respublik Principles of Civil Procedure of the USSR and Union Republics
Osnovy Grazhdanskogo Zakonodatel'stva Principles of Civil Legislation of the USSR and Union Republics
Osnovy Ispravitel'no-trudovogo Zakonodatel'stva Soiuza SSR i Soiuznykh Respublik Principles of Corrective Labor Legislation of the USSR and Union Republics
Osnovy Zakonodatel'stva Soiuza SSR i Soiuznykh Respublik o Brake i Sem'i Principles of Legislation of the USSR and Union Republics on Marriage and the Family
Osnovy Zhilishchnogo Zakonodatel'stva Soiuza SSR i Soiuznykh Respublik Principles of Housing Legislation of the USSR and Union Republics
osobennost' characteristic
osparivat' to contest
osushchestvlenie realization
osushchestvlenie prava the exercise of a right
osushchestvlenie prinuzhdeniia the carrying out of enforcement
osushchestvlenie roditel'skikh prav exercise of parental rights
osvobodit' to free
osvobodit' pomeshchenie to vacate an accommodation
osvobozhdenie avoidance, release
osvobozhdenie ot otvetstvennosti release from liability
otchim stepfather
otchuzhdat' to alienate
otchuzhdatel' alienator
otchuzhdenie alienation
vozmezdnoe otchuzhdenie reciprocal alienation
otdelenie separation
otdelenie tserkvi ot gosudarstva separation of church from state
otkrytie discovery
otmena repeal
otmena usynovleniia revocation of an adoption
otnoshenie relationship
imushchestvennye otnosheniia property relationships
neimushchestvennye otnosheniia non-property relationships
obshchestvennye otnosheniia social relationships
otrasl' branch
otrasl' prava branch of law
otstuplenie deviation
otsutstvie absence
otsutstvie po uvazhitel'nym prichinam absence on valid grounds

ottsovstvo paternity
otvetchik defendant
otvetstvennost' accountability, liability, responsibility
  otvetstvennyi rabotnik executive
  grazhdanskaia otvetstvennost' civil liability
  otvetstvennost' prodavtsa za nenadlezhashchee kachestvo veshchi responsibility of the seller for the defective quality of an article
  otvetstvennost' storon za narushenie dogovora obligation of parties for a breach of contract
  otvetstvennost' za narushenie obiazatel'stv liability for breach of obligations
  otvetstvennost' za povrezhdenie zdorov'ia i smert' grazhdanina liability for injury to the health or for the death of a citizen
  otvetstvennost' za vred, prichenennyi deistviiami dolzhnostnykh lits liability for harm caused by the actions of officials
  otvetstvennost' za vred, prichinennyi istochnikom povyshennoi opasnosti liability for harm caused on the grounds of increased danger
  otvetstvennost' za vred, prichinennyi nesovershennoletnimi i nedeesposobnymi liability for harm caused by minors and incapable parties
  samostoiatel'naia imushchestvennaia otvetstvennost' independent property liability
  solidarnaia imushchestvennaia otvetstvennost po obiazatelstvam joint and several property responsibility for obligations
ozorstvo mischief

padcheritsa stepdaughter
passport passport; identity card
pasynok stepson
patent patent
patronat foster care
perekhod transition, transfer
  perekhod nasledstva k gosudarstvu escheat
  perekhod prav i obiazannostei transfer of rights and obligations
perevozchik carrier
  perevozka carriage
  priamaia perevozka direct carriage
pismennyi written
  pis'mennaia sdelka written transaction
  pis'mennoe dokazatel'stvo written proof
plan plan
  planovaia distsiplina plan discipline
  planovoe zadanie task set by the plan
  planovye osnovaniia dogovora kontraktatsii plan basis of the procurement contract
  planovyi akt na postavku plan act of supply
  planovyi akt raspredeleniia plan act of distribution
  planovyi dogovor plan contract
  planovyi dogovor postavki plan contract of supply
plata payment
  plata za kommunal'nye uslugi payment for communal services
  platel'shchik alimentov maintenance payer
  platezhnoe poruchenie payment order, payment authorization
  platezhnoe trebovanie payment demand
po by, per
  po doverennosti per procuration

po neostorozhnosti negligently
po zaiavleniiu zainteresovannykh lits by petition of interested parties
podat' to file
podacha filing
podacha zhaloby filing a complaint
podat' zaiavlenie to file a declaration
podchinenie subordination
podnanimatel' sublessee
podnaznachenie naslednika substitution of an heir
podopechnyi ward
podriadchik contractor
general'nyi podriadchik general contractor
podsudnost' jurisdiction
podsudnost' grazhdanskikh del jurisdiction of civil cases
podtverzhdenie confirmation
podzakonnyi substatutory
podzakonnye akty substatutory enactments
pogashat' dolg liquidate a debt
pokazaniia testimony
poklazhedatel' depositor
pokupat' buy
pokupatel' buyer
pokupnaia tsena purchase price
pol sex
polozhenie status
imushchestvennoe polozhenie property situation, property status
sotsial'noe polozhenie social status
polozhenie statute
Polozhenie o Poriadke Naznacheniia i Vyplaty Gosudarstvennykh Pensii Statute on the Procedure for Granting and Paying State Pensions
Polozhenie o Sotsialisticheskom Gosudarstvennom Proizvodstvennom Predpriiatii Statute on the Socialist State Production Enterprise
Polozhenie o Tovarishcheskikh Sudakh Statute on Comrades' Courts
polozheniia grazhdanskogo prava provisions of civil law
pol'zovat'sia to use; to enjoy
pomeshchenie lodging
bronirovat' pomeshchenie to reserve an accommodation
dachnoe pomeshchenie vacation accommodation
pol'zovanie pomeshcheniiami use of accommodations
zhiloe pomeshchenie housing, accommodation
popechitel' warder
popechitel'stvo guardianship, trusteeship
popustitel'stvo connivance
poriadok procedure
poriadok predostavleniia zhilykh pomeshchenii procedure for allocating living accommodations
poriadok vzyskaniia i uplaty alimentov procedure for collecting and paying maintenance
porochashchii defamatory
poruchenie-obiazatel'stvo agreement-obligation
platezhnoe poruchenie payment order
poruchitel' surety
poruchitel'stvo suretyship
poshlina fees

postanovlenie decree
postavka supply, delivery
  planovyi akt na postavku plan act of supply
  polozhenie o postavkakh statute on deliveries
  postavshchik supplier
postupok conduct
  nepravomernyi postupok unlawful conduct
  pravovoi postupok legal conduct
potrebitel' consumer, user
povedenie behavior
  nedostoinoe povedenie undignified behavior
  protivopravnoe povedenie unlawful behavior
povelitel'nyi mandatory
  povelitel'naia norma mandatory norm
poverennyi agent
povyshennaia opasnost' increased danger
pozhiznennoe soderzhanie lifelong maintenance
pravilo rule
pravitel'stvo government
pravo law, right
  avtorskoe pravo copyright
  finansovoe pravo financial law
  grazhdanskoe pravo civil law
  izobretatel'skoe pravo the law of inventions
  kolkhoznoe pravo collective farm law
  konstitutsionnoe pravo constitutional law
  nasledstvennoe pravo law of succession
  obiazatel'stvennoe pravo law of obligations
  prava i obiazannosti storon rights and obligations of the parties
  pravo lichnoi sobstvennosti the right to personal property
  pravo na chast' nazhitogo v techenie braka imushchestva the right to part of the property acquired during the marriage
  pravo na nasledstvo right to succession
  pravo na otkrytie law of discovery
  pravo na pensiiu right to a pension
  pravo na poluchenie alimentov the right to maintenance
  pravo na vosmeshchenie vreda the right to compensation for harm
  pravo nasledovaniia the right to inherit; heirship
  pravo otkazat'sia ot nasledstva the right to renounce an estate
  pravo pokupatelia the right of the buyer
  pravo pol'zovat'sia zhilym pomeshcheniem right to use living accommodations
  pravo rasporiazhat'sia obshchei sobstvennost'iu the right to dispose of common property
  pravo sobstvennosti law of ownership
  pravo sobstvennosti right of ownership
  semeinoe pravo family law
  trudovoe pravo labor law
pravomernyi lawful
  pravomernoe deistvie lawful action
pravoobrazuiushchii right-forming
pravookhranitel'nye organy organs for the defense of rights
pravootnoshenie legal relationship
predpolozhenie assumption

zakonnoe predpolozhenie legal presumption
predposylka premise
predpriiatie enterprise
predstavitel'stvo representation
predstavliaemoe the represented party
predusmotrennyi zakonom stipulated by law
prekrashchenie termination
prekrashchenie braka termination of marriage
prekrashchenie dogovora termination of a contract
prekrashchenie obiazatel'stva termination of an obligation
prekrashchenie pravootnosheniia termination of a legal relationship
presechenie prevention
presechenie deistvii of actions
prezumptsiia presumption
priamaia perevozka direct carriage
priamoi umysel deliberate intent
pribyl' profit
prichina cause
prichinenie vreda the causing of harm
prichinitel' vreda tortfeasor
prichinnaia sviaz' causal relationship
primenenie application
primeniat' zakonodatel'stvo to apply legislation
primernyi ustav model charter
prinadlezhnost' membership
national'naia prinadlezhnost' nationality
rasovaia prinadlezhnost' race
priniatie acceptance
priniatie detei na vospitanie acceptance of children for rearing
priniatie nasledstva acceptance of an estate
prinuzhdenie enforcement
prinuditel'naia peredacha veshchi enforced transfer of an article
prinuditel'noe ispolnenie enforced execution
prinuditel'nyi obmen enforced exchange
priobretatel' purchaser
prirashchenie accrual
prirashchenie dolei accrual of shares
priroda nature
prirodnye ob"ekty natural objects
prisoedinenie accession
privlekat' v kachestve sootvetchikov to join as co-defendants
priznak sign
rodovoi priznak generic trait
priznaki iuridicheskogo litsa characteristics of a juridical person
priznanie declaration, admission
priznanie braka nedeistvitel'nym annulment of a marriage
priznanie grazhdanina bezvestno otsutstvuiushchim i ob"iavlenie ego umershim declaration that a citizen is missing and declared dead
priznanie prava declaration of a right
prodavets seller
proiskhozhdenie origin
proizvodstvo production
proizvodstvenno-khoziaistvennaia deiatel'nost' production-economic activity
proizvodstvennoe ob"edinenie production association

prokuror procurator
propiska registration
  razreshenie na propisku residence permit
prosrochka delay
  prosrochka ispolneniia delay of performance
  prosrochka platezha kvartplaty late payment of rent
prostaia neostorozhnost' simple negligence
protivopravnost' unlawfulness
  protivopravnoe deistvie wrongful action
  protivopravnoe povedenie unlawful behavior
  protivopravnost' deistvii unlawfulness of actions
protokol record
  protokol raznoglasii list of disagreements
psikhicheskoe rasstroistvo psychological disturbance

raschet calculation
  raschet za okazannye uslugi bill for services rendered
  khoziaistvennyi raschet economic accountability
  raschetno-kassovoe obsluzhivanie accounting/banking services
  raschetnoe pravootnoshenie accounting relationship
  raschety financial operations
  raschety za produktsiiu payments for products
rasovaia prinadlezhnost' race
raspiska receipt
  dolgovaia raspiska promissory note
rasporiazhat'sia dispose of
  rasporiazhenie disposition; order
rasprostranitel'noe tolkovanie broad interpretation
rasshiritel'noe tolkovanie extended interpretation
rassmotrenie examination, scrutiny
  rassmotrenie dela po sushchestvu examination of a case on its merits
rassrochka intallment system
  rassrochka platezha payment by installments
rastorzhenie cancellation, dissolution, annulment, abrogation, recission
  rastorzhenie braka termination of a marriage
  rastorzhenie dogovora recission of a contract
razdelenie division
raznariadka counter-warrant
razreshenie permission, authorization, resolution, settlement
  razreshenie na propisku residence permit
  razreshenie spora resolution of a dispute
  razreshitel'nyi poriadok procedure by permission
razvod divorce
raz"iasnenie explanation
  rukovodiashchie raz"iasneniia guiding explanations
real'nyi real, realizable, realistic
  real'naia sdelka executed transaction
  real'noe ispolnenie obiazatel'stva actual performance of an obligation
  real'nyi ushcherb material damage
rebenok child
  vnebrachnyi rebenok child born out of wedlock
retortsiia retortion
rezhim regime

rezhim obshchei sobstvennosti regime of common property
risk risk
risk sluchainoi gibeli risk of accidental ruin
roditel' parent
nedostoinyi roditel' unfit parent
roditel' s kem prozhivaet rebenok custodial parent
rodstvo kinship
rodstvennik po proiskhozhdeniiu relative by birth

samostoiatel'nyi independent
samostoiatel'naia imushchestvennaia otvetstvennost' independent property liability
samostoiatel'naia smeta independent account
samostoiatel'nyi balans independent balance
sanktsiia sanction
sdelka transaction
bezvozmezdnaia sdelka gratuitous transaction
deistvitel'naia sdelka valid transaction
grazhdansko-pravovaia sdelka civil law transaction
konsensual'naia sdelka consensual transaction
mnimaia sdelka sham transaction
odnostoronniaia sdelka unilateral transaction
osporimaia (otnositel'no nedeistvitel'naia) sdelka voidable (relatively invalid) transaction
pis'mennaia sdelka written transaction
real'naia sdelka executed transaction
sovershenie sdelki conclusion of a transaction
uchastniki sdelki participants in a transaction
uslovnaia sdelka conditional transaction
ustnaia sdelka oral transaction
vozmezdnaia sdelka compensatory transaction
zakliuchat' sdelku to conclude a transaction
sebestoimost' cost price
sem'ia family
semeinoe pravo family law
shtraf fine
shtraf za postavku nekomplektnoi produktsii fine for supplying an incomplete product
shtrafnaia neustoika liquidated damages penalty
sila force
nepreodolimaia sila force majeure
obratnaia sila retroactive force
sistema system
sistema pervogo riska system of first risk
sistema proportsional'noi otvetstvennosti system of proportionate liability
sistematicheskoe tolkovanie systematic interpretation
slaboumie mental retardation
sliianie merger
sluchai event, incident
strakhovyi sluchai insured/insurance event
smeshannoe strakhovanie mixed insurance
smezhnye otrasli prava contiguous branches of law
sobranie collection, assembly
sobranie postanovlenii collection of decrees

sobranie upolnomochennykh assembly of representatives
sobstvennik owner
sobstvennost' property, ownership
lichnaia razdel'naia sobstvennost' personal, separate property
lichnaia sobstvennost' personal property
obshchaia sobstvennost' common property
obshchaia dolevaia sobstvennost' common shared property
obshchaia sovmestnaia sobstvennost' common joint property
sotsialisticheskaia sobstvennost' socialist ownership
sobytie event
soglashenie agreement
soglashenie o neustoike agreement on the liquidation of damages
soglashenie o poruchitel'stve agreement on suretyship
soglashenie o zaloge agreement on pledge
soglashenie storon mutual agreement of the parties
soglasie consent
soiuz union
brachnyi soiuz marital union
sokhrannost' safekeeping
solidarnyi solidary
solidarnaia imushchestvennaia otvetstvennost po obiazatel'stvam joint and several property responsibility for obligations
solidarnye dolzhniki joint and several debtors
sosobstvennik co-owner
sostavliat' to constitute
sostavliat' doverennost' to draw up power of attorney
sostavliat' zaveshchanie to draw up a will
sotsialisticheskii socialist
sotsialisticheskaia moral' socialist morality
sotsialisticheskaia sobstvennost' socialist ownership
sotsialisticheskaia zakonnost' socialist legality
sotsialisticheskoe obshchezhitie socialist communal life
sovershennoletie adulthood
spetsifikatsiia specifications
spor dispute
spornoe otnoshenie disputed relationship
sposobnost' ability, capacity
spravka information
srok period
srok dogovora the period of a contract
srok ispolneniia period of performance
garantiinyi srok guarantee period
ssylka exile
stikhiinyi elemental, natural
stikhiinoe bedstvie natural calamity; natural catastrophe
stikhinnoe iavlenie natural calamity
storona party
storony v dogovore parties to a contract
strakhovanie insurance
dobrovol'noe strakhovanie voluntary insurance
gosudarstvennoe strakhovanie state insurance
imushchestvennoe strakhovanie property insurance
lichnoe strakhovanie personal insurance

obiazatel'noe strakhovanie obligatory insurance
smeshannoe strakhovanie mixed insurance
strakhovanie domashnego imushchestva household property insurance
strakhovanie sel'skokhoziaistvennykh kul'tur agricultural crop insurance
strakhovanie stroenii building insurance
strakhovanie trudosposobnosti employment insurance
strakhovanie zhizni life insurance
strakhovatel' the insurer
strakhovaia otsenka insurance evaluation
strakhovaia premiia insurance premium
strakhovaia summa insurance sum
strakhovoe pravootnoshenie insurance legal relationship
strakhovshchik the assured [insurance]
strakhovyi interes insurance interest
strakhovyi risk insurance risk
strakhovyi sluchai insured event
strakhovyi vznos insurance fee
stroitel'stvo construction
kapital'noe stroitel'stvo capital construction
khoziaistvennoe stroitel'stvo economic development
sub"ekt subject
sub"ekt avtorskogo prava subject of copyright
sub"ekt grazhdanskogo prava subject of civil law
sub"ekt otvetstvennosti subject of liability
sub"ekt prava subject of a right
sub"ektivnoe pravo subjective right
sud court
grazhdanskoe sudoproizvodstvo civil procedure
sud otmeniaet reshenie court annuls a decision
sudebnaia praktika court practice
sudebno-meditsinskaia ekspertiza forensic medicine
sudebnyi ispolnitel' bailiff of the court
sudoproizvodstvo judicial organization; judicial proceedings
treteiskii sud arbitration tribunal
v sudebnom poriadke in a judicial proceeding
suprug spouse (male)
supruga spouse (female)
sushchestvennyi substantive
sushchestvennoe zabluzhdenie substantive mistake
svidetel'stvo evidence, certificate
svidetel'skoe pokazanie testimony of witnesses
svidetel'stvo o rozhdenii birth certificate
avtorskoe svidetel'stvo inventor's certificate
Svod Zakonov Digest of Laws
svodnyi zakon o rekvizitsii i konfiskatsii imushchestva comprehensive law on the regulation and confiscation of property
svoevremennyi timely, prompt
svoevremennaia uplata prompt payment
svoistvo affinity
syr'e i materialy raw materials and supplies

tekhnikum technical school
tipovoi model, standard

tipovoi dogovor model contract
tolkovanie interpretation
autentichnoe tolkovanie authentic interpretation
bukval'noe tolkovanie literal interpretation
tolkovanie grazhdansko-pravovykh norm interpretation of civil law norms
istoricheskoe tolkovanie historical interpretation
logicheskoe tolkovanie logical interpretation
rasprostranitel'noe tolkovanie broad interpretation
rasshiritel'noe tolkovanie extended interpretation
sistematicheskoe tolkovanie systematic interpretation
torgovlia trade
torgovlia posylochnaia mail-order trade
torgovaia organizatsiia trading organization
tovar goods, wares
tovarno-denezhnaia forma commodity-money form
tovarnoe obrashchenie commodity circulation
tovarnyi znak trademark
transportnyi kodeks transport code
transportnyi ustav transport charter
trebovanie demand
obiazatel'stvennye trebovaniia promissory obligations
platezhnoe trebovanie payment demand
treteiskii arbitration
treteiskii sud arbitration tribunal
trud labor
trudovoe pravo labor law
trudovoi dogovor labor contract
trudovoi dokhod earned income
tsena price
dogovornaia tsena contractual price
pokupnaia tsena purchase price
ustanovlennaia tsena set price
tsennost' value
material'nye tsennosti goods of material value
ob"iavlennaia tsennost' declared value

ubytki losses
uchastnik participant
udovletvorenie satisfaction
udovletvorenie pretenzii satisfaction of claim(s)
vstrechnoe udovletvorenie mutual satisfaction
ugolovnaia otvetstvennost' za dachu lozhnykh svedenii criminal liability for the falsification of information
ugolovnyi zakon criminal law
ukaz edict
ukazanie instruction
instruktivnye ukazaniia instructive orders
ukhod care
uklonenie deviation, evasion
uklonenie prodavtsa ot peredachi veshchi refusal of the seller to transfer an article
umershii decedent; the deceased
umysel intention
umyshlenno deliberately

uplata payment
  uplata alimentov payment of alimony
  uplata neustoiki payment of liquidated damages
upravlenie administration
  gosudarstvennoe upravlenie state management
  operativnoe upravlenie operative management
upravomochennoe litso person empowered by law to act
ushcherb damage
  real'nyi ushcherb material damage
  ushcherb, prichinennyi prestupnymi deistviiami damages caused by criminal actions
uslovie condition, clause, term
  usloviia dogovora podriada terms of a work contract
  uslovnaia sdelka conditional transaction
  uslovnoe osuzhdenie k lisheniiu svobody conditional sentence to deprivation of freedom
usmotrenie discretion
ustanovlenie establishmnent
  ustanovlennaia tsena set price
  ustanovlennye zakonom prepiatstviia legally established obstacles
ustav, statute, charter, set of rules and regulations
  transportnyi ustav transport charter
usynovitel' adoptive parent
  usynovlenie adoption
  usynovlennyi adoptee
uvazhenie lichnosti respect for the individual

valiuta (foreign) currency
  valiutnye tsennosti foreign currency valuables
vedomstvo department
velenie command
  obiazatel'noe velenie obligatory order
Verkhovnyi sovet Supreme Soviet
  Verkhovnyi sud Supreme Court
vina fault
  vina dolzhnika fault of the debtor
  vinovnoe deistvie culpable action
vklad deposit
  vkladchik depositor
  vklady v sberegatel'nye kassy deposits in savings banks
vladet' to possess
vlast' authority
vnebrachnyi rebenok child born out of wedlock
vnuchka granddaughter
  vnuk grandson
voditel'skoe pravo driver's license
volevoi volitional
volia will
voskhodiashchaia liniia ascendant lineage
vospolnitel'nyi discretionary
  vospolnitel'naia norma discretionary norm
vosproizvedenie reproduction
vosstanovlenie restoration
  vosstanovlenie polozheniia, sushchestvovavshego do narusheniia prava resto-

ration of the situation existing prior to the violation of a right
vozmeshchenie compensation
vozmeshchenie poter' compensation for losses
vozmeshchenie prichinennykh ubytkov compensation for losses caused
vozmeshchenie ubytkov compensation of damages
vozmezdnost' reciprocity
vozmezdnyi reciprocal
vozmezdnaia sdelka compensatory transaction
vozmezdnoe otchuzhdenie reciprocal alienation
vozmezdnyi dogovor compensatory contract
vozmezdnykh nachalakh, na reciprocity
voznagrazhdenie remuneration
voznikat' to arise
vozniknovenie beginning, origin
vozniknovenie i prekrashchenie iuridicheskikh lits origin and termination of juridical persons
vozniknovenie obiazatel'stva origin of an obligation
vremia time
vremennoe otsutstvie temporary absence
vspomogatel'nyi auxiliary
vspomogatel'noe pomeshchenie auxiliary premises
vstrechnyi mutual, counter
vstrechnoe obiazatel'stvo mutual obligation
vstrechnoe trebovanie counterclaim
vstrechnoe udovletvorenie mutual satisfaction
vstrechnyi isk counterclaim
vstupat' to enter
vstupat' v obiazatel'stvo to enter into an obligation
vstupit' v deistvie to take effect
vydeliat' single out; define
vyselenie eviction
vysylka banishment
vzaimnyi mutual
vzaimnoe pogashenie vstrechnykh trebovanii mutual liquidation of counterclaims
vzaimnoe zaiavlenie mutual declaration
vzaimnye prava i obiazannosti mutual rights and obligations
vzaimnyi dogovor mutual contract
vzyskanie penalty; exaction
vzyskanie po ispolnitel'nomu listu recovery under a court judgment
vzyskanie ubytkov recovery of damages

zabluzhdenie mistake
sushchestvennoe zabluzhdenie substantive mistake
zachet setoff
zachetnaia neustoika calculated liquidated damages
zadatok earnest
zaem loan
zaemnye otnosheniia loan relationships
zaemshchik borrower
ZAGS (Zapis' aktov grazhdanskogo sostoianiia) registry of acts of civil status
zaiavka claim, demand
zaiavlenie statement, declaration, application
zaiavlenie zainteresovannykh lits petition of interested parties

iskovoe zaiavlenie complaint
zaimodavets lender
zaimoobrazno as a loan
zakaz order
zakaz pokupatelia order of the buyer
zakazchik client
zakonnaia neustoika lawful liquidated damages
zakonnoe pravo legal right
zakonnoe predpolozhenie legal presumption
zakonnost' legality
zakonnyi interes lawful interest
predusmotrennyi zakonom stipulated by law
zakonodatel'stvo legislation
zakonodatel'stvo o nedrakh legislation on minerals
zakonodatel'stvo, grazhdansko-pravovoe civil law legislation
zakonodatel'stvo, lesnoe forestry legislation
zakonodatel'stvo, vodnoe water legislation
zakonodatel'stvo, zemel'noe land legislation
zakonodatel'stvo, zhilishchnoe housing legislation
zalog pledge
zalog, osnovannyi na dogovore pledge based on contract
zalogodatel' pledgor
zalogoderzhatel' pledgee
zalogovyi bilet pawnticket
zalozhennoe imushchestvo pledged property
zapis' registration
zapis' materi rebenka registration of a child's mother
zapis' ottsa rebenka registration of a child's father
zashchita chesti i dostoinstva grazhdan i organizatsii defense of the honor and dignity of citizens and organizations
zashchita chesti i dostoinstva protection of honor and dignity
zashchita interesov lichnosti protection of personal interests
zashchita prav avtora protection of author's rights
zastrakhovannoe imushchestvo insured property
zastroishchik, individual'nyi individual builder
zaveshchanie will
zaveshchatel' testator
zaveshchatel'nyi otkaz legacy
zemel'noe zakonodatel'stvo land legislation
zhaloba complaint
zhenikh bridegroom
zhilishchno-ekspluatatsionnaia kontora housing-management office
zhilishchno-stroitel'nyi kooperativ housing construction cooperative
zhilishchnoe zakonodatel'stvo housing legislation
zhilishchnyi fond housing fund
zhilishchnyi fond zhilishchno-stroitel'nykh kooperativov housing fund of housing construction cooperatives
zhilishchnyi kodeks housing code
zhiteiskii obychai social custom
zhitel'stvo residence
postoiannoe zhitel'stvo permanent residence
zloupotreblenie abuse, misuse
zloupotreblenie narkoticheskimi veshchestvami abuse of narcotic substances
zloupotreblenie spirtnymi napitkami abuse of alcoholic beverages

English-Russian Glossary

ability sposobnost′ (*see also* capacity)
absence otsutstvie
  absence on valid grounds otsutstvie po uvazhitel′nym prichinam
abuse, misuse zloupotreblenie
  abuse of alcoholic beverages zloupotreblenie spirtnymi napitkami
  abuse of narcotic substances zloupotreblenie narkoticheskimi veshchestvami
acceptance priniatie
  acceptance form of payments aktseptnaia forma raschetov
  acceptance of an estate priniatie nasledstva
  acceptance of children for rearing priniatie detei na vospitanie
accession prisoedinenie
accident avariia; krushenie
account (record) schet; (description) otchet
  accounting (bookkeeping) bukhgalteriia, schetovodstvo; (calculation, reckoning) raschet
  accounting relationship raschetnoe pravootnoshenie
  accounting/banking services raschetno-kassovoe obsluzhivanie;
accountable otvetstvennyi (*see also* responsibility, liability)
  accountability podotchetnost′
  economic accountability khoziaistvennyi raschet (khozraschet)
accrual prirashchenie
  accrual of shares prirashchenie dolei
acquiescence molchanie
act (legal) akt; (deed) deianie; postupok
  acts of codified nature akty kodifikatsionnogo kharaktera
  action deistvie
  actual deistvitel′nyi; fakticheskii
  actual performance of an obligation real′noe ispolnenie obiazatel′stva
addition dopolnenie
administration upravlenie; administratsiia
  administrative administrativnyi
  administrative act administrativnyi akt
  administrative law administrativnoe pravo
  administrative law relationship administrativnoe pravootnoshenie
  administrative management administrativnoe upravlenie
  administrative organ organ upravleniia
adoption usynovlenie
  adoptee usynovlennyi
  adoptive parent usynovitel′
adulthood sovershennoletie
advance avans
affinity svoistvo
age vozrast

age of marriage brachnyi vozrast
attainment of the age of marriage dostizhenie brachnogo vozrasta
agent poverennyi
agreement soglashenie
agreement of the parties soglashenie storon
agreement on pledge soglashenie o zaloge
agreement on suretyship soglashenie o poruchitel′stve
agreement on the liquidation of damages soglashenie o neustoike
agreement-obligation poruchenie-obiazatel′stvo
alienate otchuzhdat′
alienation of an article iz″iatie veshchi
alienator otchuzhdatel′
alimony alimenty
alternative al′ternativnyi
alternative liquidated damages al′ternativnaia neustoika
analogy analogiia
analogy of law analogiia zakona
analogy of law and legislation analogiia prava i zakona
annul annulirovat′
annulment of a contract rastorzhenie dogovora
annulment of a marriage nedeistvitel′nost′ braka; priznanie braka nedeistvitel′nym
appearance iavka
application (use) primenenie
apply legislation primeniat′ zakonodatel′stvo
application (request) zaiavlenie
application for carriage zaiavka na perevozku
arbitrate reshat′ v arbitrazhnom poriadke
arbitration tribunal treteiskii sud
arbitrazh arbitrazh
area okrug
arise voznikat′
article (item) predmet; (thing) veshch′; (paragraph) stat′ia
article of everyday use predmet obikhoda
article of personal use predmet lichnogo potrebleniia
articles of domestic and everyday use predmety domashnego obikhoda i byta
assembly of representatives sobranie upolnomochennykh
authentic interpretation autentichnoe tolkovanie
author's contract avtorskii dogovor
authority vlast′
auxiliary obligation dopolnitel′noe obiazatel′stvo
auxiliary premises vspomogatel′noe pomeshchenie
average adjuster dispasher
average statement dispasha

bailiff of the court sudebnyi ispolnitel′
banishment vysylka
basis osnovanie
bearer nositel′
bearer of a subjective right nositel′ sub″ektivnogo prava
bilateral dvustoronnii
bilateral contract dvustoronnii dogovor
bilateral restitution dvustoronniaia restitutsiia
bill for services rendered raschet za okazannye uslugi

bill of lading konosament
birth rozhdenie
  birth certificate svidetel′stvo o rozhdenii
  birth register kniga zapisei rozhdenii
blood test ekspertiza krovi
borrower zaemshchik
branch of law otrasl′ prava
breadwinner kormilets
bride nevesta
  bridegroom zhenikh
bringing a suit in court pred″iavlenie iska v sude
broad interpretation rasprostranitel′noe tolkovanie
building insurance strakhovanie stroenii
buyer pokupatel′
by petition of interested parties po zaiavleniiu zainteresovannykh lits

calculated liquidated damages zachetnaia neustoika
capacity sposobnost′ (*see also* ability)
capital kapital′nyi
  capital construction kapital′noe stroitel′stvo
  capital repair of property kapital′nyi remont imushchestva
care ukhod
carrier perevozchik
  carriage contract dogovor perevozki
carrying out of the enforcement osushchestvlenie prinuzhdeniia
cash compensation denezhnaia kompensatsiia
cash payment denezhnyi raschet
cause prichina
  causal relationship prichinnaia sviaz′
  causer of harm prichinitel′ vreda
  causing of harm prichinenie vreda
change izmenenie
characteristic osobennost′
  characteristics of a juridical person priznaki iuridicheskogo litsa
charterer frakhtovatel′
check chek
  check giver chekodatel′
  check holder chekoderzhatel′
child rebenok
  children deti
  child(ren) born in a marriage rebenok (deti), rozhdennyi(ye) v brake
  child(ren) born out of wedlock vnebrachnyi(ye) rebenok (deti)
  child(ren) conceived in a marriage rebenok (deti), zachatyi(ye) v brake
  childcare institution detskoe uchrezhdenie
circumstance obstoiatel′stvo
  circumstances excluding liability of the debtor obstoiatel′stva, iskliuchaiushchie otvetstvennost′ dolzhnika
citizen grazhdanin
civil grazhdanskii
  civil code grazhdanskii kodeks
  civil activity grazhdanskii oborot
  civil law grazhdanskoe pravo
  civil law act grazhdansko-pravovoi akt
  civil law dispute grazhdansko-pravovoi spor

civil law legislation grazhdansko-pravovoe zakonodatel'stvo
civil law method of regulation grazhdansko-pravovoi metod regulirovaniia
civil law regulation grazhdansko-pravovoe regulirovanie
civil law relationship grazhdanskoe pravootnoshenie
civil law transaction grazhdansko-pravovaia sdelka
civil legal capacity grazhdanskaia pravosposobnost'
civil legal relationship grazhdanskoe pravootnoshenie
civil liability grazhdanskaia otvetstvennost'
civil procedure grazhdanskii protsess
civil proceedings grazhdanskoe sudoproizvodstvo
claim for damages vzyskanie ubytkov
clarification vyiasnenie
client doveritel'; zakazchik
code kodeks; ustav
co-owner sosobstvennik
coincidence of the persons of the debtor and the creditor sovpadenie dolzhnika i kreditora v odnom litse
collective farm kolkhoz
collective farm law kolkhoznoe pravo
commission agent komissioner
commission store komissionnyi magazin
commodity circulation tovarnoe obrashchenie
commodity-money form tovarno-denezhnaia forma
common household obshchee khoziaistvo; sovmestnoe khoziaistvo
common joint ownership obshchaia sovmestnaia sobstvennost'
common joint property obshchaia sovmestnaia sobstvennost'
common property obshchaia sobstvennost'
common shared property obshchaia dolevaia sobstvennost'
communal life (*see also* hostel) obshchezhitie
communist morality kommunisticheskaia moral'
compensation vozmeshchenie
compensation for losses vozmeshchenie poter'
compensation for losses caused vozmeshchenie prichinennykh ubytkov
compensation of damages vozmeshchenie ubytkov
compensatory contract vozmezdnyi dogovor
compensatory transaction vozmezdnaia sdelka
competence kompetentsiia
competent organ kompetentnyi organ
complaint iskovoe zaiavlenie; zhaloba
comprehensive law svodnyi zakon
comprehensive law on the regulation and confiscation of property svodnyi zakon o rekvizitsii i konfiskatsii imushchestva
conclude a transaction zakliuchat' sdelku
conclusion vyvod
condition uslovie
conditional uslovnyi
conditional sentence to deprivation of freedom uslovnoe osuzhdenie k lisheniiu svobody
conditional transaction uslovnaia sdelka
conditions of a work contract usloviia dogovora podriada
confirmation podtverzhdenie
confiscation of property konfiskatsiia imushchestva
connivance popustitel'stvo
consensual konsensual'nyi

consensual contract konsensual'nyi dogovor
consensual transaction konsensual'naia sdelka
consent soglasie
consideration rassmotrenie
constitution konstitutsiia
constitutional law konstitutsionnoe pravo
consumer services bytovoe obsluzhivanie
contest osparivat'
contiguous branches of law smezhnye otrasli prava
contract dogovor
contract for carriage of goods dogovor perevozki gruza
contract for the carriage of passengers and baggage dogovor perevozki passazhirov i bagazha
contract for the pledge of a house dogovor zaloga doma
contract for the rental of housing dogovor zhilishchnogo naima
contract for the rental of living accommodations dogovor naima zhilogo pomeshcheniia
contract for the sale of a house with the condition that the seller is maintained for life dogovor prodazhi doma s usloviem pozhiznennogo soderzhaniia prodavtsa
contract for use of living space in houses belonging to citizens by right of personal ownership dogovor za pol'zovanie zhiloi ploshchadi v domakh, prinadlezhashchikh grazhdanam na prave lichnoi sobstvennosti
contract in favor of a third party dogovor v pol'zu tret'ego litsa
contract of agency dogovor porucheniia
contract of bank loan dogovor bankovskoy ssudy
contract of barter dogovor meny
contract of carriage by sea dogovor morskoi perevozki
contract of carriage for local carriage dogovor perevozki v mestnom soobshchenii
contract of commission agency dogovor komissii
contract of credit sale dogovor prodazhi v kredit
contract of delivery dogovor postavki
contract of deposit for safe custody dogovor khraneniia
contract of deposit of unidentified things dogovor khraneniia s obezlicheniem veshchei
contract of domestic hire dogovor bytovogo prokata
contract of gift dogovor dareniia
contract of insurance dogovor strakhovaniia
contract of lease of employment living accommodations dogovor naima sluzhebnogo zhilogo pomeshcheniia
contract of loan dogovor zaima
contract of maintenance for life dogovor pozhiznennogo soderzhaniia
contract of pledge dogovor o zaloge
contract of property lease dogovor arendy imushchestva
contract of property rent dogovor prokata imushchestva
contract of public competition dogovor o konkurse
contract of purchase and sale of dwellings dogovor kupli-prodazhi zhilykh domov
contract of retail sale dogovor roznichnoi kupli-prodazhi
contract of supply dogovor postavki
contract of the gratuitous use of property dogovor bezvozmezdnogo pol'zovaniia imushchestvom
contract of time-charter dogovor taim-chartera

contract of towage dogovor buksirovki
contractor podriadchik
contractual dogovornyi
contractual liquidated damages dogovornaia neustoika
contractual obligation dogovornoe obiazatel'stvo
contractual price dogovornaia tsena
contractual relationship dogovornoe otnoshenie
copyright avtorskoe pravo
cost stoimost'
cost accounting khozraschet
cost price sebestoimost'
counter vstrechnyi
counter obligation vstrechnoe obiazatel'stvo
counter satisfaction vstrechnoe udovletvorenie
counter-warrant raznariadka
counterclaim vstrechnoe trebovanie; vstrechnyi isk
court sud
court annuls a decision sud otmeniaet reshenie
court judgment ispolnitel'nyi list
court practice sudebnaia praktika
crafts and trades kustarno-remeslennye promysly
credit kredit
credit legal relationships kreditnye pravootnosheniia
credit relationship kreditnoe pravootnoshenie
creditor kreditor
criminal ugolovnyi
criminal law ugolovnyi zakon
criminal liability ugolovnaia otvetstvennost'
culpable action vinovnoe deistvie
curatorship popechitel'stvo
custodian khranitel'
custodial parent roditel', s kem prozhivaet rebenok
custom obychai

damage ushcherb
damage caused by criminal actions ushcherb, prichinennyi prestupnymi deistviiami
liquidated damages neustoika
debt dolg
debtor dolzhnik
debtor under an obligation dolzhnik po obiazatel'stvu
deceased; decedent umershii
declaration priznanie
declaration of a citizen as missing and declaring him dead priznanie grazhdanina bezvestno otsutstvuiushchim i obiavlenie ego umershim
declaration of a right priznanie prava
declared value of baggage ob"iavlennaia tsennost' bagazha
decree postanovlenie
defamed person oporochivaemoe litso
defective products zabrakovannaia produktsiia
defendant otvetchik
defense of the honor and dignity of citizens and organizations zashchita chesti i dostoinstva grazhdan i organizatsii

delay of performance prosrochka ispolneniia
deliberate intent priamoi umysel
  deliberately umyshlenno
delimitation otgranichenie
demurrage demeredzh
department vedomstvo
dependents izhdiventsy
deposit for safe custody khranenie
  depositor poklazhedatel'; vkladchik
  deposits in savings banks vklady v sberegatel'nye kassy
  deposits of money by citizens denezhnye vklady grazhdan
deprivation of freedom lishenie svobody
descendant lineage niskhodiashchaia liniia
deviation otstuplenie
digest of laws svod zakonov
direct carriage priamaia perevozka
discovery obnaruzhenie, otkrytie
discretion usmotrenie
  discretionary norm vospolnitel'naia norma
dispose (of) rasporiazhat'sia
  disposition rasporiazhenie
  dispositive capacity deesposobnost'
  dispositive norm dispozitivnaia norma
  dispositivity dizpozitivnost'
dispute spor
  disputed relationship spornoe otnoshenie
dissemination rasprostranenie
division razdelenie
divorce razvod
  divorce lawsuit delo o rastorzhenii braka
draw up a will sostavliat' zaveshchanie
  draw up power of attorney sostavliat' doverennost'
driver's license voditel'skoe pravo

earned income trudovoi dokhod
earnest zadatok
economic accountability khoziaistvennyi raschet; khozraschet
  economic activity ekonomicheskii oborot; khoziaistvennaia deiatel'nost'
  economic development khoziaistvennoe stroitel'stvo;
  economic legislation khoziaistvennoe zakonodatel'stvo
  economic organization khoziaistvennaia organizatsiia
  economic relationships ekonomicheskie otnosheniia
edict ukaz
employment contract trudovoi dogovor
  employment insurance strakhovanie trudosposobnosti
enforcement prinuzhdenie
  enforced exchange prinuditel'nyi obmen
  enforced execution prinuditel'noe ispolnenie
  enforced transfer of an article prinuditel'naia peredacha veshchi
enter into obligations vstupat' v obiazatel'stva
enterprise predpriiatie
equivalent-reciprocal character ekvivalentno-vozmezdnyi kharakter
escheat perekhod nasledstva k gosudarstvu
estate nasledstvo

estate-leaver nasledodatel'
event sobytie
everyday needs bytovye nuzhdy
eviction vyselenie
examination of a case on its merits rassmotrenie dela po sushchestvu
exchange obmen
exchange of a living accommodation obmen zhilogo pomeshcheniia
exclusionary liquidated damages iskliuchitel'naia neustoika
executive otvetstvennyi rabotnik
exercise of a right osushchestvlenie prava
exercise of parental rights osushchestvlenie roditel'skikh prav
exile ssylka
existing code deistvuiushchii kodeks
existing legislation deistvuiushchee zakonodatel'stvo
expert examination; expert testimony ekspertiza
expiration of the period of validity of a contract istechenie sroka deistviia dogovora
explanation raz"iasnenie
extended interpretation rasshiritel'noe tolkovanie
extent of legal liquidated damages razmer zakonnoi neustoiki
extra charge of penalty nachislenie peni

factual composition fakticheskii sostav
failure to observe nesobludenie
failure to perform an obligation neispolnenie obiazatel'stva
family law semeinoe pravo
fault vina
fault of the debtor vina dolzhnika
fees poshlina
file a declaration podat' zaiavlenie
filing of a complaint podacha zhaloby
filing of claims and suits pred"iavlenie pretenzii i iskov
financial law finansovoe pravo
financial operations raschety
fine shtraf (*see also*, penalty)
fined liquidated damages shtrafnaia neustoika
food program prodovol'stvennaia programma
force majeure nepreodolimaia sila
foreign currency valuables valiutnye tsennosti
foreign state insurance ingosstrakh
forensic medicine sudebno-meditsinskaia ekspertiza
forestry legislation lesnoe zakonodatel'stvo
formalize documents oformliat' dokumenty
foster care patronat
fraud obman
fulfill obligations ispolniat' obiazannosti
function of securing obespechitel'naia funktsiia
fundamental law osnovnoi zakon
fundholder fondoderzhatel'

general contractor general'nyi podriadchik
general guardian opekun nad imushchestvom
general provisions, basic positions osnovnye polozheniia
General Statute on Ministries of the USSR Obshchee polozhenie o ministerstvakh SSSR

generic trait rodovoi priznak
gift darenie
give a promissory note vydavat' dolgovuiu raspisku
giver daritel'
goods of material value material'nye tsennosti
government pravitel'stvo
grammatical interpretation grammaticheskoe tolkovanie
grand cabotage bol'shoi kabotazh
granddaughter vnuchka
  grandfather dedushka
  grandmother babushka
  grandson vnuk
gratuitous transaction bezvozmezdnaia sdelka
gross negligence grubaia neostorozhnost'
grounds osnovaniia
  grounds for development osnovaniia vozniknoveniia
group warrant gruppovyi nariad
guarantee garantiia
  guarantee period garantiinyi srok
guardian opekun
  guardianship opeka, popechitel'stvo
guiding explanations rukovodiashchie raz"iasneniia
guilt vina

heir naslednik
  heirs ab intestato nasledniki po zakonu
  heirs of the first (second) category nasledniki pervoi (vtoroi) ocheredi
hidden defect skrytyi nedostatok
historical interpretation istoricheskoe tolkovanie
holder of an order orderoderzhatel'
hooligan behavior khuliganskie deistviia
hostel obshchezhitie
house dom
  house and lot domovladenie
  household property insurance strakhovanie domashnego imushchestva
  housing code zhilishchnyi kodeks
  housing construction cooperative zhilishchno-stroitel'nyi kooperativ
  housing fund zhilishchnyi fond
  housing fund of housing construction cooperatives zhilishchnyi fond zhilishchno-stroitel'nykh kooperativov
  housing legislation zhilishchnoe zakonodatel'stvo
  housing-management office zhilishchno-ekspluatatsionnaia kontora

immunity immunitet
imperative norms of law imperativnye normy zakona
implied action konkliudentnoe deistvie
impossibility of performance nevozmozhnost' ispolneniia
improper nenadlezhashchii
  improper party nenadlezhashchee litso
  improper performance of an obligation nenadlezhashchee ispolnenie obiazatel'stva
  improper time nenadlezhashchee vremia
inability to work netrudosposobnost'
indefinitely bessrochno

independent samostoiatel'nyi
  independent account samostoiatel'naia smeta
  independent balance samostoiatel'nyi balans
  independent property liability samostoiatel'naia imushchestvennaia otvetstvennost'
information spravka, informatsiia
inheritance nasledovanie
  inheritance law nasledstvennoe pravo
initiation of execution proceedings pred"iavlenie lista po vzyskaniiu
inner waterway transport vnutrennevodnyi transport
instruction ukazanie
  instructive orders instruktivnye ukazaniia
instruments of civil status akty grazhdanskogo sostoianiia
insurance strakhovanie
  insurance beneficiary litso, v pol'zu kotorogo zakliucheno strakhovanie
  insurance evaluation strakhovaia otsenka
  insurance fee strakhovyi vznos
  insurance interest strakhovyi interes
  insurance legal relationship strakhovoe pravootnoshenie
  insurance premium strakhovaia premiia
  insurance risk strakhovyi risk
  insurance sum strakhovaia summa
  insured event strakhovyi sluchai
  insured property zastrakhovannoe imushchestvo
  insurer strakhovatel'
intention umysel
interest on deposits protsenty na vklady
international treaty mezhdunarodnyi dogovor
interpretation tolkovanie
  authentic interpretation autentichnoe tolkovanie
  broad interpretation rasprostranitel'noe tolkovanie
  extended interpretation rasshiritel'noe tolkovanie
  historical interpretation istoricheskoe tolkovanie
  interpretation of civil law norms tolkovanie grazhdansko-pravovykh norm
  literal interpretation bukval'noe tolkovanie
  logical interpretation logicheskoe tolkovanie
  systematic interpretation sistematicheskoe tolkovanie
invalid nedeistvitel'nyi
  invalidity nedeistvitel'nost'
inventor's certificate avtorskoe svidetel'stvo

join as co-defendants privlekat' v kachestve sootvetchikov
joint and several debtors solidarnye dolzhniki
  joint and several property responsibility for obligations solidarnaia imushchestvennaia otvetstvennost' po obiazatel'stvam
judicial sudebnyi
  judicial proceedings sudoproizvodstvo
juridical iuridicheskii
  juridical fact iuridicheskii fakt
  juridical person iuridicheskoe litso
jurisdiction podsudnost'
  jurisdiction of civil cases podsudnost' grazhdanskikh del

keep a common household vesti obshchee khoziaistvo
kinship rodstvo

labor trud
labor contract trudovoi dogovor
labor law trudovoe pravo
lack of sufficient knowledge nedostatochnaia osvedomlennost′
land legislation zemel′noe zakonodatel′stvo
late payment for goods nesvoevremennaia oplata produktsii
late payment of rent prosrochka platezha kvartplaty
law pravo; zakon
law of discovery pravo na otkrytie
law of invention izobretatel′skoe pravo
law of obligations obiazatel′stvennoe pravo
law of ownership pravo sobstvennosti
law of succession nasledstvennoe pravo
lawful pravomernyi
lawful action pravomernoe deistvie
lawful interest zakonnyi interes
lawful marriage deistvitel′nyi brak
lawyer iurist
lease arendnyi dogovor
to lease arendovat′
legacy zaveshchatel′nyi otkaz
legality zakonnost′
legal act pravovoi akt
legal capacity pravosposobnost′
legal conduct pravovoi postupok
legal consequences pravovye posledstviia
legal custom iuridicheskii obychai
legal device pravovoi priem
legal liquidated damages zakonnaia neustoika
legal means pravovye sredstva
legal order pravoporiadok
legal presumption zakonnoe predpolozhenie
legal relationship pravootnoshenie
legal right zakonnoe pravo
legal subjectivity pravosub″ektnost′
legal succession pravopreemstvo
legally established obstacles ustanovlennye zakonom prepiatstviia
legislation zakonodatel′stvo
legislation in force deistvuiushchee zakonodatel′stvo
lender zaimodavets
lessor naimodatel′
letter of authority doverennost′
letter of credit akkreditiv
levy of judgment in execution against pledged article obrashchenie vzyskaniia na zalozhennuiu veshch′
liability otvetstvennost′
liability for breach of obligations otvetstvennost′ za narushenie obiazatel′stv
liability for harm caused by minors and incapable parties otvetstvennost′ za vred, prichinennyi nesovershennoletnimi i nedeesposobnymi
liability for harm caused by the actions of officials otvetstvennost′ za vred, prichenennyi deistviiami dolzhnostnykh lits
liability for harm caused on the grounds of increased danger otvetstvennost′ za vred, prichinennyi istochnikom povyshennoi opasnosti

liability for injury to the health or for the death of a citizen otvetstvennost′ za povrezhdenie zdorov′ia i smert′ grazhdanina
liable party obiazannaia storona
limitation of liability ogranichenie otvetstvennosti
life insurance strakhovanie zhizni
lifelong maintenance pozhiznennoe soderzhanie
lineage liniia
liquidation likvidatsiia
liquidate a debt pogashat′ dolg
liquidation of damages neustoika
liquidated damages for late payment of rent neustoika za prosrochku platezha kvartplaty
liquidation of a legal person likvidatsiia iuridicheskogo litsa
list of disagreements protokol raznoglasii
literal interpretation bukval′noe tolkovanie
loan relationships zaemnye otnosheniia
logical interpretation logicheskoe tolkovanie
long-term credit dolgosrochnyi kredit
losses ubytki

mail-order trade posylochnaia torgovlia
main administration glavnoe upravlenie
maintenance (alimony) alimenty
maintenance obligation alimentnoe obiazatel′stvo
maintenance payer platel′shchik alimentov
mandatory obiazatel′nyi; povelitel′nyi
mandatory norm povelitel′naia norma
marital union brachnyi soiuz
material damage real′nyi ushcherb
material objects material′nye predmety
material support material′noe obespechenie
material-technical supply material′no-tekhnicheskoe snabzhenie
maternity materinstvo
means of enforcement mery prinuditel′nogo vozdeistviia
means of payment sposob platezha
measure mera
measure of liability mera otvetstvennosti
measures to reconcile spouses mery k primireniiu suprugov
mental illness dushevnaia bolezn′
mental retardation slaboumie
merger sliianie
method of authority metod vlasti
minor nesovershennoletnii
mischief ozorstvo
missing person bezvestno otsutstvuiushchii
mistake zabluzhdenie
misuse zloupotreblenie
mixed insurance smeshannoe strakhovanie
model obrazets
model charter primernyi ustav
model contract tipovoi dogovor
monogamy edinobrachie
morality nravstvennost′
mutual agreement of the parties soglashenie storon
mutual assistance benefit fund kassa obshchestvennoi vzaimopomoshchi

mutual contract vzaimnyi dogovor
mutual declaration vzaimnoe zaiavlenie
mutual liquidation of counterclaims vzaimnoe pogashenie vstrechnykh trebovanii
mutual rights and obligations vzaimnye prava i obiazannosti

name imia, naimenovanie
nationality natsional'nost'; natsional'naia prinadlezhnost'
natural calamity stikhiinoe bedstvie; stikhinnoe iavlenie
natural objects prirodnye ob"ekty
necessary neobkhodimyi
necessary participant neobkhodimyi uchastnik
neediness nuzhdaemost'
negligence neostorozhnost'
negligently po neostorozhnosti
non-fulfillment nevypolnenie
non-obligatory interpretation neobiazatel'noe tolkovanie
non-property relationships neimushchestvennye otnosheniia
norm norma
normative act normativnyi akt
norms of civil law normy grazhdanskogo prava
notarial notarial'nyi

object of civil law regulation predmet grazhdansko-pravovogo regulirovaniia
object of copyright ob"ekt avtorskogo prava
object of civil law predmet grazhdanskogo prava
obligation obiazatel'stvo
obligated person obiazannoe litso
obligation involving citizens obiazatel'stvo s uchastiem grazhdan
obligation of parties for a breach of contract otvetstvennost' storon za narushenie dogovora
obligation to compensate for harm caused obiazannost' vozmestit' prichinennyi vred
obligations arising as the result of causing harm obiazatel'stva, voznikaiushchie vsledstvie pricheneniia vreda
obligations of the lessee obiazannosti nanimatelia
obligatory insurance obiazatel'noe strakhovanie
obligatory interpretation obiazatel'noe tolkovanie
obligatory order obiazatel'noe velenie
obligatory share of the estate obiazatel'naia dolia v nasledstve
obvious defect iavnyi nedostatok
office dolzhnost'
official dolzhnostnoe litso
official travel abroad komandirovanie za granitsu
operating expenses ekspluatatsionnye raskhody
operative management operativnoe upravlenie
oral ustnyi
oral transaction ustnaia sdelka
order rasporiazhenie
order for residence in an apartment order na zaselenie kvartiry
order of the buyer zakaz pokupatelia
organs for the defense of rights pravookhranitel'nye organy
origin proiskhozhdenie
origin and termination of juridical persons vozniknovenie i prekrashchenie

iuridicheskikh lits
origin of an obligation vozniknovenie obiazatel'stva
owner sobstvennik; vladelets
owner of goods gruzovladelets
ownerless property beskhoziainoe imushchestvo
ownership sobstvennost'

parent roditel'
participants in a transaction uchastniki sdelki
party storona
parties to a contract for the lease of living accommodations storony v dogovore naima zhilogo pomeshcheniia
passport
identity card passport
patent patent
paternity ottsovstvo
pawnshop lombard
pawnticket zalogovyi bilet
payment plata; uplata; platezh
payment by installments rassrochka platezha
payment demand platezhnoe trebovanie
payment for communal services plata za kommunal'nye uslugi
payment of alimony uplata alimentov
payment of liquidated damages uplata neustoiki
payment order (authorization) platezhnoe poruchenie
payments for products raschety za produktsiiu
penalty shtraf
penalty liquidated damages shtrafnaia neustoika
per procuration po doverennosti
period (of time) srok
period of a contract srok dogovora
period of limitation iskovaia davnost'
period of performance srok ispolneniia
performance of obligations ispolnenie obiazannostei
permanence bessrochnost'
permanent residence postoiannoe zhitel'stvo
person litso
person empowered by law to act upravomochennoe litso
person to whom maintenance payments must be paid alimentnoobiazannoe litso
person under obligation obiazannoe litso
personal insurance lichnoe strakhovanie
personal non-property valuable lichnoe neimushchestvennoe blago
personal property lichnaia sobstvennost'
personal separate property lichnaia razdel'naia sobstvennost'
petit cabotage malyi kabotazh
petition of interested parties zaiavlenie zainteresovannykh lits
petty everyday transactions melkie bytovye sdelki
place of opening of an inheritance (of succession) mesto otkrytiia nasledstva
place of residence mesto zhitel'stva
plaintiff istets
plan plan
plan act of distribution planovyi akt raspredeleniia
plan act of supply planovyi akt na postavku; planovyi dogovor postavki

plan basis of a procurement contract planovye osnovaniia dogovora kontraktatsii
plan contract planovyi dogovor
plan discipline planovaia distsiplina
pledge zalog
pledge based on contract zalog, osnovannyi na dogovore
pledged property imushchestvo, zalozhennoe
pledged property shtraf za postavku
pledgee zalogoderzhatel'
pledgor zalogodatel'
plurality of persons mnozhestvennost' lits
possess vladet'
possessor of right of operative management obladatel' prava operativnogo upravleniia
power of attorney doverennost'
premise predposylka
presumption prezumptsiia
prevention of actions violating a right presechenie deistvii, narushaiushchikh pravo
primary obligation glavnoe obiazatel'stvo
principal doveritel'
principle printsip
principles (bases, fundamentals) osnovy
principles of civil legislation osnovy grazhdanskogo zakonodatel'stva
principles of civil procedure osnovy grazhdanskogo sudoproizvodstva
principles of corrective labor legislation osnovy ispravitel'no-trudovogo zakonodatel'stva
principles of housing legislation osnovy zhilishchnogo zakonodatel'stva
principles of legislation on marriage and the family osnovy zakonodatel'stva o brake i sem'i
procedure poriadok
civil procedure grazhdanskii protsess
procedure by permission razreshitel'nyi poriadok
procedure for allocating living accommodations poriadok predostavleniia zhilykh pomeshchenii
procedure for collecting and paying maintenance poriadok vzyskaniia i uplaty alimentov
procurator prokuror
procurement contract dogovor kontraktatsii
production association proizvodstvennoe ob"edinenie
production housing-repair association proizvodstvennoe zhilishchno-remontnoe ob"edinenie
production housing-repair trust proizvodstvennyi zhilishchno-remontnyi trest
production-economic activity proizvodstvenno-khoziaistvennaia deiatel'nost'
profession professiia
professional tools predmety professional'nogo truda
profit pribyl'
promissory note dolgovaia raspiska
promissory obligations obiazatel'stvennye trebovaniia
prompt payment svoevremennaia uplata
promulgation izdanie
proof dokazatel'stvo
proper quality nadlezhashchee kachestvo

property sobstvennost′
  property amassed during marriage imushchestvo, nazhitoe vo vremia braka
  property insurance imushchestvennoe strakhovanie
  property relationships imushchestvennye otnosheniia
  property situation (status) imushchestvennoe polozhenie
  property-hire contract dogovor imushchestvennogo naima
protection of an estate okhrana nasledstvennogo imushchestv
  protection of author's rights zashchita prav avtora
  protection of honor and dignity zashchita chesti i dostoinstva
  protection of personal interests zashchita interesov lichnosti
  protection of social interests okhrana obshchestvennykh interesov
provisions of civil law polozheniia grazhdanskogo prava
psychological disturbance psikhicheskoe rasstroistvo
publication opublikovani, vypusk v svet
purchase price pokupnaia tsena
  purchase-sale contract dogovor kupli-prodazhi
  purchaser priobretatel′
purposeful tselenapravlennyi

race rasa; rasovaia prinadlezhnost′
range of persons krug lits
rates of payment for transport stavki provoznykh platezhei
raw materials and supplies syr′e i materialy
real transaction real′naia sdelka
reciprocal vozmezdnyi
  reciprocal alienation vozmezdnoe otchuzhdenie
  reciprocity vozmezdnoe nachalo
  reciprocity vozmezdnost′; na vozmezdnykh nachalakh
recission of a contract rastorzhenie dogovora
recovery under a court judgment vzyskanie po ispolnitel′nomu listu
refraining from action vozderzhanie ot sovershenia deistviia
refusal of the seller to transfer an article uklonenie prodavtsa ot peredachi veshchi
regime of common property rezhim obshchei sobstvennosti
registration of a child's father (mother) zapis′ ottsa (materi) rebenka
registry of acts of civil status ZAGS
relative by birth rodstvennik po proiskhozhdeniiu
release from liability osvobozhdenie ot otvetstvennosti
remuneration voznagrazhdenie
renew a contract vozobnovit′ dogovor
repeal otmena
representation predstavitel′stvo
represented party predstavliaemoe
reserve an accommodation bronirovat′ pomeshchenie
residence permit razreshenie na propisku
resolution of a dispute razreshenie spora
respect for the individual uvazhenie lichnosti
responsibility (*see also* accountability; liability) otvetstvennost′
restoration of the situation existing prior to the violation of a right vosstanovlenie polozheniia, sushchestvovshego do narusheniia prava
restrictive interpretation ogranichitel′noe tolkovanie
retortion retortsiia
retraction oproverzhenie
  retraction of defamatory information oproverzhenie porochashchikh svedenii
  retraction of false information oproverzhenie nepravil′nykh svedenii

retroactive force obratnaia sila
revision of legislation obnovlenie zakonodatel'stva
revocation of an adoption otmena usynovleniia
right pravo
  right of ownership pravo sobstvennosti
  right of the buyer pravo pokupatelia
  right to a pension pravo na pensiiu
  right to compensation for harm pravo na vosmeshchenie vreda
  right to dispose of common property pravo rasporiazhat'sia obshchei sobstvennost'iu
  right to inherit pravo nasledovaniia
  right to maintenance pravo na poluchenie alimentov
  right to part of the property acquired during the marriage pravo na chast' nazhitogo v techenie braka imushchestva
  right to personal property pravo lichnoi sobstvennosti
  right to renounce an estate pravo otkazat'sia ot nasledstva
  right to succession pravo na nasledstvo
  right to use living accommodations pravo pol'zovat'sia zhilym pomeshcheniem
  right-forming pravoobrazuiushchii
  rights and obligations of the parties prava i obiazannosti storon
risk risk
  risk of accidental ruin risk sluchainoi gibeli
rule pravilo
  ruling opredelenie

safeguarding the rights of consumers okhrana prav potrebitelei
safekeeping sokhrannost'
sale sbyt
sanction sanktsiia
satisfaction of claims udovletvorenie pretenzii
savings nakoplenie
scholarly interpretation nauchnoe tolkovanie
scientific-technical expertise nauchno-tekhnicheskaia ekspertiza
search for the judgment debtor rozysk dolzhnika
seasonal goods sezonnye tovary
securing performance of obligations obespechenie ispolneniia obiazatel'stv
  securing warrant okhrannoe svidetel'stvo
seller prodavets
separate property obosoblennoe imushchestvo
  separateness of an accommodation izolirovannost' pomeshcheniia
  separation of church from state otdelenie tserkvi ot gosudarstva
service apartment sluzhebnaia kvartira
set price ustanovlennaia tsena
setoff zachet
sex pol
sham marriage fiktivnyi brak
  sham transaction pritvornaia sdelka
share dolia
  share co-owner dolevoi sosobstvennik
  share of the succession nasledstvennaia dolia
shipper gruzootpravitel'
  shipping invoice gruzovaia nakladnaia
short-term credit kratkosrochnyi kredit
simple negligence prostaia neostorozhnost'

single out vydeliat′
single person odinokoe litso
single-form order order edinoi formy
situs of a legal person mesto nakhozhdeniia iuridicheskogo litsa
social custom zhiteiskii obychai
social housing fund obshchestvennyi zhilishchnyi fond
social organization obshchestvennaia organizatsiia
social relationship obshchestvennoe otnoshenie
social status sotsial′noe polozhenie
socialist sotsialisticheskii
socialist community life sotsialisticheskoe obshchezhitie
socialist legality sotsialisticheskaia zakonnost′
socialist morality sotsialisticheskaia moral′
socialist ownership sotsialisticheskaia sobstvennost′
source of increased danger istochnik povyshennoi opasnosti
specific performance of an obligation ispolneniie obiazannosti v nature
specification spetsifikatsiia
spouse suprug (m.); supruga (f.)
standard etalon
state gosudarstvo
state administration gosudarstvennoe upravlenie
state housing fund gosudarstvennyi zhilishchnyi fond
state insurance gosudarstvennoe strakhovanie
state organ gosudarstvennyi organ
state plan discipline gosudarstvennaia planovaia distsiplina
state workers' savings bank gosudarstvennaia trudovaia sberegatel′naia kassa
stateless person litso bez grazhdanstva
status polozhenie
statute polozhenie
statute on comrades' courts polozhenie o tovarishcheskikh sudakh
statute on deliveries polozhenie o postavkakh
statute on the procedure for granting and paying state pensions polozhenie o poriadke naznacheniia i vyplaty gosudarstvennykh pensii
statute on the socialist state production enterprise polozhenie o sotsialisticheskom gosudarstvennom proizvodstvennom predpriiatii
stepdaughter padcheritsa
stepfather otchim
stepmother machekha
stepson pasynok
stipulated by law predusmotrennyi zakonom
stipulated condition ogovorennoe uslovie
structural part sostavnaia chast′
subcontractor's contract dogovor subpodriada
subject of a civil legal relationship predmet grazhdanskogo pravootnosheniia
subject of a contract of sale predmet dogovora kupli-prodazhi
subject of a right sub″ekt prava
subject of civil law sub″ekt grazhdanskogo prava
subject of copyright sub″ekt avtorskogo prava
subject of delivery predmet postavki
subject of liability sub″ekt otvetstvennosti
subjective right sub″ektivnoe pravo
sublessee podnanimatel′
subordination podchinenie
subsidiary (branch) filial

subsidiary podsobnyi
  subsidiary household podsobnoe khoziaistvo
  subsidiary household plot podsobnoe domashnee khoziaistvo
substantive mistake sushchestvennoe zabluzhdenie
substatutory enactments podzakonnye akty
substitution of an heir podnaznachenie naslednika
succession nasledstvennoe imushchestvo
  succession on intestacy nasledovanie po zakonu
  succession under a will nasledovanie po zaveshchaniiu
supplementary dopolnitel'nyi
  supplementary debtor dopolnitel'nyi dolzhnik
  supplementary living space dopolnitel'naia zhilaia ploshchad'
  supplementary services dopolnitel'nye uslugi
supplier postavshchik
Supreme Court Verkhovnyi sud
  Supreme Soviet Verkhovnyi sovet
surety poruchitel'
  suretyship poruchitel'stvo
surname familiia
system sistema
  system of first risk sistema pervogo riska
  system of proportionate liability sistema proportsional'noi otvetstvennosti
  systematic interpretation sistematicheskoe tolkovanie

tacit acceptance molchalivyi aktsept
take effect vstupit' v deistvie
task zadanie
  task set by the plan planovoe zadanie
technical school tekhnikum
temporary vremennoe
  temporary absence vremennoe otsutstvie
terminate a contract prekratit' dogovor
  termination of actions presechenie deistvii
  termination of a legal relationship prekrashchenie pravootnosheniia
  termination of a marriage rastorzhenie braka; prekrashchenie braka
  termination of an obligation prekrashchenie obiazatel'stva
testator zaveshchatel'
testimony pokazaniia
  testimony of witnesses svidetel'skoe pokazanie
things of economic value ekonomicheskie tsennosti
time vremia
tort delikt
  tortfeasor prichinitel' vreda
trademark tovarnyi znak
trading organization torgovaia organizatsiia
transaction sdelka
transfer of rights and obligations perekhod prav i obiazannostei
transport charter transportnyi ustav
  transport code transportnyi kodeks
type of occupation rod zaniatii
  types of earnings (income) subject to withholding for support payments vidy zarabotka (dokhoda), podlezhashchikh uchetu pri uderzhanii alimentov

unclaimed finds nevostrebovannye nakhodki
under article v silu stati

undignified behavior nedostoinoe povedenie
unfair nespravedlivo
unfit parent nedostoinyi roditel′
unilateral transaction odnostoronniaia sdelka
unjust acquisition of property neosnovatel′noe preobretenie imushchestva
  unjust retention of property neosnovatel′noe sberezhenie imushchestva
unlawful act nepravomernyi akt
  unlawful action nepravomernoe deistvie
  unlawful behavior protivopravnoe povedenie
  unlawful conduct nepravomernyi postupok
  unlawfulness of actions protivopravnost′ deistvii
unshared bezdolevoi
unsolemnized marriage fakticheskii brak
usable thing veshch′ prigodna k upotrebleniiu
use pol′zovanie
  to use pol′zovat′sia
  use of accommodations pol′zovanie pomeshcheniami
  user potrebitel′

vacate an accommodation osvobodit′ pomeshchenie
vacation-home construction cooperative dachno-stroitel′nyi kooperativ
valid deistvitel′nyi
  valid transaction deistvitel′naia sdelka
  validity of a marriage deistvitel′nost′ braka
  validity of an act deistvitel′nost′ akta
verbal formulation slovesnaia formulirovka
vestiges perezhitki
violation narushenie
  violation of contractual discipline narushenie dogovornoi distsipliny
  violator narushitel′
void transaction nichtozhnaia sdelka
  void (totally invalid) transaction nichtozhnaia (polnostiu nedeistvitel′naia) sdelka
  voidable (relatively invalid) transaction osporimaia (otnositel′no nedeistvitel′naia) sdelka
volitional volevoi
voluntary insurance dobrovol′noe strakhovanie
  voluntary marriage dobrovol′nost′ vstupleniia v brak
  voluntary procedure dobrovol′nyi poriadok
volunteer society dobrovol′noe obshchestvo

ward podopechnyi
  warder popechitel′
warrant nariad
water legislation vodnoe zakonodatel′stvo
wedding palace dvorets brakosochetaniia
whereabouts mestoprebyvanie
wide range shirokii krug
will volia; (document) zaveshchanie
withholding the transfer of an article zaderzhka peredachi veshchi
work contract dogovor podriada
written pismennyi
  written proof pis′mennoe dokazatel′stvo
  written transaction pis′mennaia sdelka
wrongful action protivopravnoe deistvie

Index

About the Editor

Oleg Nikolaevich Sadikov is a Distinguished Scientist of the RSFSR, Doctor of Juridical Sciences, and head of the Civil Law Department of the All-Union Scientific-Research Institute of Soviet Legislation. He is a member of the editorial boards of the periodicals *Soviet State and Law* and the *Soviet Yearbook of International Law* and has published a number of works on private international law, the legal regulation of international transportation, and civil law of the USSR.